Sandwich Structural Composites

Sandwich Structural Composites
Theory and Practice

Wenguang Ma and Russell Elkin

CRC Press is an imprint of the
Taylor & Francis Group, an **informa** business

First edition published 2022
by CRC Press
6000 Broken Sound Parkway NW, Suite 300, Boca Raton, FL 33487-2742

and by CRC Press
2 Park Square, Milton Park, Abingdon, Oxon, OX14 4RN

© 2022 Taylor & Francis Group, LLC

CRC Press is an imprint of Taylor & Francis Group, LLC

Reasonable efforts have been made to publish reliable data and information, but the author and publisher cannot assume responsibility for the validity of all materials or the consequences of their use. The authors and publishers have attempted to trace the copyright holders of all material reproduced in this publication and apologize to copyright holders if permission to publish in this form has not been obtained. If any copyright material has not been acknowledged, please write and let us know so we may rectify in any future reprint.

Except as permitted under U.S. Copyright Law, no part of this book may be reprinted, reproduced, transmitted, or utilized in any form by any electronic, mechanical, or other means, now known or hereafter invented, including photocopying, microfilming, and recording, or in any information storage or retrieval system, without written permission from the publishers.

For permission to photocopy or use material electronically from this work, access www.copyright. com or contact the Copyright Clearance Center, Inc. (CCC), 222 Rosewood Drive, Danvers, MA 01923, 978-750-8400. For works that are not available on CCC, please contact mpkbookspermissions@tandf.co.uk

Trademark notice: Product or corporate names may be trademarks or registered trademarks and are used only for identification and explanation without intent to infringe.

Library of Congress Cataloging-in-Publication Data
Names: Ma, Wenguang, editor. | Elkin, Russell, editor.
Title: Sandwich structural composites : theories and practices /
[edited by] Wenguang Ma and Russell Elkin.
Description: First edition. | Boca Raton, FL : CRC Press, 2022. | Includes
bibliographical references and index. | Summary: "This book offers comprehensive coverage sandwich structural composites. It describes structure, properties, characterization, and testing of raw materials. In addition, it discusses design and process methods and applications of sandwich structural composites. The book will be of benefit to industrial practitioners, researchers, academic faculty, and advanced students looking to advance their understanding of these increasingly important materials"— Provided by publisher.
Identifiers: LCCN 2021036440 (print) | LCCN 2021036441 (ebook) |
ISBN 9780367441722 (hbk) | ISBN 9781032079073 (pbk) |
ISBN 9781003035374 (ebk)
Subjects: LCSH: Laminated materials. | Sandwich construction.
Classification: LCC TA418.9.L3 S36 2022 (print) | LCC TA418.9.L3 (ebook)|
DDC 620.1/18—dc23/eng/20211006
LC record available at https://lccn.loc.gov/2021036440
LC ebook record available at https://lccn.loc.gov/2021036441

ISBN: 978-0-367-44172-2 (hbk)
ISBN: 978-1-032-07907-3 (pbk)
ISBN: 978-1-003-03537-4 (ebk)

DOI: 10.1201/9781003035374

Typeset in Times
by codeMantra

Contents

Preface vii
Biographies xi
Introduction xiii

Chapter 1 Sandwich Structural Core Materials and Properties 1
Wenguang Ma and Russell Elkin

Chapter 2 Special Properties and Characterization Methods of
Core Materials 73
Wenguang Ma and Russell Elkin

Chapter 3 Face Sheet Materials for Sandwich Composites 85
Trevor A. Gundberg

Chapter 4 Laminating Processes of Thermoset Sandwich Composites 125
Wenguang Ma and Russell Elkin

Chapter 5 All-Thermoplastic Sandwich Composites 159
Wenguang Ma and Russell Elkin

Chapter 6 Characterizations of Sandwich Structures 185
Wenguang Ma and Russell Elkin

Chapter 7 Sandwich Structure Design and Mechanical Property
Analysis 253
Wenguang Ma and Russell Elkin

Chapter 8 Sandwich Composite Structure Modeling by Finite
Element Method 295
Guohua Zhou

Chapter 9 Application of Sandwich Structural Composites 333
Wenguang Ma and Russell Elkin

Chapter 10 Sandwich Composite Damage Assessment and Repairing 425
Wenguang Ma and Russell Elkin

Index 449

Preface

Wenguang Ma – as a co-author of the book *Sandwich Structural Composites – Theory and Practice*, earned his Ph.D. degree in plastics engineering from University of Massachusetts Lowell. For more than 20 years, he has been working in the sandwich composites industry area from developing polyethylene terephthalate (PET) foam core materials by extrusion, as well as using balsa wood, plastic foams, and honeycomb core materials for creating sandwich structures for a wide range of applications and developing one of the world's largest pultrusion line for production of sandwich composite panels. Currently, Dr. Ma is Director of Technical Operation at Vixen Composites USA.

Dr. Ma has published about 20 research articles including PET foam core development and application. His other research subjects include balsa and plastic foam cores, and sandwich composite structures. He has given a series of technical seminars to international composites conferences on an extensive set of topics including foam core material properties, applications, structural design and characterization, all thermoplastic sandwich construction, and quality control/repairing of sandwich laminates.

He contributed one chapter to the book *Rigid Structural Foam and Foam-Cored Sandwich Composites* in the series on Polymeric Foams published by CRC/Taylor & Francis Group. After that book was published, the editor of CRC, Allision Shatkin, invited him to write a book about the developments in last 30 years on the theories and practices of the sandwich composites. He gladly accepted the invitation, and invited Russell Elkin, his previous colleague, as a co-author to complete the book. He also invited Trevor Gundberg and Guohua Zhou to contribute one chapter each, respectively.

Russell Elkin, co-author of the book, is the Product Development Manager with Baltek, Inc. of 3A Composites. He has worked for Baltek and its predecessor companies since 1994 after graduating from Carnegie Mellon University with a degree in mechanical engineering. In addition to working peripherally alongside Dr. Ma on the development of PET foam cores, he has designed and assisted on the design of hundreds of sandwich laminates across all market segments. He has also managed dozens of projects converting customers from open molding to vacuum infusion in both structural and process design, training, and implementation.

Trevor Gundberg, contributor of Chapter 3, is Vice President of Composite Engineering of Vectorply. He joined Vectorply in 2006 as a Composite Materials Engineer, previously holding technical positions within the composites industry at both Toray Carbon Fibers America and DIAB, Inc. He has been Director of Composites Engineering. Trevor holds a Bachelor of Science in Mechanical Engineering from North Dakota State University (with a minor in Chemistry) and a Master of Science in Mechanical Engineering from the University of Delaware and is a registered Professional Engineer in the state of Alabama.

Contributor of Chapter 8, Dr. Guohua Zhou, is a research scientist at Optimal, Inc. working on finite element simulation method development for engineering application, and he is also an experienced computer-aided engineering (CAE) engineer. Dr. Zhou earned his Ph.D. degree in structural engineering at the University of California, San Diego (UCSD), with a focus on computational mechanics. He has solid background in linear elasticity, composite elasticity, vibration, nonlinear solid mechanics, micromechanics, fracture mechanics, and damage mechanics.

"Sandwich" is a special construction of individual materials or structural elements laminated together, where a thick, low-density core material separates thin sheets of stiff, strong, and relatively dense material. The core is typically light-density wood, honeycomb, or polymeric foam. The skins are usually metal, fiberglass, or carbon fiber reinforcements, laminated using thermosetting resins, or embedded in a thermoplastic resin matrix. Faces and core are joined by any variety of lamination and/or bonding methods. The key aspect of sandwich construction is the bond between skin and core, which must ensure an adequate load transfer from one constituent component to the other. The outcome is an extraordinary product. High bending stiffness, superior strength-to-weight ratios, and good buckling resistance are just a few of the advantages. Sandwich construction also permits function integration. Insulation and fire protection become parts of the structure, for example. Separate materials to provide these traits do not need to be added later. Highly efficient structures may be designed for each specific application by the proper selection and assembly process of integrating the core and facing materials together.

The earliest patent for sandwich construction was filed in the late 19th century, but sandwich has been used in modern industry since World War II. Balsa wood was the first core material used for building the de Havilland DH.98 Mosquito bomber's fuselage. The first foam core material was polyvinyl chloride (PVC) rigid foam, first commercialized in Germany from the 1930s to the 1940s. The first successful structural adhesive bonding of honeycomb sandwich structures was achieved in 1938. Today, there is a very wide assortment of materials commercially produced as sandwich cores. Many tailored for some specific performance or application. Metal, plastics, hardwood, especially fiber-reinforced composite and other facing materials have also been developed in the last 50 years for meeting all applications.

Sandwich structural composites are being used successfully for a variety of applications such as spacecraft, aircraft, train and car structures, wind turbine blades, boat/ship superstructures, boat/ship hulls, and many other areas. The largest volume of sandwich structure composites has been used in the wind energy industry since the 1980s. The balsa wood, rigid foam, honeycomb, and other core or their combination in cored sandwich constructions are used for building rotor blades and nacelle enclosures of wind energy generators. Since the 1990s, the raw materials, lamination processes, design, and testing methods of the sandwich composites industries have been advanced into a new era.

While many still consider high-performance composites and sandwich to be a recent development, the concept and application have been around for more than 75 years. Most engineering students are introduced to composites in an introductory to materials class. Due to the breath of material types (there are material libraries now), FRP and sandwich receive only a brief mention.

A similar situation exists in texts about composites. There are a large number of excellent books dedicated to design and analysis, but nothing with a primary focus on materials. Furthermore, the sections in many of these texts mainly cover honeycomb and targets the aircraft and aerospace markets. Books on processing are few and those tend to be high on theory, which is important of course, but any discussion of the individual raw materials and how material and process selection affect the resulting laminate is often brief and unsophisticated. Manuals on boat construction and building aircraft often provide more relevant information than many textbooks. Since the turn of the 21st century, the number of published works about composites has grown significantly, but these still cite old and what the authors know to be outdated and inaccurate references. Current texts continue to contain a great number of assumptions, poorly supported by data, especially about wood-based sandwich cores. We hope this book will fill in a number of those gaps and provide information to the reader which is not available in a product selector app or website.

This volume begins with an introduction about the development and significance of core material. The reader may be introduced to a few new core materials, including polyethylene terephthalate (PET) foam core developed in the last 30 years. Chapters 1 and 2 cover in-depth core types, format, typical properties, and characterization methods. Chapter 3 details various facing materials, reinforcing fibers, and matrix resins. Chapter 4 introduces various laminating processes, adhesive bonding of premade skins to different core materials, and wet laminating processes including hand lay-up, prepreg and autoclave curing, vacuum infusion, and resin transfer molding and pultrusion. All thermoplastic sandwich composites are a new development in the last 20 years. Laminating core and skin sheets using heat and pressure is reviewed in Chapter 5.

The structural design, property prediction, computer software simulation, and testing are necessary for developing a new product, optimizing the lamination process, and performing quality control and property confirmation in the sandwich composite industry. Chapters 6–8 present characterizations of sandwich structures, structural design, mechanical property analysis, and finite element analysis (FEA) for mechanical and dynamic predictions. Applications of sandwich composites in different industries are given in Chapter 9, and damage detection, repairing, and nondestructive inspection after repair are presented in Chapter 10.

This book is intended to present the latest worldwide developments in sandwich composites over the past 30 years, to benefit the industrial practitioner, researcher, academic faculty, and both undergraduate and graduate school students, and hopefully direct more interest and dedication into the current industrial activity for an even more prosperous future.

A final note: a book like this is never finished. There is still so much to learn.

Biographies

Wenguang Ma (0000-0001-6946-985X), guest speaker invited, holds a Ph. D. degree in plastics engineering from University of Massachusetts Lowell, and is working as a Technical Director in Vixen Composites LLC, and ever worked in a couple of companies as Technical Manager and Senior Engineer on the research and development of the composites. His experiences and contributions include structure polyethylene terephthalate (PET) foam core development and application for sandwich composites, and sandwich composite design/ development by pultrusion and other processes. He is the author or co-author of over 20 technical papers and book chapters in the areas of the sandwich composites, polymer foam, polymer blending and modification, biodegradable polymers, etc.

Russell Elkin (0000-0002-1886-9420) is the Product Development Manager with Baltek, Inc., USA. He has worked for Baltek and its predecessor companies since 1994 after graduating from Carnegie Mellon University with a degree in mechanical engineering. At Baltek, he works with customers around the world on composite and sandwich laminate design. He also gathered vast knowledge and experience in manufacturing processes, including advanced training in vacuum infusion and Lite-RTM where he oversees the development of new products.

Trevor A. Gundberg (0000-0003-1435-3248) joined Vectorply in 2006 as a Composite Materials Engineer, previously holding technical positions within the composites industry at both Toray Carbon Fibers America and DIAB, Inc. He was Director of Composites Engineering from 2010 to 2020 and is now Vice President of Composite Engineering. Trevor holds a Bachelor of Science in Mechanical Engineering from North Dakota State University (with a minor in Chemistry) and a Master of Science in Mechanical Engineering from the University of Delaware, is an ASQ-certified Six Sigma Green Belt, and is a registered Professional Engineer in the state of Alabama.

Guohua Zhou (0000-0002-5484-6955) is a research scientist at Optimal, Inc. working on finite element simulation method development for engineering application, and he is also an experienced computer-aided engineering (CAE) engineer. He has been leading the sandwich composite testing and modeling work for the side and roof panels for a series of electrified bus products at Optimal, Inc.; the work involves evaluation of noise vibration and harshness (NVH), stiffness, strength, and durability of sandwich composite. Guohua earned his Ph.D. degree in structural engineering at the University of California, San Diego (UCSD), with a focus on computational mechanics. He has solid background in linear elasticity, composite elasticity, vibration, nonlinear solid mechanics, micromechanics, fracture mechanics, and damage mechanics. He is interested in solving mechanics problems in both engineering and science fields. Through years of academic training and working in the industry,

Guohua has a deep understanding of theoretical knowledge and industrial applications, and he has developed a wide range of problem-solving capabilities, including deriving the computational algorithm based on the physics of the problem and writing the finite element source code, conducting tests, and developing modeling approaches based on commercial finite element software. He is eager to solve real-world problems by filling the gaps between state-of-the-art research and full-scale industry implementation. Guohua loves learning new things and communicating new ideas, and welcomes any future potential collaborations.

Introduction

I.1 SANDWICH COMPOSITES AND THEIR SIGNIFICANCES IN MODERN INDUSTRIES

A sandwich composite is a special construction of laminated individual materials or structural elements together, where a relatively lightweight, thick, and compatible core material separates stiff, strong, and relatively dense thin faces. The core can be mostly any material, provided that it meets a few requirements. It should be of low density. It should have an elastic modulus far lower (<100×) than the skins, but it must have sufficient strength and modulus in shear and compression to do the job. The skins are usually metallic or FRP composites, but wood or unreinforced plastic may also be used. The faces and the core are joined by adhesive bonding, or by wet molding processes such as hand lay-up laminating, resin infusion process, pultrusion process, or in the case of thermoplastic composite facings, melt-bonded to the core. The resulting bond ensures an adequate load transfer from one sandwich constituent component to the other.

Most often, there are two faces, identical in material, fiber orientation, and thickness, which primarily resist the in-plane and lateral bending loads. However, in many cases, the faces may differ in thickness, materials, fiber orientation, or any combination of these three. This may be due to the fact that in use, one face is an external face, while the other is an internal face. The former sandwich is regarded as a mid-plane symmetric sandwich and the latter a mid-plane asymmetric sandwich.

Composite designers determined early on that sandwich construction can dramatically increase a laminate's stiffness with little added weight. A sandwich structure is very cost-effective because the relatively low-cost core replaces more expensive composite reinforcement material. And the stiffer but lighter sandwich panel requires less supporting structure than a solid laminate.[1]

In a sandwich panel, the core functions like the connecting web of an I-beam, separating the face skins at a constant distance, while the skins themselves function as the I-beam flanges as seen in Figure I.1. The sandwich panel's bending stiffness is

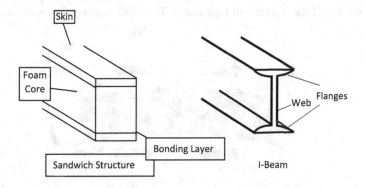

FIGURE I.1 I-beam and sandwich structure comparison.

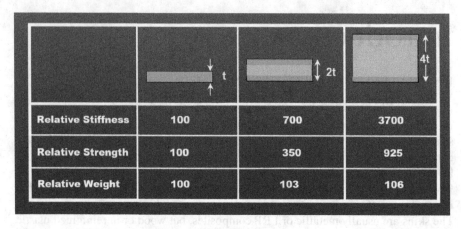

FIGURE I.2 Comparison of solid and sandwich laminates for out-of-plane loads.

roughly proportional to the core thickness cubed in the same way an I-beam is stiffer as the distance between the flanges increases. Doubling the panel thickness by using light-density core between the solid lamination yields a panel three and half times stronger and seven times stiffer; in the same way, doubling the panel thickness again yields the panel nine times stronger and 37 times stiffer, and with very little weight increase as described in Figure I.2.

When a bending force works on a sandwich panel as shown in Figure I.3, the core transfers the force from one skin to the other and bears the shear and compression stress. A core also helps distribute loads and stresses on the skins, which makes a cored sandwich an excellent design for absorbing impacts. A core's compressive modulus prevents the thin skins from wrinkling (buckling) failure, while its shear modulus keeps the skins from sliding independent of each other when subjected to bending loads.

The core of a sandwich structure can be of almost any material or architecture, but in general, cores fall into four main types: (a) foam or cellular solid core (also light wood), (b) honeycomb core, (c) truss core, and (d) web core.

The two most common honeycomb types are the hexagonally shaped cell structure (hexcell) and the square cell (egg-crate). Web core construction is analogous to

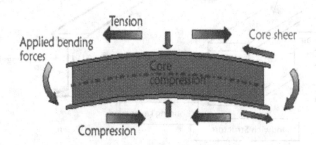

FIGURE I.3 Bending force works on a sandwich structure.

a group of I-beams with their flanges welded together. Truss or triangulated core construction is being widely used for the bridge constructions discussed above. In most foam core and honeycomb core sandwich constructions, one can assume for all practical purposes that the in-plane and lateral bending loads are carried by the faces only. However, with some truss core and web core designs, a portion of these loads are carried by the core.

The face sheets can be mostly any material in the form of thin sheet. The properties of primary interest for the facings are high axial stiffness giving high flexural rigidity, high tensile and compressive strength, impact and temperature resistance, environmental resistance, wear resistance, and optionally an attractive surface finish.

The facing-core interface is often the most vulnerable part of the sandwich structure. This interface is often bonded (e.g., graphite epoxy facings joined to an aluminum honeycomb core). Alternatively, the facing-core interface can be blended or functionally graded as is sometimes suggested for ceramic-metal sandwich structures.

The facing can also be divided into another two groups: rigid sheets before laminating on to the cores, which will be constructed with the cores by using adhesives and by a dry consolidation process; and facings formed during the laminating process on to the cores by a wet process. Most large sandwich products and most fiber-reinforced sandwich composites belong to the second group that are made by using thermosetting liquid resin, the dry fiber reinforcement materials, and the different wet laminating processes.

The dry fabrication process is completed by using three separate materials: face sheets, core, and adhesive. The faces are metallic or nonmetallic rigid sheets; the adhesives could be liquid resins, or thermosetting or thermoplastic adhesive films. The fabrication processes are hot pressing, vacuum bagging, pinch roller pressing, and continuous pressing, with many variations for all of these.

Wet lamination usually is used for making fiber-reinforcing composites in which the facing is fabricated during production process. The facing consists of dry fibrous materials, such as glass, carbon, synthetic, metallic, or natural fibers. The matrix materials are primarily thermosetting resins, and to a lesser extent, thermoplastic resins that are gaining attention because of their ability to be recycled. The wet processes can be completed by using liquid resin and different fabrication methods, such as vacuum infusion, resin transfer molding, chop and spray lamination, hand layup, hot press, and pultrusion.

Advantages of the sandwich construction are considerably higher stiffness and strength-to-weight ratios than an equivalent beam made of only the core material or the face-sheet material. This leads to a high bending stiffness and strength-to-weight ratios for the composite. Designers now have the flexibility to choose facing material, core type, optimize the thickness & density as well as utilize the core sheet to improve the lamination process and minimize the total part cost. As an added benefit, sandwich panels offer excellent heat insulation, sound abatement, kind of fire proofing and vibration damping, and the appropriate selection of the core material. However, localized loads are one of the major causes of failure in sandwich constructions, because the faces are significantly thinner than the same material used in a monocoque construction to resist the same loads. The point load on the structures should be avoided in application.

I.2 DEVELOPMENT OF SANDWICH COMPOSITES

The first patent for honeycomb core was filed in 1905 in Germany. The first aircraft sandwich panel with thin mahogany facings and an end-grain balsa wood core was made in 1919. In the late 1930s, hardwood facings were bonded to relatively thick slice of paper honeycomb used in manufacture of furniture in USA. In 1938, a plywood skin/cork wood core sandwich wing monoplane was displayed in France.

The sandwich construction has been used in modern industry since World War II. In England, sandwich construction was first used by de Havilland in the Mosquito bomber of World War II which employed plywood sandwich construction with balsa as the core. Also the concept of sandwich construction in the United States originated with composite faces and a low-density core. In 1942, Vidal Research Corp designed and fabricated the fuselage for a prototype version of the Vultee VB-13 (XBT-16) using fiberglass-reinforced polyester as the face material using both a glass-fabric honeycomb and a balsa wood core. Figure I.4 shows that the fuselage was made in the workshop.

The first use of honeycomb structures for structural applications had been independently proposed for building application and published already in 1914. In 1934, Edward G. Budd patented a welded steel honeycomb sandwich panel from corrugated metal sheets, and Claude Dornier aimed in 1937 to solve the core-skin bonding problem by rolling or pressing a skin which is in a plastic state into the core cell walls. The first successful structural adhesive bonding of honeycomb sandwich structures was achieved by Norman de Bruyne of Aero Research Limited, who patented an adhesive with the right viscosity to form resin fillets on the honeycomb core in 1938. The North American XB-70 Valkyrie made extensive use of stainless-steel honeycomb panels using a brazing process they developed.

Also in the late 1940s, two young World War II veterans formed Hexcel Corporation, which over the decades has played the most important role of any firm in the growth of sandwich structure. Starting with honeycomb core, Hexcel makes well big of the world's honeycomb core materials even today. In 1992, Bitzer of Hexcel gave an excellent overview of honeycomb core materials and their applications. Bitzer states that every two (or more) engine aircraft in the western world utilizes some honeycomb core sandwich, and that while only 8% of the wetted surface

FIGURE I.4 Balsa core was used for making the fuselage of Mosquito Air Bomber during World War II in Australian plants.

Introduction

of the Boeing 707 is sandwich, 46% of wetted surface of the newer Boeing 757/767 is honeycomb sandwich. In the Boeing 747, the fuselage cylindrical shell is primarily Nomex honeycomb sandwich, and the floors, side-panels, overhead bins, and ceiling are also of sandwich construction.[2]

The honeycomb cores made by stainless and aluminum sheets, and synthetic composite sheets such as Nomex and Kevlar have been used in aircraft fuselage, parts such flooring panels, interior walls, storage bins, exterior control surfaces, engine nacelles, and helicopter blades and tail booms. The Airbus A380 airliner uses honeycomb extensively in the wing flaps and the doors. The rotor blades of the helicopter have honeycomb or foam cores. More recent developments show that honeycomb structures are also advantageous in applications involving nanohole arrays in anodized alumina, microporous arrays in polymer thin films, and activated carbon honeycombs.

In the 1970s, tremendous activity began in Sweden regarding the use of composite sandwich construction for naval ship hulls, which led Swedish Royal Navy to switch from continuing to use steel hulls to fiberglass composite sandwich constructions. This effort involved analysis, optimization, small-scale tests, full-scale tests for both underwater explosions and air explosions, etc. They were able to show that a properly designed composite sandwich hull could be as structurally sound as a steel hull. Figure I.5 shows a boat hull built by the sandwich structure. As a result, from some date in the 1980s, all Swedish Royal Navy ship hulls were made of sandwich construction. Then, these works turned attentions to the rest of the Scandinavian naval ship hulls that are sandwich composites as well as hundreds of the ferry boats that traverse the waters in and among the Scandinavian countries.[2]

Since 1980, composite front cabs of locomotives have been built for the XPI locomotives in Australia, the ETR 500 locomotives in Italy, the French TGV, and the Swiss locomotive 2000. Interestingly, the major design criteria are the pressure waves occurring during the crossing of two high-speed trains in a tunnel. In Japan, the new Nozomi 500 bullet trains use honeycomb sandwich for the primary

FIGURE I.5 Boat hull built by foam core sandwich laminates.[3]

structure. In 1995, the sandwich construction is now being used in double-decker buses and bridges in Europe.

After applications in the marine industry, sandwich has been used in wind turbine blade construction on a large scale since the 1970s. The wind energy business is unique as fiber-reinforced composites and sandwich created the opportunity for development and innovation. While small-scale turbine blades may be designed with wood or aluminum, utility-scale blades cannot. Wind blades are typically manufactured as two half-shells, which are secondarily assembled along with a shear web into a whole blade. Each half blade is a sandwich structure, often featuring a combination of different core materials, each used in a particular location. The other components, such as shear webs as shown in Figure I.6, are also cored sandwich constructions. The wind energy industry has seen phenomenal growth in the past 25 years, and blades are getting longer to increase turbine efficiency. Offshore turbine blades see the highest loads due to their length and are more expensive to maintain; these massive structures are now built in greater volume. The growth rate of wind energy is 20% per annum.[4]

The first foam core material was polyvinyl chloride (PVC) rigid foam that was made by mixing isocyanate blend with PVC, first commercialized in Germany by Dr. Lindemann in 1930s–1940s. It is rumored that this early version of PVC foam was used in the German E-boats and even in the famous "Bismarck" battleship. After World War II, PVC foam production continued, and its development continued for new applications. The primary application for PVC foam core is in the construction of recreational boats and yachts. Foam sandwich structure can be used for building

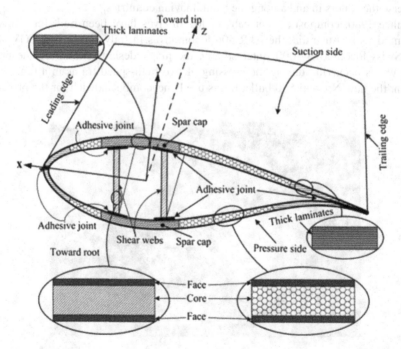

FIGURE I.6 Wind blade construction.[5]

Introduction

the hull, bulkhead, stringer, deck, superstructure, and even furniture. The advantage of foam in a marine environment is no water absorption.[6]

Polyurethane rigid foam was used as the core of sandwich composites starting in the 1950s. More expensive than balsa but cheaper than PVC foam, polyurethane foam has been used historically in the construction of the more expensive boats.[7]

In early 1990s, styrene-acrylonitrile copolymer (SAN) foam was introduced into the market. It is claimed to have good thermal stability and no outgassing problem during resin curing of sandwich facings at elevating temperature.

The first thermoplastic rigid foam core, polyethylene terephthalate (PET) foam, emerged to the composite market in 1998. A special format of PET "strand" foam was introduced to the market in 2000. This was the first type of structural core product made by a conventional foaming extrusion process, by a special die design, and is recyclable by thermal melt processing. The PET foam core can be used with all types of liquid thermosetting resins and is compatible with vacuum infusion, adhesive bonding, prepreg, RTM, compression molding, and thermoforming.[8]

Other high-performance foam cores are polyamide (PA), polymethacrylimide (PMI), polyethersulfone (PES), and polyetherimide (PEI) foams. PMI foam is a closed-cell rigid foam based on PMI. The field of application of PMI foam is aircraft, rail vehicle, and racing car construction, which need processing temperatures up to 130°C (266°F).[9] PEI foam is a closed-cell, thermoplastic polymer foam that combines fire resistance with low smoke and toxicity, along with good dielectric properties. It has a good strength-to-weight ratio and very low moisture absorption, and is a core material for the structural lightweight applications that demand high fire resistance, radar transparency or operation in extremely hot or cold environments.[10] PES foam is made from extrusion of thermoplastics and has a unique high service temperature. It offers excellent fire, smoke, and toxicity properties. Furthermore, it is nonhygroscopic and has superior damage/impact performance and improved dielectric properties.[11]

The polystyrene (PS) foam is used for insulation in most applications. However, in recent years, PS foam with a greater density has been promoted as a structural core material in the composite industry for wind turbine blade construction using epoxy resin systems. Its low chemical resistance limits its usage in laminating with unsaturated polyester and vinyl ester resins.

The commonly used face materials can be divided into two main groups: metallic and nonmetallic materials. The former group contains steel, stainless steel, and aluminum alloys. There is a vast variety of alloys with different strength properties, whereas the stiffness variation is very limited. Non-metallic materials include materials such as plywood, cement, wood veneer, plastic laminates, and fiber-reinforced composites. Between 1870 and 1890, the first synthetic resins were developed. These polymer resins are transformed from their liquid state to solid state by crosslinking molecules. Early synthetic plastics include celluloid, melamine, and Bakelite. In the early 1930s, American Cyanamid and DuPont independently formulated polyester resin for the first time. In the same time period, Owens-Illinois Glass Company began weaving glass fiber into a textile fabric on a commercial basis. Between 1934 and 1936, experimenter Ray Green combined these two new products and began molding small boats. During World War II, the development of radar required nonmetallic housings and the U.S. military advanced the fledgling composites technology with many research projects. Following World War II, composite materials emerged as a major engineering material.

The sandwich composites industry began to take off in the 1940s and grew rapidly in the 1950s. By 1955, most of the composites processing methods used today had been developed. Open molding, hand lay-up, chopping, compression molding, filament winding, resin transfer molding, vacuum bagging, and vacuum infusion were all developed between 1946 and 1955. Since then, many auxiliary materials and assistive technologies have been developed to complement the various processing methods.

Methods to characterize and evaluate composites and sandwich core materials were also developed during this period – during World War II and the next two decades. The various test methods are used for evaluating the properties of the cores, faces, and sandwich structures. The core in the sandwich structure suffers the shear, compression, and tension stress during the different time and in the different location in most applications. The shear stress on the core is parallel to the face, but the compression and tension stress are perpendicular to the faces. In other words, the flat shear, flatwise tension, and compression properties are important behaviors of the core materials. Some testing methods focus on checking bonding strength between face and core, and some focus on the properties of whole sandwich structures.

Testing methods for evaluating sandwich materials and structures were derived from methods for evaluating solid materials and then evolved to the ASTM, ISO, and other industrial test methods for evaluating sandwich structure's performance. These methods measure static or dynamic properties on small sample coupons or entire structures, such as wind turbine blades. New technologies, such as ultrasonic, thermography, X-ray, and shearography, are used for nondestructive inspection (NDI) techniques for production in-line quality control and off-market application fatigue and damage inspection.

Thousands of research papers have been published, and systems for theoretical analysis have been built since the original sandwich structures were used in modern applications. These are important for optimizing part design and predicting failure conditions. The first research paper concerning sandwich construction was published in Germany in 1944 dealing with sandwich panel subjected to in-plane compressive loads. In 1948, the differential equations and boundary conditions for the bending and buckling of sandwich plates using the Principle of Virtual Displacement were derived but pursued on the bulking problem. In the same year, a small deflection theory for sandwich was published.[2]

In 1956, sandwich plate optimization was discussed in a chapter of the book *Minimum Weigh Analysis of Compression Structures*. In 1960, a paper on the correlation among and an extension of the existing theories for flat sandwich panels subjected to lengthwise compression, including optimum design, was published. In 1966, the first book on sandwich structures was published in the Netherlands, followed by another book on sandwich structures in England in 1969. These books remained the "bibles" for sandwich structures until the mid-1990s.

Also in the mid-1960s, the US Naval Air Engineering Center sponsored research to develop fiberglass composite sandwich constructions to compete in weight with conventional aluminum aircraft construction for aircraft. This effort was directed toward achieving a stealth aircraft, although that war had not yet been coined. Much of this research effort was in the development of minimum weigh optimization methods.

For buckling, elastic instability or buckling was discussed by some textbooks published earlier. The solutions can be used for dealing with structures and sandwich elements. By using the appropriated flexural stiffness and some equations, the overall instability of honeycomb and solid core sandwich panel subjected to in-plane compressive loads or in-plane shear loads, for all boundary conditions, can also predict core shear instability. There are also two equations that can resolve face wrinkling. For truss core and web core sandwich panels, care must be taken to ensure that both the core and the face do not buckle. Sandwich shell structures behave significantly different from plate and beam structures; the shell buckle is usually at a fraction of the load predicted by standard methods of analysis because shells are very imperfection-sensitive.

Then came the challenge of dynamic loading and impacts, such as the important case of wave slamming unique to ship hulls. The maximum static load expectation will reduce significantly when a sandwich structure is in contact with water on one surface. Some theoretical equations should be used when designing and analyzing the hull structure. In 1981, the methods by which to calculate the natural flexural frequencies for orthotropic and isotropic sandwich plates, including the effects of transverse shear deformation, were published. The methods provided the forced vibration response solutions for these panels subjected to dynamic sine loads, step function loads, triangular loads, exponential decay (blast) loads, and stepped triangular loads (nuclear blasts).

In 1989, K. Ha published an overview of finite element analysis applied to sandwich construction.[12] Finite element analysis techniques are developed for the solution of nonlinear problems involving sandwich composite materials. Each layer of a sandwich panel is represented explicitly so that multilayer panels and other configurations are easily considered. The finite element discretization is performed using arbitrary-shaped, fully compatible shell and continuum elements, as well as special elements representing full-depth and/or face sheet stiffeners. Each of these elements is capable of representing arbitrarily large displacements and rotations, since no simplifying assumptions are made in the theoretical formulation. Elastic-plastic effects may also be considered. An associated computer program is described which may be used to perform linear or nonlinear solutions for mechanical and thermal loading, linear natural frequency calculations.

Recent developments of the sandwich structural composites include the following:

Created new progressively sophisticated methods for modeling of sandwich structures, which require accounting for the three-dimensional effects, physical and geometric nonlinearities, and constitutive relations for the newly developed materials.

Investigated and introduced new sandwich designs and concepts into practice. Many such concepts are developed for the core aiming at improving its functionality both transferring and distributing applied loads among the facings as well as enhancing toughness of the structure. In addition to widely used honeycomb, cellular, and balsa cores, various truss-core designs are extensively investigated.

Used multifunctionality of sandwich structures for a natural design objective. Such features as heat transfer management, radar wave absorption, noise, and fire insulation are considered in diverse industrial settings.

Investigated a new process of joining additive manufactured (AM) lattice structures and carbon fiber-reinforced plastics (CFRPs) to manufacture hybrid lattice sandwich structures without secondary bonding. Multiple variations of lattice structures are designed and 3D-printed using Digital Light Synthesis (DLS) and a two-stage (B-stage) epoxy resin system. The resulting lattice structures are only partially cured and subsequently thermally co-cured with preimpregnated carbon fiber reinforcement.[13]

Studied new material concepts for sandwich structure applications. Examples of materials incorporated into new sandwich designs include, but are not limited to, nanotubes, shape memory alloy, and piezoelectrics, while the aims may vary from enhanced strength, stiffness, and toughness to sensing internal damage.

Considered intensely environmental effects in design and production, including fire, due to their growing significance in numerous applications. While sandwich structures can be adopted to incorporate thermal protection layers, internal damage caused by such phenomena is not always easily detected. Residual properties of damaged sandwich structures are also of interest both after environmental exposure and after excessive mechanical loading.[14]

Although it is a mature market, sandwich construction has major growth potential and a bright future. Already in the primary structure for satellites and aircraft, sandwich construction will be increasingly used particularly for large aircraft. More countries are now using composite sandwich constructions for their navy's ship hulls and other type ship hulls. One of the largest uses next will be for bridge construction because the sandwich composite structures have so many advantages especially anticorrosive behavior. More applications will be in building construction and various vehicles in land transportation. With growing need for alternative sources of energy, wind energy mill systems are being developed, all of which rely heavily on composite sandwich constructions.

I.3 TYPES OF STRUCTURAL SANDWICH COMPOSITES

Structural sandwich composites may be divided into different categories based on raw materials, such as metallic and nonmetallic, cores, and face materials, thermoplastic and thermoset; product type, like beam, plate, shell and special designed shape; applications, for example in aerospace, military, civil, marine, transportation, building construction, electronics, and biomedical areas; and processing methods, for instance intermittent and continuous production, open and close mold, and dry and wet lamination.

I.3.1 DIFFERENT CORE MATERIALS FOR SANDWICH STRUCTURES

Sandwich composites should be first categorized by raw materials. Then divided into different types based on the core materials. The global core materials market is projected to grow to USD 1.92 billion by 2022, at a compound annual growth rate of

8.77% between 2017 and 2022. The core types can be lattice trusses and web core, honeycomb, solid cores including cellular structural foams and woods, microsphere mats and 3D woven fabric cores, etc.

Lattice truss and web cores are mostly used in metal sandwich structures, which are made by welding the core and skin together. The lattice materials are basically consisting of a three-dimensional (3D) network of fully triangulated solid (or hollow) rods. Generally, these are used to reduce the mass and increase the stiffness/strength of metal panel and beam, which are used in bridge and building construction. Potential applications also include blast resistant structures, multifunctional materials, and replacement for the structurally efficient but more expensive honeycombs.

There are many different honeycomb core materials. Basically, they are metallic and nonmetallic honeycombs. The metallic can be stainless steel and aluminum sheet honeycomb; the nonmetallic include those made by glass fiber-reinforced plastic, carbon fiber-reinforced plastic, synthetic fiber cloth saturated by phenolic resin, thermal plastics, and paper saturated by resin. The cores of aerospace and satellite structures including fuselage, wing, cabin linings, ceiling panels, air ducts, overhead compartments, winglets, and fins, are often aluminum, stainless steel, or Nomex honeycomb which are made by aramid and Kevlar fiber-reinforced clothes.

Most plastic honeycombs mostly are made from polypropylene (PP), and some are polycarbonate (PC) and polyethylene terephthalate (PET), which can be processed by block and tube extrusion process, sheet thermal forming and corrugation process. Plastic honeycomb is mostly used in civil industries including the application of the hulls of boats and yachts, passenger ferries, catamarans, snowboards, truck bodies, and public transit buses. Most honeycomb cores are used by adhesive lamination or prepreg lamination. However, plastic honeycomb core can be laminated by a layer of film and veil by hot press, and then can be used for wet resin lamination, such as vacuum infusion and pultrusion process.

Honeycombs of craft paper and cardboard are typically light density. Paper honeycomb impregnated by phenolic resin can be used for hot press applications. During World War II, paper honeycomb was introduced in the airplane industry. After the war ended, paper honeycomb was used primarily as a structural material in the reconstruction of Europe. The shortage of building materials (wood, bricks, etc.) made paper honeycomb an excellent cost-effective available alternative. Now, most paper-based honeycomb composites are used for making panels for furniture, cabinetry and partitions of office cubicles, RVs, and boats.

Solid cores such as plastic foam and light wood are actually microcellular materials. Foams are rigid plastics and have closed micro cells with a cell size of 50–200 μm for most foam cores, and some microcellular foams have a cell size below 50 μm. The cell wall thickness is 2–5 μm. Balsa wood has a longitude cell structure with a length of about 600 μm, a diameter of about 30 μm, and a cell thickness of about 2 μm. The general-purpose foam cores for civil industries are polyethylene terephthalate (PET), polyvinyl chloride (PVC), polyurethane (PUR), styrene acrylonitrile (SAN), and polystyrene (PS) plastic foams. The foams for using in advance industries are polyetherimide (PEI), polyethersulfone (PES), and polymethacrylimide (PMI) foams.

A few special foams for sandwich are carbon, aluminum, ceramic, and syntactic. Carbon, aluminum, and ceramic foam cores are used with metal or carbon fiber composite faces to apply sandwich structures at high temperatures. Syntactic foam are composite materials synthesized by filling a metal, polymer, or ceramic matrix with hollow spheres called microballoons, which cure to a closed-cell foam-like matrix. These are typically used in applications that require very high fire resistance or extreme compression such as deep-sea vessels or undersea petroleum infrastructure.

Three-dimensional (3D) woven fabrics have been introduced and provide a viable alternative in sandwich construction. A basic common definition of 3D fabric is that these types of fabrics have a third dimension in the thickness layer. Fibers or yarns are intertwined, interlaced, or intermeshed in the X (longitudinal), Y (cross), and Z (vertical) directions. During laminating, the liquid resin will wet top and bottom face layers, and vertical direction glass fiber at the same time to generate a single-fabric system sandwich structure. When compared with materials such as foam and honeycomb, a 3D fabric core may provide many significant manufacturing advantages. As the material is provided as a fabric, it is initially very flexible and can be used easily as a core in nonconventional applications such as curved surfaces and sandwich pipes. Furthermore, as the FRP skins and core are cured at the same time, they will ideally have an improved structural unity, and this may eliminate the risk of delamination. Moreover, since the 3D fabric comes in a roll, it is a very easy to transport material and long lengths of composite beams can be produced without any seams or overlap in the core. These products have broad application prospects in automobile, locomotive, aerospace, marine, windmills, building, and other industries.

There are a number of thin, low-density "fabric-like" materials which are made by glass or polyester fiber mat full of density-reducing hollow plastic microspheres. The hollow spheres displace resin, and so the resultant middle layer, although much heavier than a foam or honeycomb core, is lower in density than the equivalent thickness of glass fiber laminate. They are usually only 1–5 mm in thickness and can be used as core to lower the density of a thin-wall sandwich lamination. They also are used like another layer of a laminate, being designed to "wet out" with the laminating resin during construction for improving the surface finish. Being so thin, they can also conform easily to 2D curvature, and so are quick and easy to use.

Hybrid core concept is to use different core materials for making one product based on different property requirements in different area and location of the product. Considerations for using different cores are the core properties including weight, stiffness, strength, damage tolerance, heat resistance, etc., the costs including raw material costs, resin uptake, finishing options, ease of processing, life cycle costs and the sustainability, such as CO_2 footprint, recyclability, and life cycle analysis, etc. For example, to make a wind turbine blade using sandwich structural composite, it is considered to combine balsa core with high stiffness and strength in the blade's root section along with low-density foam core for saving weight in the blade tip.

I.3.2 Overview of Different Facing Materials and Matrixes for Sandwich Structures

As stated before, the facings resist nearly all of the applied edgewise (in-plane) loads and flatwise bending moments. For low-density cores, the thin, separated facings

Introduction

provide nearly all of the bending rigidity to the construction. The core also provides most of the shear rigidity of the sandwich construction. By proper choice of materials for facings and core, constructions with high ratios of stiffness to weight can be achieved.

The facing materials can be almost any structural materials which are available in the form of thin sheet and can be used to form the facings of a sandwich panel. The properties of primary interest for the facings are high axial stiffness giving high flexural rigidity; high tensile and vertical compressive strength; impact and temperature resistance; environmental resistance; wear resistance; and optionally an attractive surface finish.

The commonly used facing materials can be divided into two main groups: metallic and nonmetallic materials. The metallic group is comprised of primarily steel, stainless steel, and aluminum alloys. Within each type of metal, there are a vast variety of alloys with different strength properties, whereas the stiffness variation is very limited. The second group is the larger in the two groups, including plywood, cement, veneer, reinforced thermal plastics, fiber-reinforced thermoplastic, and thermoset composites.

The facing can also be divided into another two groups: rigid sheets before laminating on to the cores including metal sheets, plywood and wood veneer, fiberglass reinforcement plastics (FRP) sheet, cement, and ceramic plates, which will be constructed with the cores by using adhesives and by a dry consolidation process, such as compression, vacuum bagging, and pinch roller pressing process; and facings formed during the laminating process on to the cores by a wet process. Most large sandwich products and most fiber-reinforced sandwich composites belong to the second group that are made by using thermosetting liquid resin, the dry fiber reinforcement materials, and the different wet laminating processes.

Wet lamination usually is used for making fiber-reinforcing composites in which the facing is fabricated during production process. The facing consists of dry fibrous materials, such as glass, carbon, synthetic, basalt, boron and ceramic, metallic and natural fibers. The matrix materials are the liquid thermosetting resins primarily, and to a lesser extent, thermoplastic resins that are gaining attention because of their recyclable properties. In the wet laminating process, three main materials are used for making the sandwich composite products: the core, reinforcement material, and liquid resin. Fabrication of the facing and bonding the facing to the core are completed in a one-shot process, such as vacuum infusion, resin transfer molding, chop and spray lamination, hand layup, hot press, and pultrusion. The fiber-reinforcing fabrics are the major materials used in the wet laminating process because they have a high Young's modulus thus high stiffness and are flexible and air permeable, and so the liquid resin can penetrate them and bond them to the core after curing to make sandwich composite products with a complicated shape, high stiffness, and unique mechanical properties.

The most commonly used fiber-reinforcing materials are glass, carbon, mineral, synthetic, and natural fibers. Of these, "E-glass," or electrical glass, fibers enjoy the greatest utilization as a reinforcement material in modern composite industries. Fiberglass in its current style was first made in the middle of the 1930s, and was used as a reinforcement material with unsaturated polyester for making composites in the early 1940s. Following that, many modifications have been accomplished for making new grades of glass fibers with a higher modulus, a wide variety of construction formats and many more applications. Currently, fiberglass is classified by chemical composition as E-glass, S-glass, A-glass, R-glass, and others. Glass-epoxy or glass-vinyl ester/unsaturated ester is used in the facings of civil and marine structures.

Carbon fiber having a high tensile strength and modulus was first discovered in the late 1950s and is produced by slowly carbonizing synthetic fiber filaments, such polyacrylonitrile fiber, at a high temperature in a controlled atmosphere. Carbon fibers have a much higher tensile modulus and much lower density than fiberglass. Due to their premium cost, carbon fiber reinforcements are mainly used for making high-performance composite products in applications for aerospace, sport goods, and military and medical equipment applications.

The greatest usage synthetic fibers in the composite industry are aramid fibers, made from aromatic polyamide, and SPECTRA® fiber, made from ultra-oriented, ultra-high-molecular-weight polyethylene (UHMWPE). These have properties of light weight, high strength, and high toughness. The key properties of fibrous aramid make it an important material in many different composite markets, such as bullet proof vests, aircraft body parts, and other products in advance industries.

Natural fibers are enjoying greater usage recently for reducing energy consumption and applications desiring the use of sustainable natural materials in the composite industries. Natural fibers include flax, hemp, jute, and kenaf, which are well adapted for making door panels, seat backs, and trunk liners of automobiles.

Other fibers used as reinforcement materials are mineral fibers, such as basalt fiber, ceramic fiber, and boron fiber, all of which have a higher tensile modulus and strength, and compressive strength, than fiberglass. Metal fibers are also used as reinforcement materials, such as fibers made by ferrous and nonferrous alloys including steel, stainless steel, bronze, aluminum, brass, and copper. Applications for metal fibers in sandwich composites include those requiring electrical conductivity and providing electromagnetic interference shielding.

There are different fiber formats mentioned above, but one widely used fabric is woven roving that is produced by interlacing continuous fiber roving into relatively heavy weight fabrics. Woven roving is generally compatible with most resin systems and used primarily to increase the flexural and impact strength of sandwich structural laminates. Multiaxial nonwoven stitched fabrics are ideally suited for selective reinforcement. Fibers are bundled and stitched together, rather than woven, so there is no crimp in the yarns. As a result, these reinforcements offer optimized directional strength for their finished composite parts. For this reason, they are popular in the wind energy industry. These fabrics are ideal for wet lay-up, pultrusion and vacuum infusion processing, to create strong, stiff sandwich composite parts, and are compatible with liquid resins.

The purpose of the resin system, or "matrix," is to bind the dry fibers of the reinforcement together into integrated composite facings, to transfer load from fiber to fiber, and to carry the load from the composite facings to the core. The choice of resin will depend upon many factors: the structural performance, lamination method, the cost, processing, the service environment expected, surface cosmetics, fire resistance, and the expected service life.

Introduction

Almost all of the resins used in sandwich composites are a combination of several resins and various additives. Hence, the term resin system is often used instead of resin. Such additives can raise or lower the viscosity, change the resistance to ultraviolet radiation, improve the inter-laminar shear strength as well as the toughness, increase the strength of resin rich areas in the laminate, increase or reduce translucence, and change the surface tension and or the wettability during the low viscosity period just before cure.

Unsaturated polyester resins are the most widely used in the sandwich composite wet lamination, thanks to their ease of handling, good balance of mechanical, electrical, and chemical properties, and relatively low cost. Typically coupled with E-glass fiber reinforcements, polyesters adapt well to a range of fabrication processes and are most commonly used in open-mold spray-up, compression molding, resin transfer molding (RTM), vacuum infusion, and pultrusion.

Vinyl ester is a resin produced by the esterification of an epoxy resin with an unsaturated monocarboxylic acid, most commonly methacrylic acid. The reaction product is then dissolved in a reactive solvent, such as styrene, to 35%–45% content by weight, offering a bridge between lower-cost, rapid-curing and easily processed polyesters and higher-performance epoxy resins. Since ester groups are susceptible to hydrolysis, less of these increase vinyl esters' resistance to water and chemically corrosive environments. Vinyl esters are favored in making the chemical tanks and the corrosion resistance equipment and also add value in structural laminates that require a high degree of moisture resistance (such as boat hulls and decks).

Epoxy resins have a chemical structure based on diglycidyl ether of bisphenol, cresol novolacs, or phenolic novolacs. Epoxies are not cured with a catalyst, like polyester resins, but instead use a hardener (also called a curing agent), such as polyfunctional amines, acids (and acid anhydrides), phenols, etc. The hardener (part B) and the base resin (part A) co-react in an "addition reaction" according to a fixed ratio. Epoxy resins contribute strength, durability, and chemical resistance to a sandwich composite. They offer high performance at elevated temperatures, with hot/wet service temperatures up to 121°C. Epoxy resin is known in the marine and wind energy industry for its incredible toughness and bonding strength.

Phenolic resins are based on a combination of phenol and formaldehyde or resorcinol. They find application in flame-resistant aircraft interior panels and in commercial markets that require low-cost, flame-resistant, low-smoke products, and excellent char yield and heat-absorbing characteristics. They have proven to be successful in aerospace applications, in components for offshore oil and gas platforms, and in mass transit and electronics applications.

Cyanate esters are chemical substances generally based on a bisphenol or novolac derivative, in which the hydrogen atom of the phenolic OH group is substituted by a nitrile or cyanide group. The resulting product with an -OCN group is called a cyanate ester. These are versatile matrices that provide excellent strength and toughness, allow very low moisture absorption, and possess superior electrical properties compared to other polymer matrices, although at a higher cost. Cyanate esters feature hot/wet service temperatures to 149°C and are usually toughened with thermoplastics or spherical rubber particles. They process similarly to epoxies, but their curing process is simpler because of their viscosity profile and nominal volatiles. Current applications in high-performance sandwich composites range from radomes, antennae, missiles, and nosecones to microelectronics sandwich composite products.

The choice of sandwich materials depends on the function of the structure, lifetime loading, availability, and cost. Graphite-epoxy and carbon-epoxy multilayered facings are typical in aerospace applications, while glass-epoxy or glass-vinyl ester/unsaturated esters are used in the facings of civil and marine structures. The core of aerospace structures is often aluminum or Nomex honeycomb. In civil engineering, the core is often a closed-cell foam and plastic honeycomb, while balsa of various density is a typical choice in ship and wind turbine blade manufacturing.

A major benefit of sandwich construction is the resin matrix, and fiber type and format need not be superlative. Many structures require stiffness, not strength, and do not need to be ultra-light. Sandwich permits the use of random fabrics even recycled fibers. This allows the engineer to design more cost-effective parts that are also easy to manufacture. Sandwich creates the opportunity for high-performance composites in applications where unit cost and process time had prevented broad adoption such as automotive and architecture.

I.4 DIFFERENT APPLICATIONS INCORPORATING SANDWICH

Generally speaking, the sandwich structural composites have the following applications:

Aerospace industry – because of their bending stiffness-to-weight ratio. Floorboards, composite wing, horizontal stabilizer, composite rudder, landing gear door, speed brake, flap segments, aircraft interior, and wingspans are typically made of sandwich composites.

Marine – hulls, stringers, bulkheads, decks, superstructures, interiors, and furniture. Racing boats, lifeboats, sailing boats, leisure yachts, and commercial vessels. The wide variety of marine applications utilizes the full range of composite materials and sandwich cores. Naval ships and submarines also use sandwich construction in transportation industry. The insulating, sound damping properties, and low-cost properties make them the choice materials for the constructions of walls, floors, doors, panels, and roofs for recreational vehicles, vans, trucks, trailers, and trains. Fire resistance is another important property for sandwich construction when required.

Architectural industry – excellent thermal and acoustical insulation of sandwich structural composite make it an ideal choice for it. Typical applications include structural columns, roofs, portable buildings, office partitions, countertops, and building facades. Increasingly modular construction methods are being adopted. This will further expand the use of composites and sandwich by building and construction.

Wind energy, turbine blades, shear webs nacelle housings, and spinners (center hub cover). Thanks to lightweight, low cost, and high stiffness of the sandwich structure, the wind turbine rotor can reach over 180 m.

Industrial applications, where corrosion resistance and performance in adverse environments are critical, for example, chemical tank, need corrosion-resistant structures and requiring less maintenance. The plastics based fiberglass reinforced sandwich composites are perfect for these uses.

Sandwich structure composites are now being used in infrastructures such as bridge decks, railway sleepers, bridge beams and bridge girder, and sea wall panel. And sporting goods include skis, snowboards, surfboards, canoes, and kayaks.

REFERENCES

1. S. Black, "Getting To the Core of Composite Laminates," *Composites Technology*, October 1 2003.
2. O.T. Thomsen et al, Ed., *Sandwich Structures 7: advancing with sandwich structures and materials*, Springers, 2005.
3. https://www.bavariayachts.com/.
4. J. Sloan, "Core for composites: Winds of change," *Composites Technology*, June 10, 2010.
5. https://www.sciencedirect.com/.
6. T. Gundberg, "Foam Core Materials in the Marine Industry," [Online], Available: https://www.boatdesign.net/articles/foam-core/. (Accessed Feb. 20, 2021).
7. J. Sloan, "Structural polyurethanes: Bearing bigger loads," *Composites Technology*, March 2010.
8. J. Sloan, "Wind foam sources: PET, SAN & PVC," *Composites Technology*, June 2010.
9. "Evonik Corporation Rohacell® 110 WF High Heat Grade Polymethacrylimide (PMI) Foam." http://www.matweb.com/search/datasheettext.aspx?matguid=bc292714f981423c904d2cc470191738.
10. I. P. Sabic, "Ultem foam for high-performance applications", October 2011, http://www.plasticwire.com.
11. http://www. diabgroup.com.
12. K. Ha, "Finite element analysis of sandwich plates: An overview," *Comput Struct*, 37 (4), pp. 397–403, 1990.
13. J. Austermann, J. Alec, A. Redmann, V. Dahmen, A. Quintanilla, S. Mecham and T. Osswald, "Fiber-reinforced composite sandwich structures by co-curing with additive manufactured epoxy lattices" *Journal of Composites Science*, vol. 3, no. 53, 2019.
14. V. Birmana and G. Kardomateasb, "Review of current trends in research and applications of sandwich structures", *Composites Part B*, vol. 142, pp. 221–241, 2018.

1 Sandwich Structural Core Materials and Properties

Wenguang Ma

Russell Elkin

CONTENTS

1.1 Rigid Structural Plastic Foams 2
    1.1.1 Requirements of Foam Core Materials 2
    1.1.2 Polyvinyl Chloride (PVC) Foam 3
    1.1.3 Polyethylene Terephthalate (PET) Foam 5
    1.1.4 Polyurethane (PUR) Foam 8
    1.1.5 Poly (Styrene-co-Acrylonitrile) (SAN) Foam 12
    1.1.6 Polymethacrylimide (PMI) Foam 13
    1.1.7 Polyetherimide (PEI) and Polyethersulfone (PES) Foam 15
    1.1.8 Syntactic Foam Core 17
1.2 Wood-Based Core Materials 18
    1.2.1 Balsa Wood-Based Core Material 19
        1.2.1.1 Balsa Tree 19
        1.2.1.2 Milling, Kiln Dry, and Make Block 20
        1.2.1.3 Microstructure 21
        1.2.1.4 Mechanical Properties of Balsa Lumber 21
        1.2.1.5 Mechanical Properties of Lumber-Based End-Grain Core 24
        1.2.1.6 Homologation and Density Variation 25
        1.2.1.7 Humidity, Moisture, and Its Effect on Mechanical Properties 26
        1.2.1.8 Miscellaneous Properties 28
        1.2.1.9 Product and Format 32
        1.2.1.10 Applications 33
    1.2.2 Cork-Based Sandwich Core 34
        1.2.2.1 Plantation and Harvest 34
        1.2.2.2 Mechanical Properties 35
        1.2.2.2 Other Properties and Applications 36
1.3 Honeycomb Cores 36
    1.3.1 Thermoplastics Honeycombs 38
    1.3.2 Metal Honeycombs 43
    1.3.3 Honeycombs Made from Composite Materials 48
    1.3.4 Paperboard Honeycombs 50

DOI: 10.1201/9781003035374-1

1.4 Special Foam Cores 54
    1.4.1 Metallic Foams 54
    1.4.2 Ceramic Foams 57
    1.4.3 Carbon Foams 59
1.5 Other Core Materials 61
    1.5.1 Corrugated and Lattice Truss Cores 61
        1.5.1.1 Corrugated Core 61
        1.5.1.2 Lattice Truss Core 62
    1.5.2 3D Fabric Woven Cores 64
    1.5.3 Core Mats 64
1.6 Sheet Formats of Core Materials 67
References 69

Sandwich construction consists of thin, stiff, and strong sheets of metallic or fiber composite materials separated by a thick layer of low-density material. The thick layer of low-density material, commonly known as the core, may be rigid foam, honeycomb, wood, truss, or lattice. Cellular foams are mostly polymer plastics but can also be metal and ceramics. Honeycombs consist of almost any type of material, as diverse as composite face sheets, but they are usually aluminum, thermoplastic, paper, synthetic fiber paper, and composite sheets. Natural-based core comes from balsa trees, but other light woods, such as paulownia and bamboo have been used. Other core materials, made by fabrics and textiles, have different functions in the composite industries. In this chapter, the processing method of different core materials will be introduced briefly. The special features and typical properties of each core will be discussed. The common formats of the core materials will be presented. Also, the applications of each core in the sandwich composite industries will be revealed.

After reading this chapter, the reader will know how to select the right core type, sheet format and thickness depending on the manufacturing process, and the desired performance of the overall laminate for the end-product application.

1.1 RIGID STRUCTURAL PLASTIC FOAMS

1.1.1 Requirements of Foam Core Materials

Many unique properties are required for structural sandwich cores; thus, prudent selection of the polymer from which the foam is produced is most important. Not every polymer can be foamed commercially, and this is why the polymers available for making core materials are few in number. First of all, the foam should be sufficiently stiff. As the core in a structural sandwich, its shear modulus is a representative indication of sandwich stiffness. PVC foam core with density of 40 kg/m^3, a general-purpose core material, has a shear modulus of 12 MPa, while PMI foam core, a high-performance core material, with a lower density of 30 kg/m^3 has a shear modulus of 13 MPa. However, a rigid PUR foam with a density of 70 kg/m^3 has only a shear modulus of 7.6 MPa, which is why PUR foam is a lower-performance core material.

Equally important are the surface properties of the foam core. It should have a sufficiently high surface energy so that it can be wetted out (compatible) with all the adhesives and liquid thermosetting resins that are used in composite industries, but also it should have good resistance against solvent attack so it cannot be softened by any solvents in the resin or adhesive. The close cell content of the foam core also is a critical property that should be, generally, more than 80%. If the open cell content is too high, the liquid resin will migrant into the core during laminating, increasing the weight of the sandwich laminate. In summary, the foam should have a high-enough surface energy to bond to with the adhesive or resin for laminating the facing but cannot be attacked by them physically or chemically. The bond strength must be high enough to withstand the constant bending, compressive, or tensile forces of dynamic loading, such as the forces on a boat hull or rotating turbine blade. The coefficient of thermal expansion (CTE) of the core, the laminate material, and resin or adhesive must be compatible to ensure that thermal cycling doesn't cause unequal expansions and contractions, leading to de-bonding.

The hot and cold temperature resistance is another important property of the structural core materials. During lamination, the core will endure a heat history because of the need to cure the resin at elevated temperatures and by the heat released by adhesive or resin curing (exothermic). The bond layer between facing and foam surface will shrink or swell if the foam core does not have a sufficiently high heat resistance, reducing the bond strength and possibly causing a laminate to fail. For the specific laminate application or service, the sandwich structure could be exposed to extremes in temperature environment. The foam core should maintain mechanical properties within an acceptable design range to ensure that the composite sandwich product performs the intended function.

The foam core also needs to be fatigue-resistant. A product made by using the foam cored sandwich construction, such as a boat or a wind turbine blade, will endure cyclical forces. For example, a helicopter or wind blade may be designed for over 10 million cycles in its service life. If the core cannot maintain sufficient mechanical properties for its design life in a dynamically loaded environment, the product will prematurely fail.

For specific applications, an important property is toughness or "ductility". It is an indication of the impact resistance of a core material and is reflected by the value of the foam's shear elongation at break (SEB). The rigid PVC core with a density of 40 kg/m^3 has a SEB of about 9%, which increases with increasing density. PMI has a SEB of 3% because of its high rigidity.

Additional properties to be considered for the foam core include fire resistance, water resistance, and appropriate cell structure depending upon the application. For applications in building construction and public transportation, the fire/smoke/toxic (FST) gas behavior is critical and usually must pass special tests, as required by the prevailing authority.

1.1.2 Polyvinyl Chloride (PVC) Foam

Of the various structural foam cores, one of the most commonly used is cross-linked PVC foam, which is actually a hybrid of PVC, polyurethane, and polyurea.[1-3] Cross-linked, or semi-rigid, foams are relatively stiff and strong, perform well at temperatures up to 120°C, and are resistant to styrene (except for very low density), so they can be used with unsaturated polyester and vinyl ester resins.

Another type of PVC foam is linear or made with a different polymer formulation, is more elastic than cross-linked varieties, and are widely used in marine applications, where it offers a high deformation before failure and excellent impact resistance. While linear PVC is easier to thermoform around complex curves, the tradeoff is somewhat lower mechanical properties, resistance to temperature, and styrene compared to the cross-linked version. Both offer good properties toward fatigue resistance.

Most of the cross-linked PVCs are used in applications of the wind energy and marine industries now. The flow chart in Figure 1.1 briefly shows the process of making rigid PVC foam. PVC and other materials, which include one or two different isocyanates, one or two different anhydrides, blowing agent and surfactants, are mixed thoroughly into a plastisol at room temperature, under conditions to avoid contact with moisture. The plastisol is injected into aluminum or steel molds, and the molds are heated in a multiple-opening press at 170°C–175°C under a pressure of 200–400 bar for about 20 minutes. The partially cured but unfoamed products, called embryos, are cooled and released from molds. They are then put into a chamber for expanding into a foam plank with steam heating at about 90°C–100°C for 24 hours. Finally, the foam planks are kept in a steam heating room at 40°C–70°C for several days to several weeks, depending upon target density, for finishing the expansion and curing reactions. After cutting off the outer skin and trimming off edges, the planks, called the foam buns, are ready for slicing into foam sheets that can be processed into the different core products.

The typical mechanical properties of PVC foam cores are listed in Table 1.1. Several excellent properties of the PVC foam cores are the good fatigue resistance, high ratio of stiffness and strength to weight, and resistance to styrene and other chemicals, making it suitable for processing with unsaturated polyester, vinyl ester, and epoxy resins. The limitations of PVC foam core are its modest heat resistance and toxic gases liberated during burning, related to the PVC resin. Also, PVC foam contains carbon dioxide gas under pressure, so outgassing can occur at elevated temperatures over time – that is, the gas diffuses from the closed cells and migrates to voids or unbounded areas in the laminate. In some instances, outgassing has been blamed for delamination and blistering in the sandwich construction, especially in parts made at elevated cure temperatures or finished in dark colors and in direct exposure to the sun.

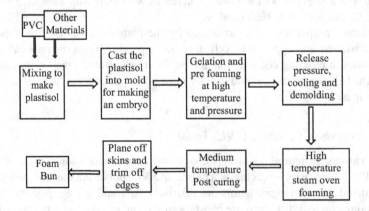

FIGURE 1.1 Production flow chart of rigid PVC foam.

TABLE 1.1
Typical Mechanical Properties of Rigid PVC Foam Cores[4]

Apparent Density	ISO 845	Kg/m³	48	60	80	100	130	200
		lb/ft³	3.0	3.8	5.0	6.3	8.1	12.5
Compressive strength	ASTM 1621B	Mpa	0.65	0.85	1.35	2.21	3.02	4.81
		Psi	94	123	196	321	438	698
Compressive modulus	ASTM 1621B	Mpa	52	71	92	136	172	242
		Psi	7,542	10,298	13,343	19,725	24,947	35,099
Tensile strength	ASTM C297	Mpa	1.42	1.61	2.21	3.52	4.78	7.15
		Psi	206	234	321	511	693	1037
Tensile modulus	ASTM C297	Mpa	55	75	95	132	175	253
		Psi	7,975	10,875	13,775	19,145	25,375	36,695
Shear strength	ASTM C273	Mpa	0.55	0.70	1.11	1.63	2.19	3.52
		Psi	80	102	161	236	318	511
Shear modulus	ASTM C273	Mpa	16	22	31	36	51	86
		Psi	2,321	3,191	4,496	5,221	7,397	12,473
Shear elongation	ASTM C273	%	12	20	30	40	40	40
Water absorption	ISO 2896	%	1.9	1.8	1.6	1.5	1.4	1.2

1.1.3 POLYETHYLENE TEREPHTHALATE (PET) FOAM

PET foam core is the first type of thermoplastic structural foam core made by the continuous extrusion foaming process. It is also the second largest volume foam used in the sandwich composite industries currently. Because of being totally made by thermoplastics, the foam's production waste, scrap, and products at end of service lifetime can be recycled by conventional extrusion melting and re-pelletizing processes. Currently, the wastes of PVC, PUR, and other semi- and fully thermosetting structural foam impose a large negative impact to the environment. PET foam is one of the most ecological solutions for reducing composite's environmental footprint.

The production route of PET foam is shown in Figure 1.2.[5] Unlike polystyrene foam, PET foam is not easily processed into thick foam planks by using a slot die extrusion method. The commercially successful process uses a multiorifice die and a simple device to shape the multistrands together into a thick foam plank. So, PET foam currently in the market is also called PET "strand" foam core. The process of using a slot die continues in the industry, but the resulting planks are of limited thickness.

The extruded coalesced-strand foam planks can be used directly as core materials after abrading or planing their facing surfaces as shown in Figure 1.3 or as components for constructing larger foam blocks by thermo-welding. The welded blocks can be cut into "end-grain" sheets (Y-Z plane) or "flat-grain" sheets (X-Z plane), as defined graphically in Figure 1.4, such that those sheets containing welds can be used as core materials. The PET strand foams, particularly at lower densities, are significantly anisotropic, meaning that they exhibit different physical properties depending upon which axis (direction) the property is measured. By way of definition, the phrase "strand direction" means the axis (X-direction in Figure 1.4) along

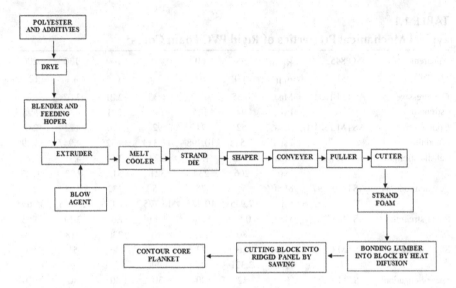

FIGURE 1.2 Production flow chart of PET foam core processing.

FIGURE 1.3 PET strand foam planks.

which the foam plank was extruded during production. The "strand direction" influences the anisotropic physical properties of the PET strand core materials.

The phrase "end-strand direction" is used in reference to structural sandwich composites wherein the sandwich facings are bonded to a Y-Z plane core sheet and thus are oriented perpendicular to the core's strand direction. In such circumstances, the "end-strand direction" incurs a shearing force during sandwich deformation that is perpendicular to the strand direction. Simply put, the strands run in the thickness direction of the core material. The phrase "longitudinal direction" is used in

Sandwich Structural Core Materials 7

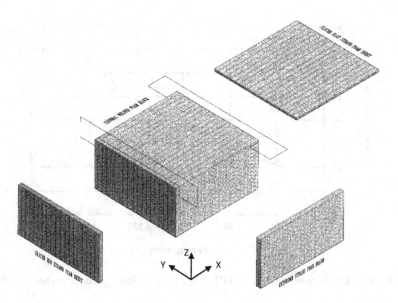

FIGURE 1.4 PET strand foam block from welding planks and sheets cutting from block.

reference to composites wherein the sandwich facings are bonded to an X-Z plane core plank or an X-Y plane core sheet such that the skins are parallel to the strand direction. In such circumstances, the "longitudinal direction" incurs a corresponding shear force direction that is parallel to the strand direction.

The phrase "transverse direction" is used in reference to the same structural sandwich as the most previous situation. In such circumstances, the "transverse direction" means a shear force direction or axis perpendicular to the strand direction, where again the strands are in the plane of the core material. As a core material, PET strand foam exhibits mechanical properties that strongly depend upon the axis along which the property was measured. For example, the shear strength for the core materials of this work in the end-strand direction differs greatly from the shear strength in the transverse direction, which in turn differs from the shear strength in the longitudinal direction. Referring to Figures 1.5 and 1.6, the PET coalesced-strand foams exhibited different compressive and shear properties depending upon the direction along which that property was measured. Although the strand character of the core materials may have not been visible for some of the higher density variants, the anisotropic character is readily apparent in the results of the mechanical property testing.

Table 1.2 lists the typical mechanical properties of the end-strand PET foam cores that are mostly available in the composite industry.[6] Because of its behavior as a semi-crystalline resin with an approximate 250°C melting temperature, the PET foam can be used with almost all thermosetting resins employed in the composite industries and can be processed at a relatively high temperature without any out-gassing. In the event of fire, no toxic gases are generated. PET foam core is a cost-effective, closed-cell foam specially formulated to provide high mechanical properties. Other advantages include carrying static and dynamic loads, resisting to fatigue superiorly, excellent chemically stability and UV resistance, and minimal water absorption.

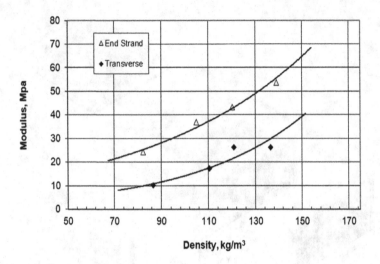

FIGURE 1.5 Compressive modulus of PET strand foam vs. density in different test directions.

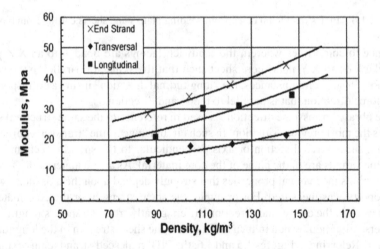

FIGURE 1.6 Shear modulus of PET strand foam vs. density in different test directions.

Because PET foam core is made by thermoplastic material, it can be thermoformed by heating the foam sheet to about 180°C for 10 minutes, and then using conventional shaping methods such as vacuum or pressure forming over tooling, to produce a contoured core with desired shape. The photo in Figure 1.7 illustrates some thermoformed PET foam sheets; the core's cell structure and configuration remain in good shape after forming.

1.1.4 Polyurethane (PUR) Foam

Rigid PUR foams have been used as structural foam cores for many years in the marine industry. There are two types of rigid PUR foam: polyisocyanurate (PIR) formulations and polyurethane (PUR). These designations are misleading, however,

Sandwich Structural Core Materials

TABLE 1.2
Typical Mechanical Properties of PET Strand Foam

Density	ISO 845	kg/m²	Average Type	65	110	145	210
			Range	60–70	105–115	140–150	200–220
Compressive strength	ISO 844	N/mm²	Average	0.80	1.4	2.2	3.5
perpendicular to the plane			Minimum	0.7	1.2	2.0	3.2
Compressive modulus	DIN	N/mm²	Average	50	85	115	170
perpendicular to the plane	53421		Minimum	35	75	100	145
Tensile strength	ASTM	N/mm²	Average	1.5	2.2	2.7	3.0
perpendicular to the plane	C297		Minimum	1.2	1.6	2.2	2.4
Tensile modulus	ASTM	N/mm²	Average	85	120	170	225
perpendicular to the plane	C297		Minimum	70	90	140	180
Shear strength	ISO	N/mm²	Average	0.46	0.8	1.2	1.85
	1922		Minimum	0.4	0.7	1.1	1.5
Shear modulus	ISO	N/mm²	Average	12	20	30	50
	1922		Minimum	10.5	18	26	44
Shear elongation at break	ISO	%	Average	25	10	8	5
	1922		Minimum	15	5	4	3
Thermal conductivity at	ISO	W/mK	Average	0.033	0.033	0.036	0.041
room temperature	8301						

FIGURE 1.7 Thermoformed PET foam sheet.

because both are polyurethanes – that is, they are polymers formed by reacting a monomer containing at least two isocyanate functional groups with another monomer having at least two hydroxyl groups. The resulting polymer is a chain of organic, monomeric units joined by urethane (carbamate) linkages. In PIR foam, the proportion of methylene diphenyl diisocyanate (MDI) is greater than for PUR. The PIR reaction also uses a polyester-derived polyol in place of a polyether polyol. PIR foam typically has an MDI/polyol ratio (also called its index) of between 200 and 500, whereas PUR indices are normally about 100. The net result is that PIR foams are more cross-linked and, therefore, are stiffer but more friable than PUR foams.[7]

PIR foams, sometimes called "trimmer" foams, generally are low density (30–100 kg/m^3) with high insulating values and good compressive strength. Friability, however, limits their utility in sandwich panel applications, because this lack of toughness can result in failures at the foam-to-facing bond especially under dynamic loads. As a result, structural use of these foams is usually limited to two roles: (a) as core-carrier material for glass-fiber reinforcement or (b) to provide an internal mold shape for laminate overlays.

Considerably tougher (less friable) than PIR foams, PUR foams are produced in densities ranging from 30 to 500 kg/m^3, depending upon the formulation, and can retain a substantial strength and toughness up to 135°C, which allows their use with high-temperature curing prepreg in ovens and autoclaves. Rigid PUR foam is also made using polyester-derived polyols for increased rigidity. PUR foam is widely used in structural sandwich shapes, such as boat transoms and stringers, in foam cored RTM parts and composite tooling, and as an edge close-out in honeycomb cored aircraft sandwich interior panels.

A third category, a blend of PIR and PUR foams, attempts to get the best of both worlds. These foams provide some improvement in strength compared to PUR foams, with reduced friability relative to PIR foams.

PUR foam can be produced by batch and continuous process to produce foam buns that will be sliced into any thickness. The continuous process can also be used for making sandwich panels by using coiled sheets of the skin materials. A batch process is shown in Figure 1.8 by using hand mixing and a mold with the lid.[8] The mold also

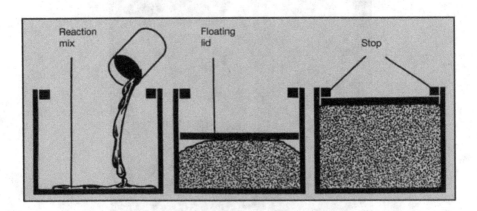

FIGURE 1.8 Batch process for making PUR bun by using a close lid mold.

Sandwich Structural Core Materials

may have an open top, and the walls of mold can be hinged together, or the hinges are released for allowing the foam to expand freely in three direction. If a completely closed mold is used, a rectangular-shaped foam bun will be produced. On closed molds, adjusting the pressure applied to the mold controls the density of the foam.

The principle of the continuous mixing and production of foam slab stock is shown in Figure 1.9, which is the most economical method to produce large sections of foam. The side walls are necessary for controlling expansion and ensuring uniform density and cell structure. To avoid crown formation and reduce cutting losses, the flat-topped slab method is mainly used for continuous production. With this process a disposable paper web is fed above the rising foam zone and is serves as a separator to slide the continuous bun along the immobile upper platen.

The typical mechanical properties of rigid PUR foam are listed in Table 1.3.[9] In addition to the weaker polymer, urethane foams suffer from nonuniform cell structure. The performance is far lower than other polymer foams at equal density.

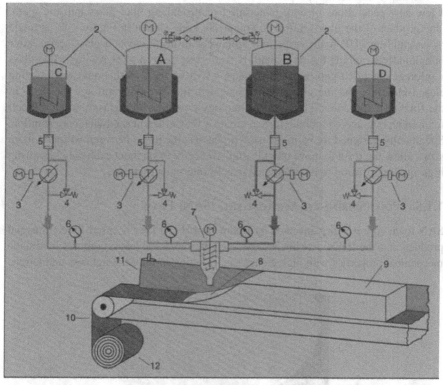

A: Polyol mix
B: Polyisocyanate
C: Blowing agent
D: Activator

1: Pressure relief valve
2: Agitator tank
3: Metering pump
4: Safety valve
5: Filter
6: Manometer

7: Agitator mixing head
8: Reaction mix
9: Reacted foam
10: Lower paper
11: Side paper
12: Paper roll

FIGURE 1.9 Diagram of continuous producing PUR rigid foam slab stock.

TABLE 1.3
Typical Mechanical Properties of PUR Rigid Foam[9]

Property	Test Method	Unit	Typical Dada								
Density	ASTM D1622	kg/m³	80	100	130	160	240	290	320	400	460
Compressive strength	ASTM D1621	MPa	0.8	1.2	2.0	2.5	5.0	6.0	9.0	12.0	14.0
Compressive modulus	ASTM D1621-B	MPa	25	40	50	65	120	160	200	250	340
Tensile strength	ASTM D1623	MPa	1.3	2	2.3	2.9	4.5	5.5	6.8	7.9	10.3
Shear strength	ASTM C273	MPa	0.6	0.9	1.2	1.8	3.4	4.7	5.3	7.1	9.3
Shear modulus	ASTM C273	MPa	9.6	16.4	30.0	32.0	43.0	61.2	74.8	90.5	143.0

However, because the PUR foam has a relatively low fatigue resistance, especially in the lower density range, only the higher density PUR foams are used for structural application, such as boat transoms, stringers, etc.

For improving certain properties of PUR foam, some modifications have created a few new generations of core materials by introducing fiber glass into the foam. One product of the fiber-reinforced composite core, shown in Figure 1.10, is made by assembling PUR foam strips into a plank with heat activated glass scrim, after individually wrapping the strips with biaxial glass or even glass roving.[10] The glass reinforcement filled gap between the strips will be filled with liquid resin during closed molding, and the glass reinforcement in the gap will reinforce the foam in the thickness direction. This core technology combines fiberglass and closed-cell foam in an engineered architecture to create a very efficient sandwich core solution. Specifically designed for resin infusion processes, the joints between wrapped foam strips allow the resin to move quickly and efficiently throughout each panel ensuring high-quality infusion of the sandwich panel's internal structure.

1.1.5 Poly (Styrene-co-Acrylonitrile) (SAN) Foam

SAN foam core was first invented by Rohm GmbH and later licensed for commercial production in the 1990s. The linear closed-cell thermoplastic foam combines good static mechanical properties with high elongation, resulting in impact toughness and fatigue

FIGURE 1.10 PUR foam reinforced by assembling foam strips wrapped with the glass roving.[10]

Sandwich Structural Core Materials

resistance – in essence, delivering the improved performance over both cross-linked and linear PVC. Due to its high acrylonitrile content, SAN has quite good chemical resistance and thus can be used with most of the resin systems used in composite construction.[1]

SAN foam can be processed at temperatures up to 85°C, which accommodates lower temperature curing epoxy prepreg systems. SAN foam also has excellent strength, buoyancy, and insulating values in subsea structural laminates subjected to hydrostatic loads, such as remotely operated vehicles for the offshore oil industry. The high-temperature-resistant version of SAN foam offers high rigidity and thermal stability for process temperatures as great as 110°C.

SAN foam is made by first polymerizing between two, tempered glass sheets a blend of styrene, acrylonitrile with polyurethane chemistries and blowing agent shown in Figure 1.11.[11] After removal of the glass sheets, the clear, unfoamed sheet is expanded at a temperature in the range of 50°C–70°C. Typical properties of SAN foam are listed in Table 1.4. Sandwiches cored with SAN foam give the highest heat resistance when used with epoxy prepreg and vacuum infusion lamination. However, since the foam is a copolymer of styrene monomer, it has a limitation to use with the unsaturated polyester and vinyl ester at elevated temperatures, such as for pultrusion and heated RTM processing. Due to its partial polyurethane structure, SAN foam is not recyclable by traditional extrusion melting processes into postconsumption products.

1.1.6 Polymethacrylimide (PMI) Foam

PMI foam cores, commercially available in 11 different grades, exhibit the highest mechanical properties of any polymer foam core, at a comparable density. For example, the shear modulus and tensile modulus are 20% higher for PMI foam than PVC foam. Heat distortion temperature ranges from 177°C to 235°C, the highest for any foam core product, which makes PMI foam suitable for laminates with high cure temperatures as well as high, sustained temperature applications.[1]

Because of its high-temperature performance, low-smoke evolution during burning, and halogen-free chemistry, PMI foams have been used in public transportation applications such as high-speed marine vessels and trains, where fire performance is

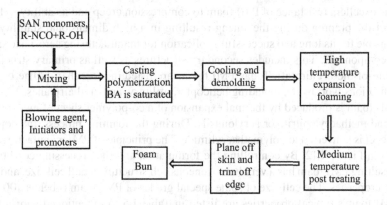

FIGURE 1.11 SAN foam manufacturing processes.

TABLE 1.4
Typical Mechanical Properties of SAN Foam[12]

Nominal density		Kg/m^2	71	94	104	115	143
		lb/ft^4	4.4	5.9	6.5	7.2	8.9
Density range		Kg/m^2	66–75	89–99	100–107	108–122	133–153
		lb/ft^4	4.1–4.7	5.6–6.2	6.2–6.7	6.7–7.6	8.3–9.6
Compressive strength	ASTM D1621	MPa	0.88	1.41	1.67	1.98	2.85
		Psi	128	205	242	287	413
Compressive modulus	ASTM D1621-1973	MPa	62	101	120	143	209
		Psi	8,992	14,649	17,405	20,740	30,313
	ASTM 1621-2004	MPa	45	59	79	90	119
		Psi	6,527	10,008	11,458	13,053	17,259
Shear strength	ASTM C273	MPa	0.81	1.15	1.30	1.47	1.93
		Psi	117	167	189	213	280
Shear modulus	ASTM C273	MPa	26	40	46	52	70
		Psi	4,061	5,802	6,672	7,542	10,153
Shear elongation at break	ASTM C273	%	24%	17%	15%	13%	10%
Tensile strength	ASTM D1623	MPa	1.30	1.72	1.91	2.11	2.62
		Psi	189	249	277	306	380
Tensile modulus	ASTM D1623	MPa	85	118	134	151	196
		Psi	12,328	17,114	19,435	21,901	28,427
Thermal conductivity	ASTM C518	W/mK	0.03	0.04	0.04	0.04	0.04
HDT	DIN 53424	°C	100	100	100	100	100
		°F	212	212	212	212	212

regulated by the authorities. These foams have been incorporated into a number of intercity trams and newer high-speed trains, in both interior and structural exterior applications. PMI foam can be thermoformed to conform to the complex part shapes and can be used with epoxy prepreg that cure up to 125°C–175°C for the train's construction. The entire thermoforming and lay-up can be vacuum-bagged and co-cured in concert.

The excellent resistance of PMI foam to compression creep and its ability to fully support the prepreg during the curing resulting in a high dimensional stability are responsible for its long and successful application for manufacturing aerospace sandwich components. This includes secondary structures as well as primary structures for state-of-the-art aircraft. Today PMI foams are listed in more than 170 aerospace specifications worldwide, including helicopter, military, and civil airplanes.[13]

PMI foam is produced by thermal expansion of a co-polymer sheet of methacrylic acid and methacrylonitrile or acrylonitrile. During the foaming process, the copolymer sheet is converted to polymethacrylimide. The principle of PMI manufacturing is shown in Figure 1.12. By controlling the formulation and the processing conditions, the resulting PMI foam has a very homogeneous cell structure, small cell size, and isotropic properties. The cell size of some special grades of PMI foam is below 100 μm.

PMI foam's typical properties are listed in Table 1.5. As mentioned above, PMI foam can offer very low-density foam but still has good mechanical properties.

Sandwich Structural Core Materials

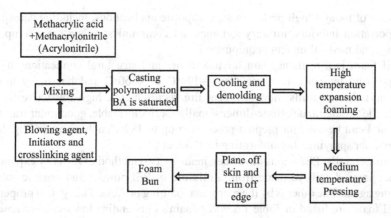

FIGURE 1.12 PMI foam manufacturing processes.

TABLE 1.5
Typical Mechanical Properties of PMI Foam[14]

Density	kg/m^3	32	52	75	110	ISO 845
	lbs./cu.ft.	2.00	3.25	4.68	6.87	ASTM D 1622
Compressive strength	MPa	0.4	0.9	1.5	3.0	ISO 844
	Psi	58	130	217	435	ASTM D 1621
Tensile strength	MPa	1.0	1.9	2.8	3.5	ISO 527-2
	Psi	1.45	275	406	507	ASTM D 638
Shear strength	MPa	0.4	0.8	1.3	2.4	DIN 53294
	Psi	58	116	188	348	ASTM C 273
Elastic modulus	MPa	36	70	92	160	ISO 527-2
	Psi	5,220	10,150	13,340	23,200	ASTM D 638
Shear modulus	MPa	13	19	29	50	DIN 53294
	Psi	1,885	2,755	4,205	7,250	ASTM C 273
Strain at break	%	3	3	3	3	ISO 527-2
						ASTM D 638

However, because of its high price and low productivity, PMI foam applications are limited to those requiring a very low weight, high mechanical properties, high temperature resistance, and low cost sensitivity.

1.1.7 POLYETHERIMIDE (PEI) AND POLYETHERSULFONE (PES) FOAM

PEI foam is a rigid lightweight, low flame, smoke, and toxicity foam core based on the polyetherimide polymer. The material combines a high strength-to-weight ratio with extremely low moisture absorption. The foam also possesses excellent dielectric properties. The foam is targeted at applications where structural fire performance, radar transparency, or performance under extreme hot or cold environments are required. PEI foam is thermoformable and compatible with phenolic and bismaleimide prepregs. These properties make it an excellent candidate as a sandwich core for many of today's high-performance composite applications, including aerospace/transportation interiors, military radomes, telecommunication systems, composite tooling, and medical imaging equipment.[15]

PEI foam is a resilient foam for use in fire and structural applications within an operating temperature range from −194°C to +160°C and has many unique characteristics − fulfills most stringent fire requirements, high impact resistance (nonbrittle failure mode), three-dimensionally thermoformable, good radar transparency, and can be used for prepreg processing up to 180°C, adhesive bonding, hand lay-up, fiber spraying, thermoforming method, etc.

Commercially, PEI foams have been made by two methods. One is by expanding the hot-pressed embryo made from a mixture of PEI powder and organic solvent blowing agent. The other is by the extrusion foaming process. The typical properties of PEI foam are listed in Table 1.6.[16] PEI foam's outstanding low moisture absorption gives a decisive advantage compared to PMI foam during high temperature processing. Often, PMI foam boards must be conditioned (dried and/or stored in a special area) before they can be machined, compression molded or thermoformed. This extra step adds time, costs, and overhead to the process. PEI foam avoids this scenario. Further, PMI foam may require a multistep annealing process. As one of its detractors, the closed cell percentage is quite low for PEI foam, rendering it unsuitable for vacuum infusion processing unless it is first sealed with a coating.

Low moisture absorption combined with the proven, excellent flame-smoke-toxicity (FST), dielectric, acoustic, and thermal performance of PEI foam is an ideal thermoplastic foam solution for the aircraft industry. Applications include luggage bins, galleys, and lower wall panels.

TABLE 1.6
Typical Mechanical Properties of PEI Foam[16]

Density	ISO 845	kg/m^3	Average	60	80	110
			Typ. range	54–69	72–95	99–126
Compressive strength perpendicular to the plane	ISO 844	N/mm^2	Average	0.70	1.1	1.4
			Minimum	0.60	0.9	1.2
Compressive modulus perpendicular to the plane	DIN 53421	N/mm^2	Average	46	62	83
			Minimum	40	56	60
Tensile strength in the plane	ISO 527 1-2	N/mm^2	Average	1.7	2.0	2.2
			Minimum	1.2	1.7	1.9
Tensile modulus in the plane	ISO 527 1-2	N/mm^2	Average	45	54	64
			Minimum	35	50	54
Shear strength	ISO 1922	N/mm^2	Average	0.80	1.1	1.4
			Minimum	0.65	0.9	1.15
Shear modulus	ASTM C393	N/mm^2	Average	18	23	30
			Minimum	15	20	25
Shear elongation at break	ISO 1922	%	Average	25	23	18
			Minimum	15	15	10
Impact strength	DIN 53453	kJ/m^2	Average	1.0	1.3	1.4
Thermal conductivity at room temperature	ISO 8301	W/mK	Average	0.036	0.037	0.040

In addition, moisture absorption itself can have a disruptive effect on electronics (interference) and may cause condensation on sensitive areas of the interior. The cycle of moisture absorption and drying that occurs as the aircraft travels through different environmental conditions (altitudes) also has the potential to cause delamination of a composite structure and can distort the dimensions of a part. Such results can lead to more frequent repairs and downtime.

Polyether sulfone (PES) foam is made from extrusion of thermoplastic polyether sulfone, which is more desirable for applications of foam from its superior thermal stability and mechanical properties. It has a unique high service temperature for thermoplastics, so that its mechanical properties can be retained at temperatures up to 200°C for thousands of hours. It offers excellent fire, smoke, and toxicity properties. Furthermore, it is nonhygroscopic, superior damage/impact performance and improved dielectric properties. PES offers a lot of benefits in several aerospace segments, which usually focus on honeycombs due to fire regulations and prepreg compatibility, and also offers other benefits such as heat and cold forms and very low water absorption. Since it is foam, it is easy to shape and trim to desired shape, which takes away the need of edge filling like honeycomb cored panel. The closed cell structure reduces moisture uptake over time considerably. Process scrap such as off-cuts in production can be recycled by typical extrusion process because PES is a thermoplastic.[17] Typical mechanical and dielectric properties are listed in Table 1.7. Extra product characteristics include excellent FST properties, low heat release rate, outstanding hot/wet performance, high temperature resistance, low water absorption, hot and cold formable, superior damage tolerance, and good chemical resistance.

1.1.8 Syntactic Foam Core

Syntactic foams are composite materials synthesized by filling a metal, polymer, or ceramic matrix with hollow spheres called microballoons or cenospheres or non-hollow low-density filler such as perlite. In this context, "syntactic" means "put together". The presence of hollow particles results in relatively low density, lower coefficient of thermal expansion, and, in some cases, radar or sonar transparency.[18]

Syntactic foams have long been used as buoyancy materials in subsea applications due to their extremely high hydrostatic strength and stiffness. Most cellular foams can

TABLE 1.7
Typical Properties of PES Foam Cores[17]

Property	Test Method	Unit	Typical Dada			
Density	ASTM D1622	kg/m^3	40	50	90	130
Compressive strength	ASTM D1621	Mpa	0.4	0.6	1.2	1.7
Compressive modulus	ASTM D1621-B	Mpa	9	18	34	60
Tensile strength	ASTM D1623	Mpa	1.5	1.9	0	3.3
Shear strength	ASTM C273	Mpa	0.6	0.8	1.4	1.7
Shear modulus	ASTM C273	Mpa	8.5	13.3	24.0	30.0
Dielectric constant	ASTM D2520 A	-	1.06	1.06	1.13	N/A
Loss tangent	ASTM D2520 A	-	0.001	0.0009	0.0022	N/A

only be produced in a range of density too low for the structural requirements. Some are anisotropic, which limits their strength under hydrostatic compression. Syntactic foam products provide designers a source of lift for vehicles and structures operating in the deepest ocean environments. Manned submarines, AUVs (Autonomous Underwater Vehicles), and ROVs (Remotely Operated Vehicles) all rely on syntactic foam in performance of their missions.

Syntactic foams are excellent as low- to moderate-weight core materials for composite structures. The toughened microsphere matrix provides far greater stiffness, strength, impact, and shock properties than standard core materials at medium density range. It is also ideally suited for composite-to-metal joints (through the practice of density matching), a well-documented area of concern in all composite structures. As a closed cell structure, the core easily processes in all composite applications including vacuum bag, RTM, and pultrusion for reducing overall density.

Syntactic foam cores can be made in small scale in the laboratory for a research project and for a low volume application, can also be made in large scale in composite plant for high volume application, or can be purchased from professional core suppliers in the composite industry. A research group reported that using epoxy as binder and microglass bubble at average outer diameter of 40 μm as matrix made five different syntactic foam slabs with density range from 200 to 460 kg/m^3.[19]

Nominal mechanical properties of two syntactic foam cores are listed in Table 1.8. Usually, syntactic foam core is used in special applications for making sandwich structures with high mechanical properties but with relatively heavy weight.

1.2 WOOD-BASED CORE MATERIALS

Wood, due to its cellular structure and high strength-to-weight ratio is well-suited for use as sandwich core. To be a commercially practical and sustainable product, the tree must produce lumber with a low-enough density along with a short growing cycle. Species such as Aspen, Poplar, Redwood, Paulownia, cork, and bamboo have been used in specific applications such as skis, snowboards and impact limiters where their individual traits are a best fit. Generally their usage is very limited to these specialized applications. The most widely supplied wood for use as a sandwich core material is balsa.

Often plywood is considered to be a core material as it is used in many fiberglass parts to build thickness. Plywood has a similar flexural modulus to E-glass FRP and

TABLE 1.8
Nominal Mechanical Properties of Two Syntactic Foams on the Market[20]

Property	Nominal Density	Compressive Strength	Compressive Modulus	Shear Strength	Shear Modulus	Shear Elongation at Break	HDT
Test method	ASTM D1622	ASTM D1621	ASTM D1621	ISO 1922	ISO 1922	ISO 1922	DIN 53424
Unit	kg/m^3	Mpa	Mpa	Mpa	Mpa	%	°C
Nominal property	210	4.71	293	2.91	98	13	100
	315	9.17	515	6.21	157	7	110

would therefore not meet the definition of a sandwich core material. If plywood is incorporated into a laminate, it should be analyzed using classical lamination theory. Note the shear strength of plywood is fairly low, especially considering the density is well above 500kg/m^3. This is due to certain veneers in "rolling shear" under load.

1.2.1 BALSA WOOD-BASED CORE MATERIAL

1.2.1.1 Balsa Tree

The Balsa tree (*Ochroma lagopus*) is unique among natural, wood-based core materials due to its low-density and cellular structure. Although very soft, Balsa is a hardwood (botanically, hardwoods have seeds enclosed in the ovary of the flower[21] with the same microstructure to species such as cherry, ash, and oak.

The word balsa is Spanish for "raft", and the tree was used in the construction of rafts for thousands of years by a number of different cultures in western South America. Large-scale commercial use of balsa rafts ended in the late 1800s, though you may still be able to spot the occasional one on a river today. In the beginning of the 20th century, early uses of balsa included glider and aircraft construction, insulation, life preservers, and model wood. World War II brought vastly increased demand for balsa with the United States importing 91% of all trees harvested.[22] Post war, balsa began to be used in composite sandwich construction for boats hulls, military and commercial aircraft, and as structural insulation for liquefied natural gas tanker ships. Today, balsa is one of the most broadly used sandwich core materials in the world in an assortment of products including boats (from runabouts to Naval vessels), trains, buses, cars, tanks, cargo pallets, snowboards, and wind turbine blades. A particular balsa-based product may be designed for almost any composite application.

The balsa tree can produce lumber with a density as low as 70 kg/m^3 and as high as 350 kg/m^3. Most other trees produce wood with a far more narrow density range. The wide variety of lumber density is a benefit but can be a challenge when it comes to supply of sandwich core. A product with "soup-to-nuts" lumber increases the available supply of the core material but may reduce the design properties due to the variation in lumber density and may be more difficult to machine and fabricate laminates with uniform quality.

The geographical location of the forest or plantation and age when the tree is harvested can influence both the average density and the density range. Typically, 85–90% of harvested balsa lumber will have a density range between 100 and 250 kg/m^3 and an average density of 150–160 kg/m^3.

The balsa tree must be grown within the bands of latitude close to the equator; Tropic of Cancer and Capricorn (+23.43°). Most balsa comes from Ecuador, but it may be grown in the surrounding countries of Peru, Columbia, and Brazil, in Central America and as far north as the Yucatan peninsula. In the Eastern hemisphere, almost all balsa is grown in Papua New Guinea. Ecuador and PNG represent 99% of the total. If one thinks geographical restrictions will imperil long-term supply, the potential growing area for balsa is millions of acres, many times greater than present farming levels.

The country of origin will influence how the tree grows, affecting the volume and average lumber density. The factors are elevation (0–1,000 m above sea level), type of soil (well drained, clay and loamy), and average rainfall (500–3,000 mm per year).[23]

FIGURE 1.13 Picture of balsa nursery and mature plantation.

The particular location within the country can also influence the average lumber density as well as yield per acre. This creates opportunities for a balsa supplier to tailor (somewhat) the available volume and average density year over year but can also make it more challenging to meet demand when nature does not cooperate with customers' forecasts.

Time to harvest will also affect what is produced by the tree. Younger trees are generally lower in average density, but also smaller. Suppliers must balance the annual demand and requests for wood with a specific range of density vs. future available supply. This is best accomplished by growing the trees on plantations.

Balsa lends itself quite well as a renewable, sustainable forest product. It is a very fast-growing tree and can be harvested in as little as 3 years, and in the wild (boy-ales), full maturity will be 5–7 years. Today, seed selection and even genetic engineering is used to grow trees with lower and more consistent lumber density as well as greater yield. Soil depletion is prevented through crop rotation, compositing waste, and cover crop plantings.

The tree is quite tall, with an average height of 30 m and about 70 cm diameter at full maturity. There are few branches along the trunk. Balsa is not considered a rainforest species. In fact, the farming of balsa has received positive mention by the Forest Stewardship Council (Figure 1.13).

1.2.1.2 Milling, Kiln Dry, and Make Block

The sylviculture and milling of balsa is generally not the same as other woods. Trees that supply general lumber are milled and processed to supply the widest and longest boards possible or veneers for plywood or panels. A balsa tree that will become sandwich core is handled in a very different manner. Variation in lumber density is most important. The tree is milled in such a way to maximize yield, minimize the number of days required to kiln-dry, minimize the variation of density in a single piece of lumber, and facilitate the production of blocks for conversion into sheets of sandwich core. Figure 1.14 shows the processes of balsa Kiln drying, block bonding, and sheet slicing.

As a natural product, the lumber may contain defects such as pith, cork, stain, knots, wormholes, decay, and water heart. Most wood defects are graded on their effect of the lumber visually. For balsa, this is defined by the market for model wood. Structurally important defects that would pertain to its use as a sandwich core are either not permitted in the block or repaired in the sheet by the supplier.

Sandwich Structural Core Materials

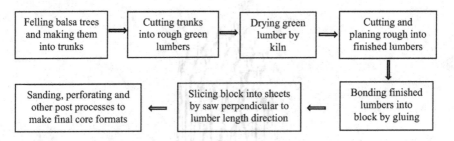

FIGURE 1.14 Balsa core sheets processing procedure.

Today, strict controls are used by most suppliers to ensure the trees are cut, milled, and put into the drying kiln in as little as 10 days. This almost eliminates the potential for the presence of decay in the lumber. A minor amount of decay will not affect the mechanical properties. Air-dried balsa lumber has been used to produce end-grain core for decades. If "time to kiln" control is lost, the potential for increased amount of fungal decay is concerning to customers using vacuum infusion. This problem will be discussed in the section about resin absorption. Today, most large suppliers have strict quality control systems to verify and certify the quality from the day the tree is felled. And the material is continuously tested as part of this certification process, often under supervision of independent approval bodies.

1.2.1.3 Microstructure

Wood itself is a composite, a combination of cellulose fiber and lignin (resin), with cellulose as the major component. In hardwoods, lignin constitutes 16%–25% of the wood substance. Hardwood fibers average about 1 mm in length. Wood may be described as a short-fiber-reinforced phenyl propanol polymer.[24] It is also the somewhat rare occurrence when the resin is stronger than the fiber. As a polymer, cellulose is fairly weak. Properties of polymers used in cellular foams are all similar or superior.

It is the microstructure of wood that creates excellent mechanical properties. The structure is essentially honeycomb with a cell length to diameter ratio of more than 25. Average cell diameter is 10 µm. This is 25–50 times smaller than cellular foam and 350× smaller than honeycomb (generally 3 mm is the smallest diameter produced). Over 90% of cells are closed (Figure 1.15).

Density is highly dependent on cell wall thickness, like all cellular materials. The vessels used to carry water and nutrients make up about ~8% of the volume (Figure 1.16).

1.2.1.4 Mechanical Properties of Balsa Lumber

The properties of wood are anisotropic, and balsa is highly so. Performance is very good along the grain, but poor across (it is far easier to break a wood board by splitting the grain). Grain is often used in reference to the annual rings and can be a synonym for fiber direction (Figure 1.17).

Looking at the microstructure, it is simple to model the grain as groups of cylinders (drinking straws). Compression perpendicular to the grain is directly pushing down on the straw's ends, trying to collapse a cylinder. The resulting strength and modulus are extremely high.

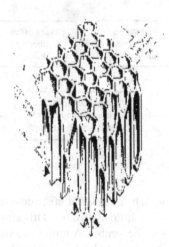

FIGURE 1.15 Microstructure drawing of wood.

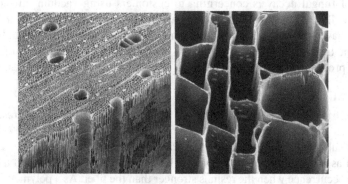

FIGURE 1.16 Microphotographs of Balsa @ 75×x and& 2,500×x.

Across the grain is like an elongated cell laying on its side. The performance of flat-train lumber in compression is generally lower than cellular foams at similar density since cellulose is a weaker polymer than those used to make foam core materials.

The first significant investigation of the mechanical properties of balsa lumber was conducted by the U.S. Forest Products Laboratory in 1944. The graphs in Figure 1.18 illustrate the mechanical properties of the end-grain sheet (longitudinal wood direction) and the lumber (radial and tangential to the growth rings) vs. density.

Compression strength in all directions are very linear as a function of density. Compression modulus parallel to the grain is more logarithmic as the failure changes from cell wall buckling to Kirkoff breaking with increasing density.

Tensile properties along the grain are also extremely high. Another way to visualize the grain is to think of "pull and peel" licorice. Pulling the candy along the piece is quite difficult, but peeling off a single strand is very, very easy. As this direction is difficult to test, and the material is not used in this manner, no significant study has been conducted to date.

Sandwich Structural Core Materials

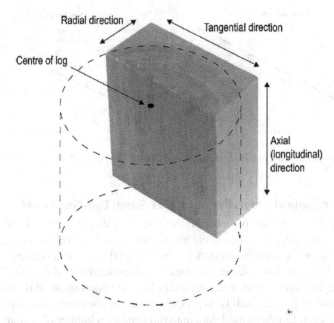

FIGURE 1.17 Balsa wood fiber directions.

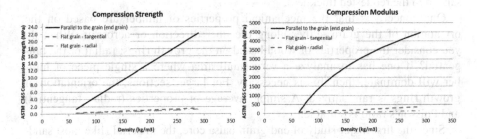

FIGURE 1.18 Compression properties of Balsa lumber in different directions.

In shear, the behaviors parallel to the grain, both longitudinal and transverse, and perpendicular to the grain (end-grain) are all quite different. Considering the "drinking straws" model, individual straws must slide over one another (one in bending (end-grain); the other in tension) parallel to the grain, and "roll" over each other perpendicular to the grain; hence, the term "rolling shear". It should be noted that the highest shear strength is end-grain, but shear modulus is best parallel to the grain in the longitudinal direction (Figure 1.19).

A sheet of balsa core may be produced in two formats: lumber-based and veneer-based. Each will be discussed in separate sections. Designers and engineers may take advantage of both density and grain orientation to create engineered "cores" that will produce progressive crush under impact or "tune" harmonic response in a ski or snowboard.

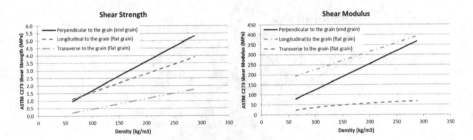

FIGURE 1.19 Shear strength & and modulus of Balsa lumber in different directions.

1.2.1.5 Mechanical Properties of Lumber-Based End-Grain Core

Looking at the properties of balsa lumber, one can theorize that the best way to design a core material for sandwich construction would be to orient the grain at 45°. This would align the grain direction with the principal stresses in bending. This may be accomplished in practice, but it comes at a considerable higher cost. The "end-grain" configuration using lumber creates the best combination of all the mechanical properties above, plus the ability to produce large volumes and consistent supply at a reasonable cost. Lumber-based core material combines lumber of various densities into a block with a low-cost wood glue, then the sheets are sliced with a saw and sanded to the requested thickness.

One concern is that the mechanical properties of lumber are based on the orientation of the grain, either along the fiber direction or across it. When do you consider the properties of wood based on the growth rings; radial or the tangential? If one has to analyze a very small area where a single piece of lumber will dominate, then this is necessary. For larger sections, the orientation of growth rings is randomized. A simple average of properties (radial, tangential) and Poisson's ratio is suitable.

Since the first major study of end-grain balsa core, the material (like most sandwich core materials) has been tested thousands of times. Differences in test results published by various balsa core suppliers may be a consequence of their specific product tested, or from their method of sample preparation or the precision of strain measurement; no different than any standard material testing.

The most recent complete recharacterization of lumber-based, end-grain balsa sheet was conducted in 2003.[25] Features of this study were as follows: (a) an update in the test method for compression; switching from a fixed compressor to a gimbaled indenter, (b) measuring the "as laminated" tensile properties, which are highly dependent of the resin, and (c) increasing the core thickness to 20mm which permitted better precision for shear strain measurement. The principal directions are defined in Figure 1.20.

Balsa is also one of few core materials where the core can be stronger than the resin or adhesive which bonds the laminate. This can make it more challenging when evaluating the performance vs. other core materials. The tensile strength with epoxy resin will be 1.5–2.3 times higher than polyester. Regression formulae for tension, compression and shear in the principle directions are shown in Table 1.9.[5]

Sandwich Structural Core Materials

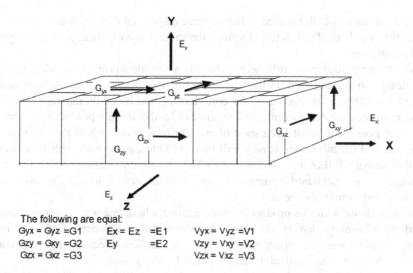

The following are equal:
Gyx = Gyz =G1   Ex = Ez =E1   Vyx = Vyz =V1
Gzy = Gxy =G2   Ey   =E2   Vzy = Vxy =V2
Gzx = Gxz =G3   Vzx = Vxz =V3

FIGURE 1.20 Coordinate system for end-grain balsa sheet (the same notation for stress (σ, τ) and & strain (ε, γ)).

TABLE 1.9
Regression Formulas for End-Grain Balsa as a Function of Density

Direction	Property	Formula (SI Units)
2	Flatwise tensile strength (epoxy)	(1402.492*LN(Density*0.0642796) − 1425.3828)/145
2	Flatwise tensile strength (ortho polyester)	(109.938*(Density*0.0624)−185.9365)/145
2	Flatwise tensile modulus	27.874*Density − 1342.6
2	Compressive strength	0.0932*Density − 4.6099
2	Compressive modulus	2910*LN(Density) − 12022
1	Shear strength (3/4" basis)	0.0193*Density − 0.2534
1	Shear modulus	1.2823*Density − 3.143

1.2.1.6 Homologation and Density Variation

It would be optimum to have all lumbers of the same density within a sheet, but this is far from necessary. All sandwich core materials have some variation, both density and properties. It has been a somewhat limited practice that one can only design with balsa based on the properties of the lowest density lumber within a sheet. The assumption becomes more pronounced with respect to local loads, since it can be possible for a connection to be located within a single piece of lumber in a sheet. If a single lumber has a density of 100 kg/m^3, then the design core shear strength must be chosen at by the lightest lumber and not based on the average sheet density. This has likely eliminated balsa as a candidate for many applications over the years, but it is easy to disprove. One example is using lumber with pith. Pith is the center of the tree and has no cellular structure. Balsa sheets with pith are often used as a lower-cost sandwich core. When a sheet of pith is tested against the same density without pith, the properties are the same.

Another way to disprove this theory is based on minimum properties. "Minimums" are design strengths based on the minimum possible density as supplied. A number of test coupons are measured for core density, and based on the variation, the minimum is calculated. Given the wide range of lumber density produced by a balsa tree, it is possible to construct a sheet of end-grain balsa core where the variation is so great, and the minimum density will be zero! Fabricate a block with alternating lumber density 100 kg/m³ and 250 kg/m³. The "minimum" density defined by the average minus 2*standard deviations is zero. The theoretical strength at this minimum density would also be zero.

If a laminate with this product is tested against a laminate with the same "standard sheet" average density, the results are statistically equivalent. Table 1.10 is a comparison between a standard product, tightly controlled, selected density, specialized product, and a theoretical min/max mixed lumber product[26]:

In general, a sheet of balsa core may be considered homologous. The distribution of lumber within the sheet is fairly randomized. Both block and sheet must meet specific tolerances, minimum and maximum lumber density, as well as average density. This also prohibits production where half is all low density and the other very high. The variation of density for end-grain balsa is similar to other core materials, and the "minimum" properties are still often greater than the average strength of other sandwich core materials at similar densities.

Special products with a narrower range of lumber density are offered, but this comes at increased cost and limits the available annual supply. It is best to contact a supplier for a specific project and obtain a custom-made product to the exact specifications.

1.2.1.7 Humidity, Moisture, and Its Effect on Mechanical Properties

Kiln-dried wood has a moisture content ~12%. Balsa like all wood will equilibrate to the environment. If very dry, after a period of time, the moisture content will be lower. If very humid, the moisture content will increase. There is a limit. In 100% humidity and room temperature, the moisture content of wood will eventually reach a maximum of 18%. The mechanical properties will generally be constant below this limit.

TABLE 1.10
Selected Lumber Density Vs. "Soup to Nuts" Performance Comparison

Product	Unit	SB.100	SL.911	SB.515 ("checkerboard")
Average density	kg/m³	149	170	163
Standard deviation		1.9	0.9	5.12
Minimum density (avg. - 2*S.D.)	kg/m³	88	141	0
Shear Strength @ avg. density	Mpa	2.6	3.0	2.9
"Minimum" shear strength	Mpa	1.6	2.2	0
Design shear strength (0.3*min)	Mpa	0.5	0.7	0
Tested core shear strength	Mpa	Not tested	2.9	2.4

When the wood is put into contact with liquid water, the moisture content will increase above 18%. The rate of absorption is completely dependent on the orientation of the grain. Water only travels along the grain, and this is how wood can be used for roofs, boats, tanks, and barrels.

The moisture content will continue to increase to 26% where the cell walls become fully saturated. This is commonly known as the fiber saturation point. Above the FSP, liquid water will begin to collect in the cells. "Green" lumber has moisture contents of 70%–90% by weight. Fully saturated wood (cells completely full of water) will have a moisture content of >300%.

The mechanical properties of wood with high moisture content will be lower by approximately 20%–30% in shear and 30%–40% in compression.[27] Note that in almost all cases, these lower strengths will still be vastly higher than most other core materials (Figure 1.21).

The only situations where the moisture content could increase to the point where the mechanical properties would be affected require the laminate to remain in contact with liquid water. High humidity alone is never the cause. A "water-logged" balsa-cored laminate is caused by (a) poor practice/quality problems during lamination; "never-bonds" and/or unfilled kerfs if flexible sheets were used. (b) Service damage such as postimpact delamination. The former may be prevented by following proper core installation guidelines or lamination with closed molding processes like resin infusion. The latter will limit incursion of water into the laminate to the damaged area and is easily repaired. If the water can be removed and the core re-dried, the mechanical properties will go back to "normal". There is no permanent loss due to contact with water unless fungal decay occurred.

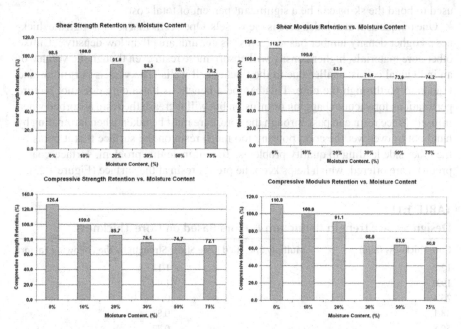

FIGURE 1.21 Properties of laminated balsa core vs. increasing moisture content.

1.2.1.8 Miscellaneous Properties

The shear strength of end-grain balsa like manufactured honeycomb will decrease with greater core thickness.[28] This is based on testing plate shear and flexural shear (beam bending). The correction factors for end-grain balsa core are shown below.[29] Even with the possible effects of moisture and core thickness, when cost is taken into consideration, balsa core provides extremely high value with often the best stiffness to price (Table 1.11).

Wood is not viscoelastic, and therefore it is easy to believe that laminates with end-grain balsa core suffer when loaded under high strain rates. This belief is undoubtedly based on how balsa fails after impact. Testing demonstrates the shear strength of balsa core does not suffer and will actually improve with increasing strain rates,[30] although the increase will be much lower than certain cellular foams.

Wood has the ability to absorb large impacts. End-grain balsa core may be crushed to 25% of its original thickness. Combinations of flat-grain and end-grain balsa are often used in impact limiters.[31]

Due to the high moduli and elevated temperature performance, creep is far less of a concern for balsa core than cellular foams or honeycomb products.[32]

Trees are nature's fatigue machines. The structure of wood has evolved to resist millions and millions of wind gusts and storms over the life of a tree. This is for "green" lumber with a MC >80%. The performance of kiln-dry wood is far higher. End-grain balsa core is no different. Figure 1.22 is a graph for the fatigue limit of various densities of end-grain balsa for $R = -1$.[32]

Besides mechanical properties, it is important for the designer to consider the total cost of the sandwich laminate. The amount of resin absorbed by the core or adhesive used to bond the skins can be a significant percent of total cost.

One unique aspect of balsa core is the vessels. Open cells that run through the thickness. Higher density lumber has more vessels per unit area than low density.[33] Unlike other core materials, higher density will absorb more resin even though the volume of an individual cell is smaller due to the thicker cell walls. The vessels permit air and liquids to go through the core. This can be a benefit as balsa core does not require perforations for vacuum bagging or infusion but will increase the resin absorbed when infused. Since resin can pass through the core, care must be taken when laminating by hand not to over-wet the core; pouring too much resin on the surface that it pools on the other side leading to quality problems. To eliminate this problem, surface-coated products are offered, which helps keep the prewet resin at the surface (Figure 1.23).

TABLE 1.11
Design Shear Strength Reduction Factors Based on Core Thickness

Core Thickness (mm)	Multiply Specified Core Shear Strength by the Following Factor:
12.7	1.11
19.0	1.00
25.4	0.92
38.1	0.83
50.8	0.77

Sandwich Structural Core Materials

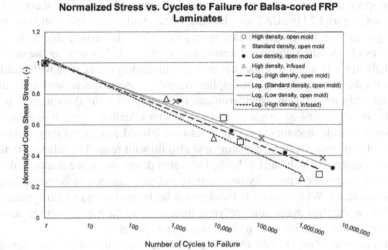

FIGURE 1.22 Fatigue limit of balsa cored laminates for $R = -1$.

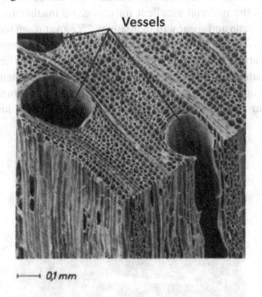

FIGURE 1.23 Capillary vessel of end-grain Balsa core.

How the end-grain sheet is produced will have an effect on the amount of resin absorbed. A sanded surface will absorb less resin than un-sanded. Coated products will lower the absorption further. Different coatings are available for specific lamination processes. Properly sanded and coated products require similar amounts of resin to prewet the surface as cellular foams; hand lay-up, vacuum bag, and prepreg. Generally, density does not affect the amount of prewet resin required for balsa core.

Resin infusion is very different even from other closed molding processes when it comes to core resin absorption. Since all of the air is removed before impregnation and the speed of resin fill is fairly slow, any and all open cells in the core sheet will fill with resin. Figure 1.24 is a CAT scan imagine of standard density balsa when infused.

As you can see, there is significant impregnation into the surface, and most or all of the vessels are filled. Resin absorption can be eliminated by coating the surface. The challenge is to maintain enough bond strength. Simply hot-coating with multiple coats of resin won't work. Certain suppliers have developed a special surface coating designed only for infusion which eliminates virtually all resin absorption into the vessels. Savings >80% on a rigid panel and ~50% on a flexible sheet can be achieved.

Sheet format also becomes more important when infused, as all perforations, grooves, kerfs will eliminate the surface coating and also fill with resin. The infused strength and resin compatibility should be checked to ensure design values are maintained.

Tests of thermal conductivity demonstrate a direct relationship between density and performance. While end-grain balsa core is fairly good as an insulating material, it will be worse than many other polymer foams due to the higher average density. Conductivity will also increase with increasing temperature.

An interesting characteristic of balsa wood is the thermal conductivity decreases at low temperatures. There is also no loss in mechanical properties, unlike polymer foams. At −184°C, bending and compression strength are 125% of room temperature.[34] This makes the material excellent for cryogenic insulation. At −184°C, the strength in compression and shear will increase by 25% over room temperature; stiffness improves by ~40%. The mechanical properties will also become lower with increasing temperature. This is vastly different than other sandwich core materials such as cellular foams which will become brittle with decreasing temperature.

Balsa core can operate and be processed at very high temperatures, up to 180°C. Like other cellular materials, the core's ability to hold heat is low so balsa can

FIGURE 1.24 CAT imagine of Balsa core after vacuum infused with resin.

Sandwich Structural Core Materials

withstand short-term exposure to temperatures far above this limit. Standard products are somewhat limited to ~120°C by the glue used to make a block, although once in a laminate any affect is minimal. Certain suppliers to offer a special product with a high temperature adhesive.

It would be easy to believe that the performance of balsa core in a fire is very poor. The opposite is true. Fire is a mass-driven reaction. The more fuel and oxygen, the greater the amount of energy released. Due to its very low density, balsa has little mass to burn, it is only 8% polymer by volume. When bare balsa core is exposed to a fire the surface burns quickly and develops an insulating layer of char. The remaining wood burns very slowly with little smoke produced. When laminated with fire resistant resin systems, balsa core can meet the most stringent fire requirements. A major benefit of sandwich laminates with balsa core is they will retain some amount of stiffness and strength post fire. If the structure must maintain its load bearing capability after a major fire event, it is best to design is a "club sandwich" with two layers of core (see illustration below). If one face sheet (skin) is consumed in a fire, the core acts as an insulator to the other skins. Special laminates and increased core thickness can provide fire resistance lasting an hour or more without the need for additional added fire barriers. A small summary of test data is shown in Table 1.12.

It is possible to increase both the temperature and fire resistance with chemical additives (a number of solutions for wood are well known) or pyrolysis; exposing

TABLE 1.12
Fire Testing Results for Unlaminated End-Grain Balsa Core

Description/Test Standard		Max Allowable	End-Grain Balsa
ASTM E-662 Smoke Density / NBS Smoke Chamber			
Nonflaming	Ds Max (min.)		74
	Ds 4 min.	200	71
Flaming	Ds Max (min.)		24
	Ds 4 min.	200	9
ASTM E-662 Toxic Gas Emission/NBS Smoke Chamber			
Nonflaming	HCN (ppm)	150	2
	CO	3500	500
	NO/NO$_2$	100	0
	SO$_2$/H$_2$S	100	5
	HF	100	4
	HCL	150	2
Flaming	HCN (ppm)	150	2
	CO	3500	600
	NO/NO$_2$	100	7
	SO2/H$_2$S	100	5
	HF	100	7
	HCL	150	2
ASTM E-84 Steiner Tunnel			
Flame Spread		Class A - 25	40
Smoke Developed		Class A - 450	50

the lumber to temperatures above 150C without oxygen. Taken to the extreme, this process may be used for making a carbon-based sandwich core.

The properties of balsa will be affected by alcohol and ammonia, the latter has been used as a way to form wood chemically as the properties are regained once the substance is removed.[35] One cannot cover a natural wood product like balsa core without discussing how fungal growth can occur and the consequences involved. A few things to keep in mind. Dry core cannot sustain either bacterial or fungal growth, but "dry" core will still have a moisture content anywhere from 4% to 18%. Fungi are different than bacteria. Bacterial growth on water-logged wood is generally a surface phenomenon, which looks bad, but scrapes away and cleans during growing, so that there is good wood underneath. Fungi is an aerobic organism only found fresh water and requires specific conditions to thrive. The wood must undergo periods of contact with water then air. Fully submerged wood will not decay as evidenced from perfect logs pulled from the bottom of lakes after decades.

In a sandwich laminate, fungal spores come into contact with balsa core from rain or freshwater lakes. As mentioned before, the wood's moisture content must go above 26% (the fiber saturation point) for fungal spores to grow. Once present, the moisture content must stay above 20%.

Some confusion can occur with respect to composite laminates with respect to moisture content. A high moisture problem in composite laminates may be considered when the moisture content is as low as 2%–3%. Kiln-dry wood has a moisture content of 12%. It can be easy to misunderstand a moisture meter reading if the meter is calibrated for fiber-reinforced composites.

If evidence of fungal growth is found, the affected core should be removed. The mechanical properties will be compromised if as little as 1% of the mass of the cell walls are consumed.

1.2.1.9 Product and Format

End-grain balsa core is typically sold in three density ranges corresponding to low, middle and heavy density lumber from the tree. Specialized products with a custom density and narrow variation of lumber density are also available.

During the slicing process, certain fibers are torn from the cells, then bent over. This can limit resin from properly penetrating the surface. Sanding the sawn sheet generally corrects this. The result is a slightly improved bond with less resin/adhesive at the bond line. Some suppliers offer special surface treatments (coatings) to limit resin absorption even further. The bond strength of coated balsa core will generally be a bit lower than un-coated but will not change the design strength or necessitate increased safety factors. This is because coating tends to make the bond more consistent; less standard deviation is observed when tested.

Format may be rigid, perforated, grooved, or flexible. Flexible sheets are supplied with a fiberglass scrim. Since the mechanical properties of end-grain balsa core are very high, the choice of scrim and the adhesive used will affect the resulting performance in a laminate. Some suppliers have optimized the scrim and application process to produce the highest possible laminate strength. Suppliers should have test data available for rigid w/scrim.

Sandwich Structural Core Materials

A second technique to create a wood-based core is to use thin veneers. A veneer is sliced, peeled or sawn from the tree. This is generally more difficult to accomplish and particularly so for balsa. The challenges are diameter and density. Only part of the tree section may be peeled; as the diameter decreases, the veneer is subjected to greater stress as it is flattened out. This leads to breakage and poor yield. A longer growing cycle is typically required as only the upper portion of the lumber density range is suited to the peeling process. These restrictions lower the potential supply and increase production costs. An alternative method of production is to use thin laminar; lumber with a thickness ~6 mm. Primary panels are fabricated, then bonded at the proper orientation to produce a block.

The benefits of fabricating sandwich core from veneers are improved homologation, and more importantly optimized fiber orientation. Combinations of 0/90, +45 and more are possible. Aligning the wood fibers in the +45 direction will create the highest shear strength possible at a specific density. All 90° (end-grain) improves density distribution vs. lumber-based core for improved compression and shear performance at high thickness (Table 1.13).

Different wood species may also be combined with balsa as well as other materials such foam. Fabrics may also be added within each bond line. This creates the opportunity design specific properties into the core material (Figure 1.25).

Not a lot of test data or research is available, but the +45 orientation overcomes one of the major problems with lumber-based balsa core, severe damage after impact. The energy to failure is 50% higher than lumber-based end-grain sheet. Pictures below capture a video during impact. Videos are available from 3A Composites Core Materials. Veneer-based wood core for sandwich is an area where much more study is needed (Figure 1.26).

1.2.1.10 Applications

Large-scale commercial use of balsa as a sandwich core material began during World War II with the British Mosquito Bomber where balsa was sandwiched between plywood face sheets for the majority of the structure. The hull of the U.S. Navy Catalina PBY flying boat was aluminum sandwich with balsa core. Most of the balsa wood supplied during the war was used for aircraft/airship construction and flotation, both life vests and rafts. The classic "end-grain" sheet was invented in the early 1960s.[36]

FIGURE 1.25 +45° veneer-based balsa core.

TABLE 1.13
Product Data Sheet of Veneer Balsa Core

All Average Values for BALTEK	VBC	Unit	Direction	0°/90° LD	±45°LD	±45°HD
Density	ISO 845	kg/m³		185	185	240
Compressive strength perpendicular to the plane	ISO 844	N/mm²	z	5.5	3.9	5.0
Compressive modulus perpendicular to the plane	ISO 844	N/mm²	Z	1000	600	800
Shear strength along bond lines	ISO 1922	N/mm²	X	1.9	3.9	5.0
Shear modulus along bond lines	ISO 1922	N/mm²	X	160	420	550
Shear strength across bond lines	ISO 1922	N/mm²	Y	1.3	1.0	1.3
Shear modulus across bond lines	ISO 1922	N/mm²	Y	100	80	100

Remark: all mechanical properties are determined at 20 mm thickness.

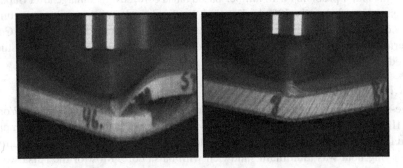

FIGURE 1.26 Impact damage of standard, lumber-based balsa core (a), and +45 veneer-based balsa core (b).

Then the end-grain balsa cores have been used for making sandwich composites in the applications of leisure boats and yachts, LNG transport vessel, airplane cargo pallet, chemical storage tanks, wind turbine blade, hopper train car, sport car floor, and road bridge decks.

1.2.2 Cork-Based Sandwich Core

1.2.2.1 Plantation and Harvest

Cork is the bark from cork trees. Cork trees are mostly found in countries surrounding the Mediterranean Sea as well as Portugal.[37] It has been utilized for more than 3,000 years. Most can guess the original application. The majority of Cork harvested

Sandwich Structural Core Materials

FIGURE 1.27 Harvesting bark from the tree.

is used for stoppers. Other applications are gaskets and vibration isolators. Only about 1% goes into composite applications (Figure 1.27).

If barks are removed from a tree on a 10-year cycle, then the tree can live for 150 years! This makes the production of cork quite sustainable. Once the bark is harvested, it must weather outside for a period of 6 months. It is then processed in a manner that permits the sheets to flatten and causes gas present in the cells to expand and produce the cellular structure.

Bark that was not thick enough and waste material not used to create stoppers is ground into granules. These are sorted and combined with different binders to create products for gaskets, vibration dampers, etc. Products suitable for sandwich composites will use polyurethane or phenolic binders. This process makes a very homogenous product. The microstructure of cork is similar to polymer foams, with a cell diameter ~40 µm. Most cells are closed; 90% of the volume is air.

1.2.2.2 Mechanical Properties

Density of cork products range from 120 to 250 kg/m^3. The most common found is ~200 kg/m3. The mechanical properties when tested as a sandwich core are very low; compression strength (2% offset) 0.29 MPa; strength at peak load = 0.39 MPa, and modulus = 9.45 MPa. Flat-wise tensile strength was measured to be 0.33 MPa with

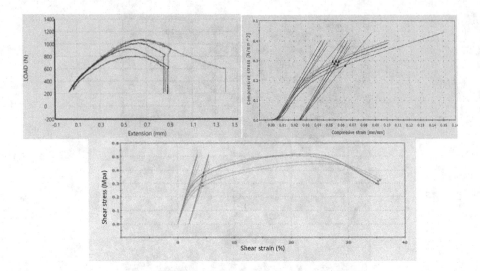

FIGURE 1.28 Tension (top left), compression (top right) and shear (bottom) behaviors of cork core during tests.[38]

a modulus 10.7 MPa. Plate shear strength (2% offset) = 0.33 MPa, strength at peak load = 0.48 MPa, and modulus = 13.5 MPa. Shear elongation at break ~35%.

Looking at the stress-strain curves in tension compression and shear, it is easy to see why the material is so well suited for applications where a high amount of damping or energy absorption is called for. Cork is a highly isotropic material. Its Poisson's ratio is approximately zero (why it works so well as a stopper) (Figure 1.28).

1.2.2.2 Other Properties and Applications

Thermal conductivity is somewhat better than end-grain balsa at similar density. Like all woods, the temperature resistance increases with decreasing temperature as shown in Figure 1.29.

Due to the low shear modulus at high relative density, cork excels as an insulator for sound.

As the product is an agglomerate of cork granules and binder there is ample porosity. This permits the material to be laminated quite easily in a variety of processes, such as vacuum bagging and prepreg with a maximum cure temperature of 180°C. Resin absorption when infused is very high, over 0.27 kg/m² per mm of thickness. The infused core density will be more than double the original.

Fire performance is excellent. Cork/phenolic has been used as an ablative layer in space applications since the 1970s. Burning creates a surface char that slowly propagates through the material. Typical formats are rolls 1.0–1.25 m wide; the thickness range is 1.5–40 mm.

1.3 HONEYCOMB CORES

Man-made honeycomb structures are manufactured by using a variety of different materials, depending on the intended application and required characteristics, from

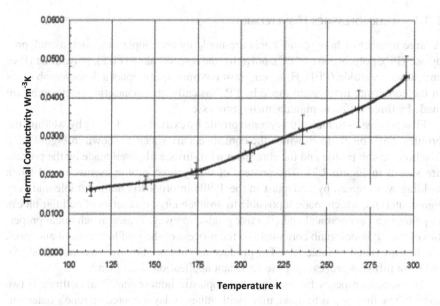

FIGURE 1.29 Thermal conductivity of cork core change with temperature.

paper or thermoplastics, used for low strength and stiffness for low-load applications, to high strength and stiffness for high-performance applications, from aluminum, stainless steel, or fiber-reinforced plastics.

The three basic techniques for honeycomb production today are molding, expansion, and corrugation. Thermoplastic honeycomb cores are usually made by extrusion molding via a block of extruded profiles or extruded tubes and other post processes from which the honeycomb sheets are sliced. Recently, a new, unique process to produce thermoplastic honeycombs has been implemented, allowing a continuous production of a honeycomb core by corrugating and thermal forming plastic sheet.

Honeycomb cores from glass fiber-reinforced plastic, carbon fiber-reinforced plastic, and synthetic aramid fabric-reinforced plastic are manufactured via the corrugation from composite materials. Honeycombs from paper cardboard or synthetic fiber fabric also are made by corrugating and postimpregnating thermosetting resin to increase mechanical properties and water resistance of the products.

Metallic honeycombs from aluminum and stainless steel are today produced by the expansion process. Continuous processes of folding honeycombs from a single aluminum sheet after cutting slits had been developed already around 1920. Continuous in-line production of metal honeycomb can be done from metal rolls by cutting and bending.

In this section, brief manufacturing process, special characteristics, and typical properties of each type of honeycomb will be introduced. Special application area of each material also will be presented.

1.3.1 THERMOPLASTICS HONEYCOMBS

A large number of honeycomb cores are made by thermoplastics, such as polypropylene (PP), polycarbonate (PC), polyethylene terephthalate (PET), polyamide (PA), and polyether imide (PEI). However, most common thermoplastic honeycomb cores in the composite market are made by PP. Basically, thermoplastic honeycombs are made by three different manufacturing processes.

First process is to make honeycomb profile by extrusion following by welding the profiles into a block, from which the core sheets are sliced as shown in Figure 1.30. The honeycomb profile and the core sheet with surface film/veil made by the process are shown in Figure 1.31. The process of honeycomb profile extrusion and block welding was created by Nidaplast in the 1980s in order to obtain durable and very light materials, which made it possible to produce large quantities of cellular blocks or plates in an economical and effective production way. Typical mechanical properties of the PP honeycomb core made by the process of the profile extrusion and block welding are listed in Table 1.14. The product also has been improved continuously for the resin infusion process and fire-retardant application since then.

The second process for making thermoplastic honeycomb is also through two steps.[40] The first step is to make tubes with adhesive layer coated on tube's outer surface by co-extrusion. Then the tubes were cut into certain length and laid in a frame mold to bond together by heating and pressing.

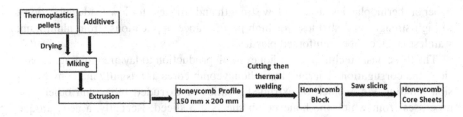

FIGURE 1.30 Process of making thermoplastic honeycomb core by profile extrusion and block welding.

FIGURE 1.31 Honeycomb profile and core sheet after laminated with surface protecting film and veil.[39]

Sandwich Structural Core Materials

TABLE 1.14
Typical Properties of Honeycomb Core Made by PP and Profile Extrusion/Welding[39]

Property	Nominal Density	Compressive Strength	Compressive Modulus	Perpendicular Tensile Strength	Shear Strength	Shear Modulus
Test method Unit	ISO 845 kg/m³	ISO 844 Mpa	ISO 844 Mpa	ASTM C297 Mpa	ISO 1922 Mpa	ISO 1922 Mpa
Nominal property	65	1.2	40	0.8	0.4	9
	100	2.6	70	0.8	07	14

In the process, the individual small tubes are produced together with an adhesively active layer in a simple co-extrusion process. Preferably, the adhesively active layer comprises a thermoplastic with lower melting temperature than that of base tube material. Following the production of the small tubes, they are lined up longitudinally alongside each other in a frame mold. By means of a subsequent thermal treatment, such hot-air treatment, the adhesive layer on the outsides of the small tubes is activated, so that the small tubes are held to each other in the desired size block. Then the honeycomb block can be cut into sheets with different thickness. Whole process is shown in a diagram of Figure 1.32. The tubes made by co-extrusion and the honeycomb sheet made by the processes are shown in Figure 1.33. As the adhesively active coating, a thermoplastic bonding agent is preferred which, in comparison with the small tube material, becomes plastic and develops its good bonding properties at lower temperatures.

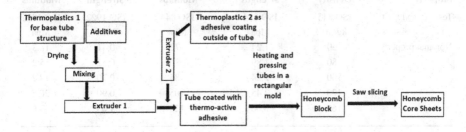

FIGURE 1.32 Process of making thermoplastic honeycomb core by tube co-extrusion and block heat bonding.

FIGURE 1.33 Tubes made by co-extrusion and honeycomb core sheet made from the tubes.[41]

Advantages of the co-extrusion and heat infuse bonding process to make honeycomb core are simple, easy control and to make the honeycomb with different densities by adjusting diameter and wall thickness of the tubes. Now the biggest volume of thermoplastic honeycombs in the composite industry are made by this process. The most common plastics used for making honeycomb core is polypropylene (PP). Polycarbonate (PC) and polyether imide (PEI) are also used for making honeycomb core for special applications. Typical mechanical properties of honeycomb cores made by PP, PC and PEI are listed in Tables 1.15–1.17.

The third method to make thermoplastic honeycomb is to fold and thermoform an uncut flat plastic sheet made by extrusion in line or made in a previous process. Accordingly, the process provides a folded honeycomb, formed from a plurality of cells arranged in rows, with the following features: the cells have lateral cell walls which adjoin one another in the form of a ring and are bounded toward two opening sides of the cell by covering-layer planes whereby the cells are each bridged or closed completely in one or other of the covering-layer planes.[43]

The process to produce the folded honeycomb core plank is shown in Figure 1.34. The folded honeycomb contains a plurality of 3D-structures, e.g. polygonally or

TABLE 1.15

Typical Properties of Honeycomb Core Made by PP and Tube Co-Extrusion/ Block Heat Bonding[42]

Property	Nominal Density	Compressive Strength	Compressive Modulus	Shear Strength	Shear Modulus
Test method Unit	ISO 845	ISO 844	ISO 844	ISO 1922	ISO 1922
	kg/m^3	Mpa	Mpa	Mpa	Mpa
Nominal property	60	1.24	72	0.41	10.3
	80	1.89	80	0.52	15.2
	100	2.24	85	0.70	17.2
	120	3.45	105	0.90	20.7
	160	5.00	120	1.03	27.6

TABLE 1.16

Typical Properties of Honeycomb Core Made by PC and Tube Co-Extrusion/ Heat Bonding[41]

Property	Nominal Density	Compressive Strength	Compressive Modulus	Shear Strength	Shear Modulus
Test Method Unit	ISO 845	ISO 844	ISO 844	ISO 1922	ISO 1922
	kg/m^3	Mpa	Mpa	Mpa	Mpa
Nominal Property	70	1.9	95	1.0	19
	80	2.2	106	1.1	21
	90	2.8	115	1.3	22
	110	3.6	155	1.5	26

Sandwich Structural Core Materials

TABLE 1.17
Typical Properties of Honeycomb Core Made by PEI and Tube Co-Extrusion/Heat Bonding[41]

Property	Nominal Density	Compressive Strength	Compressive Modulus	Shear Strength	Shear Modulus
Test Method Unit	ISO 845 kg/m^3	ISO 844 Mpa	ISO 844 Mpa	ISO 1922 Mpa	ISO 1922 Mpa
Nominal Property	56	0.83	105	0.63	12.2
	64	2.00	152	0.95	19.0
	75	3.00	179	1.35	25.2
	120	6.34	276	2.10	32.0
	144	10.67	336	3.30	40.0

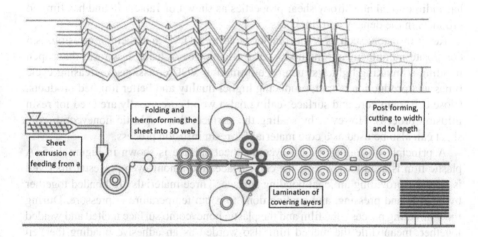

FIGURE 1.34 Folded honeycomb production steps.[44]

accurately shaped regions formed by plastic deformation and connecting areas in the covering-layer planes produced by the plastic deformation. All the cells can be closed by the application of one or more covering layers, e.g. by laminating the planar product with covering sheets. The vertical surfaces can be connected to one another, e.g. with glue or adhesive or by welding such as ultrasonic welding.

Honeycomb from a continuous thermoplastic sheet which allows production from a single continuous thermoplastic sheet by successive in-line operations by a thermoforming, a folding, and a bonding operation. The method enables the production of thermoplastic honeycombs directly from the extruder or from a roll of material. Inline post processing to panels and parts leads to further cost reductions by optimal process integration. PET nonwovens may be laminated onto the core planks to enable processing with thermoset materials.

However, making the folded honeycomb uses more material compared to making the honeycomb core sheet with the same thickness by tubing extrusion/heat bonding

TABLE 1.18
Properties of Folded Polypropylene Honeycomb Core Sheet[11]

Property	Nominal Density	Compressive Strength	Compressive Modulus	Shear Strength		Shear Modulus	
Test method Unit	kg/m³	ASTM C365 Mpa	ASTM C365 Mpa	ASTM C273 Mpa		ASTM C273 Mpa	
Nominal property	80	1.20	40	0.3 (length direction)	0.5 (width direction)	6.0 (length direction)	15.0 (width direction)

and profile extrusion/thermal welding processes, because half close cells formed on the top and bottom of the folded honeycomb planks. Also, the folded honeycomb has bidirectional anisotropy shear properties as shown in Table 1.18 and has limited products in the density and thickness.

Resin infusion, such as resin transfer molding (RTM) and vacuum infusion, is a composite manufacturing method having advantages of using less resin than open molding so making lighter structure, eliminating VOC emissions, increasing cycle times and production output, producing higher quality and better finished products. Close cell foam core and surface sealing balsa wood core usually are used for resin infusion process. However, by sealing the surface, thermoplastic honeycomb core sheet can also be used as a core material for resin infusion process.

A principle of sealing the honeycomb sheet surface is shown in Figure 1.35. A plastic film is used for covering the cell surface, and a nonwoven polyester fiber veil is used for providing an ideal bonding surface. Three materials are bonded together by heating and pressing, and cooling done to room temperature at pressure. During the laminating process, the film and the plastic honeycomb surface melted and welded together, meanwhile the melted film also worded as an adhesive bonding the veil together. This grade honeycomb core can be used in close molding process. Resin will still penetrate at the edges and through the perforations needed for a successful infusion. Other honeycombs, such as those made from metal, composite, and paper, are not able to melt and weld with the film and so cannot be made into a product suitable for closed molding.

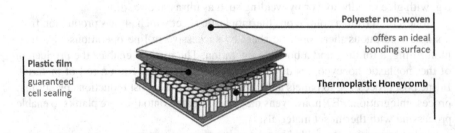

FIGURE 1.35 Resin infuse grade thermoplastic honeycomb core.[39]

1.3.2 Metal Honeycombs

Metal honeycomb, mostly made by stainless steel and aluminum sheet or foil now, is made by the expansion process and corrugating process. The honeycomb fabrication process by the expansion method begins with the stacking of sheets of the substrate material on which adhesive node lines have been printed as shown in Figure 1.36.[45] The adhesive lines are then cured to form a Hobe (honeycomb before expansion) block. The Hobe block it may be expanded after curing to give an expanded block. Slices of the expanded block may then be cut to the desired thickness. Alternately, Hobe slices can be cut from the Hobe block to the appropriate thickness and subsequently expanded. Slices can be expanded to regular hexagons, under expanded to 6-sided diamonds, and over expanded to nearly rectangular cells. The expanded sheets are trimmed to the desired length and width. The configurations of metallic honeycomb by expansion process are shown in Figure 1.37.

The expansion process requires moderately high inter-sheet bond strengths (sufficient to enable sheet stretching). The honeycomb density is controlled by cell size and thickness of foil. For low-density honeycombs with very thin webs, the required bond strengths are readily achievable with modern adhesives or by laser welding or diffusion bonding processes.[46] However, as the web (sheet) thickness to cell size ratio increases (i.e., as the relative density increases), the force needed to stretch the metal sheets eventually approaches the inter-sheet bond fracture strength. The manufacture of higher relative density hexagonal honeycombs then requires the use of corrugating manufacturing methods.

Metallic honeycomb cores have special features like elevated application temperature, high thermal conductivity, flame resistant, excellent moisture and corrosion resistance after coating with special material, fungi resistant and vary high ratio of

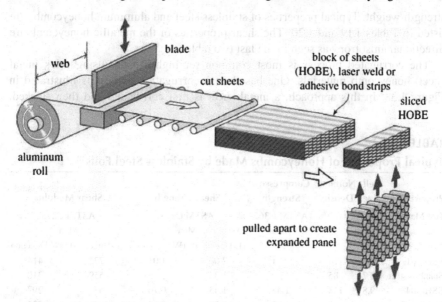

FIGURE 1.36 Process of making metallic honeycomb by expansion.[46]

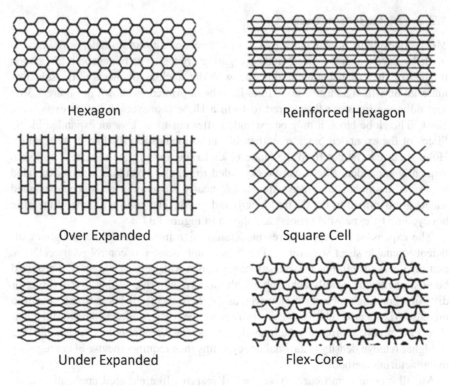

FIGURE 1.37 Cell configuration of Honeycomb honeycomb made by expansion.[45]

strength/weight. Typical properties of stainless steel and aluminum honeycombs are listed in Tables 1.19 and 1.20. The shear properties of the metallic honeycomb are directional anisotropy as seen in the last two tables.

The corrugated process is most common for high-density (using thick metal sheet) honeycomb materials. One based on a corrugation process is illustrated in Figure 1.38. In this approach, a metal sheet is first corrugated and then stacked.

TABLE 1.19
Typical Properties of Honeycombs Made by Stainless Steel Foils[48]

Property	Cell Size	Nominal Density	Compressive Strength	Shear Strength		Shear Modulus	
Test Method	mm	kg/m³	ASTM C365	ASTM C273		ASTM C273	
Unit			Mpa	Mpa		Mpa	
				(L-Direction)	(W-Direction)	(L-Direction)	(W-Direction)
SSH-301	9.8	114	2.41	2.00	1.03	772	414
Stainless steel	12.7	85	1.65	1.35	0.75	559	310
SSH-304	9.8	114	1.48	1.43	0.76	531	297
Stainless Steel	12.7	85	1.00	0.97	0.54	400	221

Sandwich Structural Core Materials

TABLE 1.20
Typical Mechanical Properties of Aluminum Honeycomb Made by Expansion Process[*,47]

Property	Nominal Density	Compressive Strength	Shear Strength		Shear Modulus	
Test Method		ASTM C365	ASTM C273		ASTM C273	
Unit	kg/m³	Mpa	Mpa		Gpa	
			(L-Direction)	(W-Direction)	(L-Direction)	(W-Direction)
Typical	26	0.74	0.61	0.31	0.17	0.08
Properties	37	1.75	0.90	0.65	0.22	0.12
	54	2.51	1.60	0.90	0.37	0.18
	69	3.81	2.12	1.23	0.48	0.21
	83	4.48	2.88	1.65	0.59	0.32
	96	6.90	4.22	2.14	0.63	0.39
	126	9.31	5.14	3.10	--	--

[*] 6.4 mm cell size 5052 aluminum foil with corrosive resistance coating honeycombs for all tests.

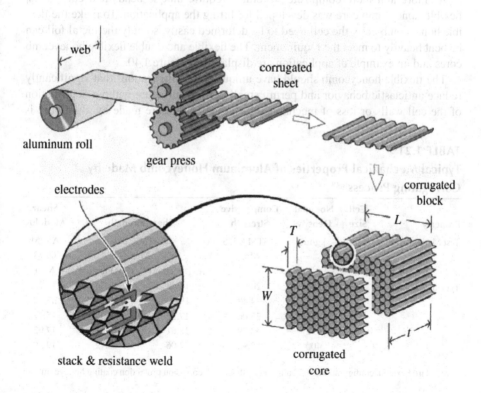

FIGURE 1.38 Corrugation manufacturing process used for making higher density honeycomb.[46]

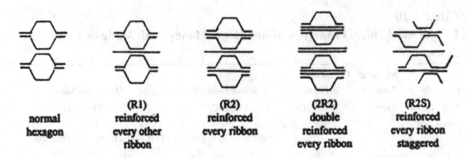

| normal hexagon | (R1) reinforced every other ribbon | (R2) reinforced every ribbon | (2R2) double reinforced every ribbon | (R2S) reinforced every ribbon staggered |

FIGURE 1.39 Core cell configurations of honeycomb made by corrugating process.[45]

The corrugated layers either adhesively bonded or welded together. Then the core is sliced to the desired thickness. Sometime, even the honeycomb made by only corrugated sheets is not strong enough, so the metal sheets are laminated between the corrugated sheets to reinforce the honeycomb as shown in Figure 1.39, in which most configurations of honeycombs made by corrugating process are shown. The honeycomb cores made by corrugating process have high density and high mechanical properties as listed in Table 1.21.

As more and more composite structures require unique bends and curvatures, flexible honeycomb core was developed for fitting the application. To make the flexible honeycomb core, the cell need to be deformed easily, so only the metal foil can be bent heavily to meet the requirement. The flexible and double flexible honeycomb cores and an example of application are displayed in Figure 1.40.

The flexible honeycomb should have unique cell configurations that significantly reduce anticlastic behavior and permit small radii of curvature without deformation of the cell walls or loss of mechanical properties. The core made by aluminum is

TABLE 1.21
Typical Mechanical Properties of Aluminum Honeycomb Made by Corrugating Process*,[48]

Property	Cell Size	Nominal Density	Compressive Strength	Shear Strength		Shear Modulus
Test Method Unit	mm	kg/m³	ASTM C365 Mpa	ASTM C273 Mpa		ASTM C273 Mpa
				(L-Direction)	(W-Direction)	
Typical Properties	3.2	232	20.00	14.48	10.34	13.79
		354	34.48	19.30	13.10	27.92
	4.8	251	22.06	17.24	10.34	15.17
		352	33.79	18.61	11.03	17.92
		400	37.92	22.06	11.72	19.30

* Standard hexagon configuration, 5052 aluminum sheet with corrosion protection coating honeycomb for all tests.

Sandwich Structural Core Materials

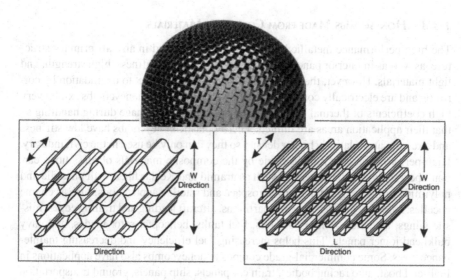

FIGURE 1.40 Flexible metallic honeycomb configuration and application.[49]

available in market, which allows the designer and fabricator freedom in the utilization of honeycomb for components requiring simple and compound curvatures. Highly contoured sandwich panels such as leading edges and flaps, nacelles, fairings, doors and access covers, and other parabolic, spherical, and cylindrical shapes are prime flexible core candidates. Duplicate die model and control tooling for aerospace use are also examples of flexible core applications. The typical mechanical properties of a few of aluminum flexible honeycombs are listed in Table 1.22.

TABLE 1.22
Typical Mechanical Properties of Flexible Honeycomb Core[49]

Property	Nominal Density	Compressive Strength	Shear Strength		Shear Modulus	
Test Method	kg/m³	ASTM C365	ASTM C273		ASTM C273	
Unit		Mpa	Mpa		Gpa	
			(L-Direction)	(W-Direction)	(L-Direction)	(W-Direction)
Typical	34	1.38	0.62	0.34	0.12	0.07
Properties	40	1.79	0.83	0.48	0.17	0.08
	50	2.41	1.14	0.66	0.22	0.09
	66	3.62	1.79	1.14	0.31	0.12
	91	6.45	2.97	1.79	0.47	0.16
	67 (double flexible)	5.24	1.93	1.31	0.21	0.12

* Honeycomb made by corrugation, 5052 aluminum foil with different thickness, 7.4 mm cell size for all tests.

1.3.3 Honeycombs Made from Composite Materials

The high-performance metallic honeycomb cores are used in aircraft primary structure, as well as in interior panels and floors requiring high stiffness, high strength, and light materials. However, the metallic honeycombs are subject to degradation by corrosion and are electrically conductive. Moreover, the metallic honeycombs exhibit very high coefficients of thermal expansion and are subject to damage during handling so that their application areas are limited. Thermoplastic honeycombs have low stiffness and strength, and relatively higher density so they can only be used in the civil industry. High-performance honeycombs made by the composite materials of glass fiber, carbon fiber, and synthetic materials, such as aramid and Kevlar fiber, with phenolic and polyimide resins are good in many aerospace and satellite applications, such as engine nacelles, radomes, galleys, flooring, partitions, aircraft leading and trailing edges, missile wings, antennas, military shelters, fuel tanks, helicopter rotor blades, and navy bulkhead joiner panels. This helps increasing fuel efficiency and decreasing maintenance costs. Some commercial-grade composite honeycombs also have applications in high-end boat, auto racing bodies, train car panels, ship panels, ground transportation structures, military shelters, ground antennas, and special-purpose structures.

The honeycombs made by composite materials have a high ratio of stiffness and strength to weight, low deformability, resistance to high temperature and to chemical corrosion, low fatigue reaction, and excellent insulation property. No-metallic composite honeycombs also have the features of good flammability, smoke and toxicity (FST) properties, excellent dielectric properties, high toughness, good thermal and moisture stability, and low densities as 24 kg/m^3 that made by Kevlar or aramid fiber.

Carbon fiber can also be used for making honeycomb core material, but the cost is very high. Structures with this type of core are often limited to space applications, for example, satellites, where weight savings is critical, and the thermal expansion needs to be kept at a minimum.

The composite honeycombs mostly are made by woven fabric or nonwoven clothes of different fibers impregnated with phenolic resin or polyimide resin, which are accomplished by expansion or corrugating process. A process made composite honeycomb by expansion is displayed in Figure 1.41. A roll of woven fabric or nonwoven paper made ether by glass fiber, carbon fiber or synthetic fiber can be used as a source for cutting individual sheets. The fabric or paper may be saturated with a low content resin that could partially cured for increasing its stiffness. Stripes of adhesive are applied on the sheets before laying them together to form a collapsed structure of sheets expandable to form a honeycomb. While still in the unexpanded form, the structure is subjected to conditions to cure the adhesive strips and adhere the layers together. The block is then expanded by pulling top and bottom faces apart from each other to yield a honeycomb block. The honeycomb block is then dipped in an uncured matrix resin bath. The dipped honeycomb with uncured matrix resin is subjected to curing heat. The dipping and curing can be repeated until the desired amount of matrix resin has been accumulated and cured to yield completed honeycomb block. Completed honeycomb block is cut or otherwise shaped into individual honeycomb sheets. The honeycomb core made with this process can achieve densities from 0.015 to 0.24 g/cc depending upon the basis weight of the un-impregnated sheet and the amount of matrix resin included in the structure.[50]

Sandwich Structural Core Materials

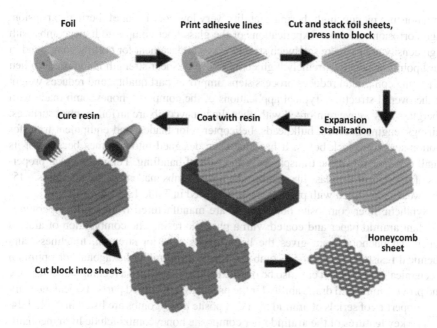

FIGURE 1.41 Expansion process for making composite honeycomb core.[51]

Another method comprises deforming the base sheets into corrugated forms, stacking, and bonding them together with bottoms of the corrugation of one sheet joining with tops of the corrugation of another sheet. The adhesive used for bonding the corrugated sheets can be cured at high temperature, so the sheets can be woven cloth, nonwoven fabric, or paper-impregnated with a certain amount of resin, such as phenolic or polyimide resin. Also, the corrugated sheets can be directly made by compressing and heating curing the glass fiber or synthetic fiber wool with binder.

The composite honeycomb with extremely low density only can be made by corrugating the nonwoven paper process because the woven cloth cannot be made very thin. If use the very thin nonwoven to make honeycomb by expansion process, the thin paper could break during expansion step. Moreover, the composite honeycomb with high temperature resistant can be only made by the corrugating process also, not by the expansion process, because a room temperature curable adhesive is required for bonding the sheets together in expansion process, which is also only can be used at low temperature.

However, a final coating could be applied to the composite honeycombs either made by the expansion or the corrugating process. The coating includes such divergent materials as plaster of cement, aluminum phosphate, sulfur, fire retardant, and others. The final coating can be accomplished in any desired manner, with some examples of techniques being powdered deposition, dispersion coating, solution coating, electrolytic deposition, chemical reaction, and chemical reduction.[52,53]

A glass fabric-reinforced honeycomb, made from fiberglass prepreg with thermosetting node adhesive, and with either a thermoplastic or thermoset resin for the web impregnations, can be used in temperatures up to 260°C. Depending on the constituent resin system, the honeycombs may be suitable for short exposures at higher temperatures. The glass fiber composite honeycombs have features of high compression strength, low-moisture pickup, high thermal stability, very low coefficient of thermal expansion, high corrosion resistance. Applications of the glass fiber composite honeycombs with high density are to make sandwich structure as a replacement for potting compound in hard points or where extremely high compressive or shear strength is required, in lieu of potting compound reduces process steps, improves part quality, and reduces weight in the overall structure. Typical applications of the composite honeycomb made with fiberglass bias weave reinforced with phenolic honeycomb are airfoil, control surfaces, fairings, engine nacelles, bulkheads, helicopter rotor blades, and equipment trays for motion-sensitive block boxes. It has also been designed into radomes because of its small cell size, dielectric transparency, and ease of handling. The mechanical properties of a serials of the glass fiber composite honeycombs made with the fiberglass ±45° bias weave reinforced with phenolic resin are listed in Table 1.23.

Synthetic fiber composite honeycombs are manufactured from high-temperature-resistant aramid paper and coated with a phenolic resin. The combination of aramid paper and phenolic resin gives the honeycombs superior strength, toughness, and chemical resistance. The honeycomb cell shape is normally hexagonal for optimum mechanical properties. It can also be over-expanded to produce a rectangular cell shape and provide improved drape ability for the production of curved parts. Typical mechanical properties of serials of aramid fiber composite honeycombs are listed in Table 1.24.

The key features of the aramid fiber composite honeycomb include high mechanical strength at low densities, outstanding resistance to corrosive attack by chemicals, excellent FST property, good dielectric properties, transparent to radio and radar waves, and compatible with most lightweight-reinforced composite materials.

1.3.4 Paperboard Honeycombs

During World War II, paper honeycomb was introduced for the airplane industry (fuel tanks for Dakotas). After the war ended, paper honeycomb was used primarily as a structural material in the reconstruction of Europe. The shortage of building material (wood, bricks, etc.) made paper honeycomb an excellent cost-effective available alternative. The development of paper honeycomb in North America was based on those postwar European applications.

The strength and rigidity are paired with extreme low aerial weight per cubic meter. These functional elements make paper honeycomb an attractive alternative with very low density (16–48 kg/m^3) for designers as the base core material for the sandwich structural products, such as doors, acoustic walls, exhibition panels, furniture, office cubic, and RV and Boat interior decoration.

The paperboard honeycombs can be made by expansion and corrugating processes mentioned in the last section and can also be made by an automatic folding process developed in recent years as shown in Figure 1.42.[55] The continuously folding process produces the paper honeycomb in triangular shape with reinforcing web as shown in Figure 1.42. The paperboard honeycombs can also be impregnated phenolic or other thermoset resin for reinforcing it as a structural core material and used for hot press lamination.[56] The special design of the paperboard honeycomb provides outstanding strength and rigidity at low cost by using natural sustainable and recyclable material and the most economical strength-to-weight ratio. By changing basis weight of paper,

TABLE 1.23
Mechanical Properties of Fiberglass Composite Honeycomb Made with ± 45° Weave Reinforced Phenolic Resin[54]

Property	Nominal Density	Compressive Modulus	Compressive Strength	Shear Strength ASTM C273 Mpa		Shear Modulus ASTM C273 Mpa	
Test method unit	kg/m³	ASTM-C365 Mpa	ASTM C365 Mpa	(L-Direction)	(W-Direction)	(L-Direction)	(W-Direction)
Typical properties	48	159	2.41	1.34	0.66	131	52
	64	317	3.86	2.17	1.03	172	83
	88	476	6.21	3.62	1.72	276	110
	128	690	12.07	4.66	3.31	310	148

TABLE 1.24
Typical Mechanical Properties of Aramid Composite Honeycombs[54]

Property	Nominal Density	Compressive Modulus	Compressive Strength	Shear Strength ASTM C273 Mpa		Shear Modulus ASTM C273 Mpa	
Test method unit	kg/m^3	ASTM C365 Mpa	ASTM C365 Mpa	(L-Direction)	(W-Direction)	(L-Direction)	(W-Direction)
Typical properties	23	38	0.60	0.50	0.25	16	11
	24	41	0.70	0.50	0.26	20	14
	29	60	0.90	0.50	0.36	25	17
	32	75	1.20	0.70	0.38	29	19
	48	138	2.40	1.25	0.73	40	25
	64	190	5.00	1.55	0.86	55	33
	72	225	5.10	2.00	1.05	66	36
	80	250	5.30	2.25	1.20	72	40
	96	400	7.70	2.60	1.50	85	50
	123	500	11.50	3.00	1.80	100	60
	139	580	14.00	3.40	1.88	113	67

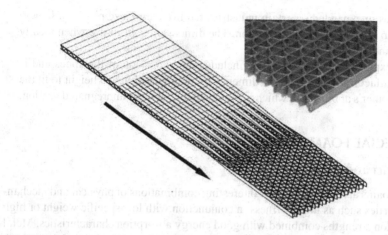

FIGURE 1.42 Automatic process for manufacturing triangle paperboard honeycomb with reinforcing web.

honeycomb thickness, cell size, and impregnating resin content, the paperboard honeycombs with a wide range of specifications are available. Physical and mechanical properties of a series of triangle-shape paperboard honeycombs are listed in Table 1.25.

Paperboard honeycomb cores are available in three physical forms:

1. Continuous unexpanded honeycomb core – the form is efficient and economical for large users. Shipped in continuous pallet load lengths, the coiled core is several hundred square meters when expanded mechanically at the point of use.

TABLE 1.25
Typical Properties of Paperboard Honeycomb with or without Impregnated Phenolic Resin[57]

Property	Nominal Density	Compressive Modulus	Compressive Strength	Shear Strength		Shear Modulus		Shear Elongation	
Test method	ASTM 271	ASTM-C365	ASTM C365	ASTM C393		ASTM C393		ASTM C393	
unit	kg/m³	Mpa	Mpa	Mpa		Mpa		%	
				L[b]	T[b]	L	T	L	T
Typical properties	25[a]	9.8	0.25	0.28	0.13	27	11	2.34	2.57
	38[a]	14.6	0.37	0.47	0.18	48	13	1.79	2.72
	47[a]	19.4	0.65	0.76	0.26	64	14	2.03	2.80
	21	3.4	0.07						
	26	4.9	0.11						
	28	4.6	0.12						
	33	7.2	0.16						
	43	9.4	0.24						

[a] Impregnated 15% phenolic resin.
[b] L, length direction; T, Transverse direction.

2. Slices unexpanded – certain industries require cut-to size pieces of honeycomb tailored to their application. The dimension is fit-to-use when locally expanded.
3. Preexpanded sheets – the type includes the expansion of the slices and is manufactured to the specific dimensions, length, width, and height to fit the customer's application, which is also available in resin impregnated version.

1.4　SPECIAL FOAM CORES

1.4.1　Metallic Foams

Metallic foams are known for their interesting combinations of physical and mechanical properties such as high stiffness in conjunction with low specific weight or high compression strengths combined with good energy absorption characteristics. Metal foam is a special class of cellular metals that originate from liquid-metal foams and, therefore, have a restricted morphology. The cells are closed, round, or polyhedral and are separated from each other by thin films.

Metal foams excellently absorb energy as vibration, impact and sound due to their cellular structure. Metal foams are in general, significantly more stable and temperature resistant than plastic foams. Moreover, they are well suited for shielding electromagnetic waves. The aluminum and zinc which exhibit a density of less than $500\,kg/m^3$, depending on the production method. Metal foams used as core materials for sandwich composite are typically made by steel, aluminum, or their alloys. Metal foam is usually offered compositely with steel and aluminum sheets realized in sandwiches, which feature a much higher bending stiffness than massive sheet metal plates with less overall weight. Due to their properties such as high stiffness and crash behavior, metal foams are highly applicable on many working fields. At the end of their lifetime, meal foams and foam composites can be recycled without any problems.

Under certain circumstances, metallic melts can be foamed by creating gas bubbles in the liquid. Normally, gas bubbles formed in a metallic melt tend to quickly rise to its surface due to the high buoyancy forces in the high-density liquid. This rise can be hampered by increasing the viscosity of the molten metal, either by adding fine ceramic powders or alloying elements to form stabilizing particles in the melt or by other means.[58]

Current methods of steel and aluminum foam manufacture may create open-cell (permeable voids) or closed-cell (sealed voids) foams with varying regularity, isotropy, and density. Three manufacturing methods are powder metallurgy, which has already been used successfully to create structural scale steel foam prototypes, hollow spheres that is in active commercial production, and Lotus-type, which has high potential for continuous casting processes necessary for low-cost steel foam production.[59]

Powder metallurgy is originally developed for aluminum foams, the powder metallurgy method was one of the first methods to be applied to steel foams and is still one of the two most popular. It produces primarily closed cell foams and is capable of developing highly anisotropic cell morphologies. The relative densities (ratio of foam density and solid material density) possible with this method are among the highest, up to 0.65, making it a strong candidate for structural engineering applications that demand that the foam retains a relatively high portion of the base material strength.

A second way for foaming melts directly is to add a blowing agent to the melt instead of injecting gas into it. Heat causes the blowing agent to decompose and release gas, which then propels the foaming process as shown in Figure 1.43. In the first step, about 1.5 wt.% calcium metal is added to an aluminum melt at 680°C. The melt is stirred for several minutes, during which its viscosity continuously increases by a factor of up to five because of the formation of calcium oxide (CaO), calcium-aluminum oxide ($CaAl_2O_4$), or perhaps even Al_4Ca intermetallic, which thicken the liquid metal. After the viscosity has reached the desired value, titanium hydride (TiH_2) is added (typically 1.6 wt%), serving as a blowing agent by releasing hydrogen gas in the hot viscous liquid. The melt soon starts to expand slowly and gradually fills the foaming vessel. The foaming takes place at constant pressure. After cooling the vessel below the melting point of the alloy, the liquid foam turns into solid aluminum foam and can be taken out of the mold for further processing. Typical densities after cutting off the sides of the cast foam blocks are between 0.18 and 0.24 g/cm³ with the average pore size ranging from 2 to 10 mm. Typical mechanical properties of a series of foam made by 316 L stainless steel are listed in Table 1.26.

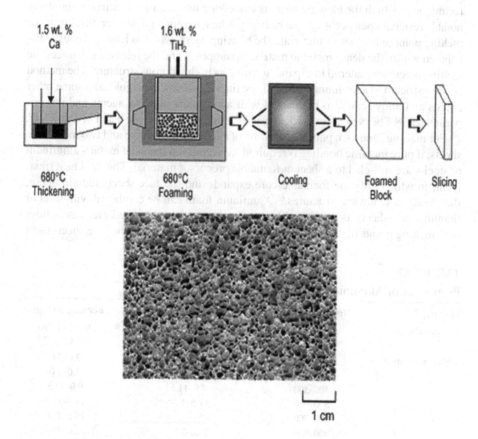

FIGURE 1.43 Direct foaming process of melts by adding gas-releasing powders and the pore structure of foam.[60]

TABLE 1.26
Properties of Foam Made by 316 L Stainless Steel[60]

Property	Nominal Density	Compressive Modulus	Compressive Yield Strength
Test method unit	ASTM 271	ASTM-C365	ASTM C365
	kg/m³	Mpa	Mpa
Typical properties	312	83	1.20
	507	196	3.00
	593	268	4.80
	772	300	6.10

Foamed metals can also be prepared from metal powders.[61] The production process begins with the mixing of metal powders – elementary metal powders, alloy powders, or metal powder blends – with a blowing agent, after which the mix is compacted to yield a dense, semi-finished product. The compaction can be achieved using any technique in which the blowing agent is embedded into the metal matrix without any notable residual open porosity. The next step is heat treatment at temperatures near the melting point of the matrix material. The blowing agent, which is homogeneously distributed within the dense metallic matrix, decomposes and the released gas forces the melting precursor material to expand, forming its highly porous structure. The method is not restricted to aluminum and its alloys; tin, zinc, brass, lead, gold, and some other metals and alloys can also be foamed with appropriate blowing agents and process parameters.[62] The typical properties of a few aluminum foams are listed in Table 1.27.

For making sandwich panels consisting of a foamed metal core and two metal face sheets, if pure metallic bonding is required, conventional sheets of metal – aluminum or steel – are roll-clad to a sheet of foamable precursor material. The final heat treatment, in which only the foamable core expands and the face sheets remain dense, then leads to sandwich structures.[65] Aluminum foam can be combined with steel or titanium face sheets as well as with aluminum face sheets. In the latter case, alloys with melting points that are different from the core material and the face sheets must

TABLE 1.27
Properties of Aluminum Foam[63]

Property	Nominal Density	Compressive Modulus	Bending Strength
Test method unit	ASTM C271	ASTM-C365	ASTM D790
	kg/m³	Mpa	Mpa
Typical properties	250–300	3.0–4.0	3.0–5.0
	300–400	4.0–7.0	5.0–9.0
	400–500	7.0–11.5	9.0–13.5
	500–600	11.5–15.0	13.5–18.5
	600–700	15.0–19.0	18.5–22.0
	700–800	19.0–21.5	22.0–25.0
	800–850	21.5–32.0	25.0–36.0

be used to avoid melting the face sheets during foaming. A large aluminum/aluminum foam sandwich was developed for a concept car in which structural aluminum foam applications were demonstrated.[64] Such sandwiches are three-dimensional, up to two meters long and about one meter wide.

Metal foams have certain appeal to the automotive industry. The aluminum-foam-steel sandwich is one example. Because these sandwich panels can reduce weight but be processed in very similar ways to steel and aluminum, they could replace conventional stamped metal parts in a car when combined with new constructional principles. At the same time, they could also reduce the number of parts in the car frame, facilitate assembly, and, therefore, reduce costs while improving performance.

1.4.2 Ceramic Foams

Ceramics, silicon carbide, and other open foams have been used as core materials in high temperature thermal protection systems owing to their light weight, high melting points, low thermal conductivities, and excellent thermal stability. Ceramic matrix composites as a shell layer can provide excellent specific strength to meet the requirements of thermal insulation/protection systems. Bulk densities of ceramic foams are 160–1280 kg/m^3; maximum use temperature is about 1,650°C in air and 2,500°C inert.

The processes of making ceramic foams can be divided into five categories: partial sintering, replica, sacrificial template, direct foaming, and bonding techniques. A wide range of processing routes have been proposed for the production of porous silicon carbide (SiC) ceramics, employing various raw materials from nature or artificial materials, leading to a variety of microstructure and pore morphologies. Each fabrication method is best suited for producing a specific range of pore sizes, pore size distribution, and overall amount of porosity.[66] Methods 1 and 2 are mostly used by industries for making the ceramic foams.

Partial sintering is the simplest, frequently used and most conventional method to fabricate porous ceramics. Particles of powder compact are bonded by partial sintering due to surface diffusion, evaporation–condensation, recrystallization, or a solution re-precipitation process. The homogeneous pore structure when sintering is terminated before fully densified.[66] Full densification is often retarded or prohibited by reducing the sintering potential. Reduced sintering potential is achieved by low sintering temperature, a constrained network of coarse powders, sintering without additives, and recrystallization. Pore size and porosity are controlled by the size of starting powders and degree of partial sintering. Generally, the powder has to be two to five times larger than the pore size required. It has been previously postulated that the elastic modulus is directly dependent on the neck radius to particle radius ratio and such neck formation allows the elastic behavior to be directly related to the sintering kinetics. Partial sintering is the easiest way to fabricate porous ceramics with porosity ranging below 65% and with pore size ranging from 0.1 to 10 um.

The replica method is based on the copy of the original foams concerning its pore shape and strut structure. Three different techniques were suggested to produce macroporous SiC ceramics: impregnation of polymer foams with SiC suspension or precursor solution, chemical vapor deposition (CVD) of SiC on polymeric foams, and infiltration of natural wood-derived or artificial carbon foams with Si sources. Macroporous SiC ceramics with open cells of high-volume porosity and with cell sizes ranging from100m to millimeters are frequently fabricated by the replica method. The original invention of using a sponge as a template was from Schwartzwalder and Somers.[67]

A polymeric sponge was used preparing ceramic cellular structure of various pore sizes, porosities, and chemical compositions firstly. Since then the sponge replica technique has become the most popular method to produce macro porous ceramics with high porosity and open cells. The most frequently used synthetic template is polymeric sponge such as polyurethane. The method is being widely used to produce lightweight structures in various applications. Porosity range of more than 90% can be produced using this method.

The method based on impregnation of polymer foams with a SiC suspension or precursor solution involves impregnating the porous or cellular structure with SiC suspension or precursor solution, removing the excess suspension by passing through rollers or centrifuges, drying the SiC or precursor-coated polymer sponge, heat treatment through careful heating to remove the polymer sponge by decomposition, and densification of the SiC coating by sintering in an appropriate temperature.[66] Typical microstructures of SiC ceramic foam via the replica technique are shown in Figure 1.44.

The replica method is the most appropriate technique to produce open-cell SiC ceramics with pore sizes ranging from 10 μm to 5 mm at porosity levels between 60% and 95%. The adhesion of an impregnating suspension on a polymeric sponge is the most crucial step in the polymer replica technique. While this technique benefits from its overall simplicity, the mechanical strength and reliability of porous structures produced with this method can be substantially degraded by the formation of

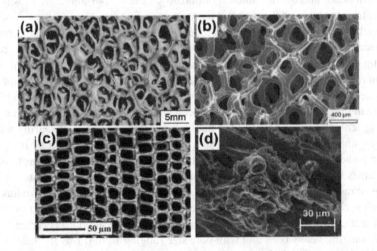

FIGURE 1.44 Typical microstructures of ceramic foams produced by the replica technique: (a) by impregnation of polyurethane sponges, (b) by CVD process, (c) by infiltration of wood-derived carbon foams with Si vapor, and (d) by infiltration of wood-derived carbon foams with Si vapor.[66]

TABLE 1.28
Typical Properties of Ceramic Foam[68]

Property	Nominal Density	Compressive Strength	Compressive Modulus	Tensile Strength	Tensile Modulus	Shear Strength
Test method unit	ASTM C271 kg/m³	ASTM C365 Mpa	ASTM C365 Mpa	ASTM C297 Mpa	ASTM C297 Gpa	ASTM C273 Mpa
Typical properties	300	8.28	3.45	3.45	1.72	3.10
	500	18.62	9.66	6.55	5.52	5.86
	700	30.34	18.97	10.34	11.38	9.31
	1,000	52.07	37.93	17.59	22.41	15.52
	1,200	67.59	53.45	23.45	31.72	19.31

hollow struts and cracked struts during pyrolysis of the polymer sponge. Typical mechanical properties of commercial ceramic foams are listed in Table 1.28 for referencing.

1.4.3 CARBON FOAMS

Carbon foam offers properties such as chemical inertness, ultra-high service temperatures, low coefficient of thermal expansion, and tailorable electrical/thermal conductivity. Carbon foams generally fall into two categories – graphitic and nongraphitic. The graphitic carbon foams offer high thermal and electrical conductivity but considerably lower mechanical strength than the nongraphitic foams. Nongraphitic carbon foams are generally stronger and cost far less to manufacture.

Graphitic foams typically are produced from petroleum, coal tar, or synthetic pitches. These precursors may be easily converted to the highly ordered graphitic crystal structure during the manufacturing process. Carbon foams produced directly from coal or organic resins are generally highly amorphous and thus do not form the graphitic microstructure. Although the highly graphitic foams offer unique properties such as high thermal and electrical conductivity and low density, they are currently not produced competitively on either a cost or a volume basis. Therefore, these foams are best suited for low-volume, high-end applications such as heat exchangers and thermal management components.

On the other hand, carbon foams made from less expensive precursor materials such as coal and wood are currently being made on a larger scale and are now competitively priced in such applications as composite cores, fire and thermal protection, composite tooling, electromagnetic shielding, and radar absorption. These applications depend on one or more critical characteristics such as weight, mechanical properties, fire resistance, low smoke toxicity, or coefficient of thermal expansion (CTE).[69]

Carbon and graphite foams are usually made from synthetic polymeric stock that is not very different from the materials used to make pencil leads or electric furnace electrodes. In another approach, coal is the feed stock. When the stock is subjected to specific pressures and temperatures, foaming gases or volatiles within the stock create carbon bubbles that impinge on each other and fracture. The result is an open cell foam. Through careful control of the gas evolution during this process, the shape, size, and number of pores can be manipulated. After the foaming process has been completed, a secondary heat treatment is applied. During this step, any remaining volatile material is removed, and properties such as electrical conductivity or electromagnetic properties can be tailored. The finished foam material is essentially an interconnected cellular network of open pores with the mechanical properties dictated by the density and by the thickness of the pore walls. Comprised almost entirely of carbon, the foams are chemically inert to almost all other materials, are noncombustible and have unique electrical and thermal properties. Foams can be made with uniform small or large pores, or a combination of large and small pores.[69] Typical mechanical properties of a series of carbon foams made by burning coal are listed in Table 1.29.

Carbon foam can also be made by converting wood into a carbon material. During this process, the wood (composed of cellulose, hemicellulose, and lignin) is heated to around 800°C in a controlled nitrogen atmosphere so that the wood does not oxidize and is converted into a carbon monolith that retains the anatomical features of the plant precursor. Since the carbonized wood retains many of the anatomical features of the precursor plant, variable properties are attainable by simply changing the species of tree. There is some shrinkage involved in the carbonization step, but now that the material is almost 100% carbon it can be further processed without any changes in dimension.

Carbon foam has shown promise for electromagnetic shielding and absorption for reasons that are both mechanical and electrical in nature. Electromagnetic shields are typically made from conductive metals, which are heavy and corrode in salt water. For instance, a carbon foam shield must be thicker than a conventional copper shield, but its lower areal density will make the weight of both shields nearly equal. In many aerospace applications harsh conditions require use of materials with, high specific strength and outstanding thermal properties. Carbon-carbon sandwich composites, which are made by using carbon foam as core and carbon fiber composite sheets as skin, have been used extensively in fulfilling this role.

These composites have the unique ability to maintain their mechanical properties such as at temperatures exceeding 2,000°C in a nonoxidizing atmosphere. This makes carbon-carbon sandwich composites prime candidates for leading edges of a spaceship or jet wings, or for use in propulsion systems which require

TABLE 1.29
Typical Mechanical Properties of Carbon Foams[70]

Property	Nominal Density	Compressive Strength	Compressive Modulus	Tensile Strength	Tensile Modulus	Shear Strength
Test method unit	ASTM D1622 kg/m^3	ASTM C365 Mpa	ASTM C365 Mpa	ASTM C297 Mpa	ASTM C297 Gpa	Torsion Shear Mpa
Nominal property	320	6.02	1.67	2.40	1.71	1.31
	400	10.31	2.12	3.78	2.00	1.69
	480	16.00	2.59	5.49	11.38	2.07

excellent thermal and mechanical properties. Carbon-carbon sandwich composites also have very good frictional and thermal properties, allowing them to be used in high-performance brake pads. These composites also have low thermal expansion coefficients and high thermal conductivity allowing them to withstand rapid temperature changes. However, carbon-carbon sandwich composites do have several drawbacks, including low oxidation resistance and high expense and difficultly to use in manufacturing.

1.5 OTHER CORE MATERIALS

1.5.1 Corrugated and Lattice Truss Cores

1.5.1.1 Corrugated Core

Metallic sandwich structures have attracted tremendous attention in recent years and become significant protective structure in engineering field ranging from aerospace structures to marine structures to transportation industry. Sandwich structures are usually based on honeycomb core and polymeric foam materials may keep air and humidity. Moisture retention is one of the problems in aircraft sandwich construction. This problem may lead to growth in the whole weight of the sandwich construction and degrading of the core properties. To overcome problems, an open channel core material such as two-dimensional prismatic core is vent able in order to avoid moisture accumulation. In addition, the suitability of the corrugated core as replacement core design structures in the sandwich construction will as well serve the concept of sustainable manufacturing.

A corrugated-core sandwich structure is embraced of a corrugation sheet between two thin surface sheets. The corrugated core keeps the face sheets apart and stabilizes them by resisting vertical deformations and also enables the whole structure to act as a single thick plate as an asset of its shearing strength. Another benefit is the same material is used for the entire structure. Unlike soft honeycomb-shaped cores, a corrugated core opposes bending and twisting as well to vertical shear. Corrugated cores when tested in the longitudinal direction offer shear strengths that are comparable with square honeycombs and significantly greater than those exhibited by traditional foam cores, which also have been verified excellent shock resistant properties, generally due to their high longitudinal stretching and shear strength. Then, corrugated-core sandwich panels, due to their extremely high flexural stiffness-to weight ratio are usually used in aeronautics, aerospace, civil engineering, and other applications.

Traditionally, the corrugated cores have three different shapes, which are trapezoidal, triangle, and sinusoidal.[71] Most corrugated cores are made by metallic materials, such as stainless steel and aluminum, but some also are made by glass fiber, carbon fiber and synthetic fiber composite sheets and cardboard paper for special applications. Figure 1.45 shows some metal sandwich panels made with corrugated core.

A novel lightweight bio-inspired double-sine corrugated (DSC) sandwich structure has been proposed to enhance the impact resistance. The out-of-plane uniform compression of the bio-inspired bi-directionally sinusoidal corrugated core sandwich panel has been studied under the quasi-static crushing load. Compared with the regular triangular and sinusoidal corrugated core sandwich panels, the bio-inspired DSC

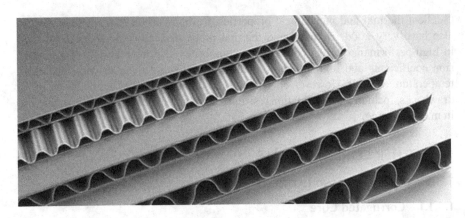

FIGURE 1.45 Metal sandwich panels with corrugated core.[72]

core sandwich panels significantly improve the structural crashworthiness as well as reducing the initial peak force greatly. The wave amplitude, wave number and corrugated core layer thickness affect the mechanical performance of the bio-inspired DSC core sandwich panel. Good application of the DSC core sandwich structure is energy absorption industry.[73]

1.5.1.2 Lattice Truss Core

Lattice truss topologies with open, three-dimensional interconnected pore networks provide opportunities for simultaneously supporting high stresses while also enabling cross flow heat exchange. These highly compressible structures also provide opportunities for the mitigation of high intensity dynamic loads created by impacts and shock waves in air or water. More recently, significant interest has emerged in lattice structures which have three-dimensional interconnected void spaces well suited.

Potential applications of the lattice truss sandwich structures include blast resistant structures multifunctional materials (e.g., simultaneously load bearing and active cooling) and replacement for the structurally efficient but expensive honeycomb sandwich composites. Optimized truss core sandwich plates offer greater weight savings and design advantages. Compared with sandwiches of hexagonal honeycomb cores, which have long been considered as the most structurally efficient systems under bending and transverse shear loads, the optimized truss core sandwich plates are found to be equally effective.[74]

Fully open cell structures can be created from slender beams (trusses) that in principle can be of any cross-sectional shape: circular, square, rectangular, I-beam, solid or hollow. The trusses can be arranged in many different configurations depending upon the intended application. Figure 1.46 shows six examples of micro truss cellular topologies used as the cores of sandwich panels.

The tetrahedral structure has three trusses each meeting at a face sheet node as shown in Figure 1.46. The pyramidal structure has four trusses meeting at a face sheet node. In both topologies, the trusses form a continuous network. Both also have directions of unobscured "easy flow". There are three of these channels in a single layer of the tetrahedral structure and two in the pyramidal system. A slightly

Sandwich Structural Core Materials

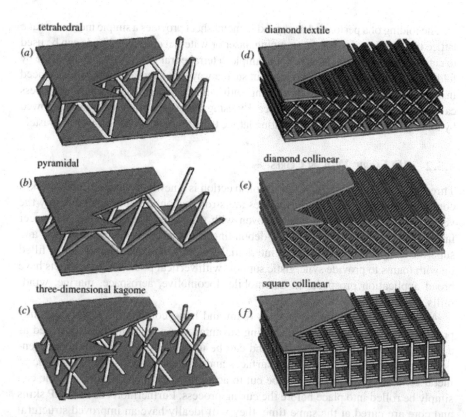

FIGURE 1.46 Examples of lattice truss topologies configured as the cores of sandwich panels.[75]

different topology is referred to as a three-dimensional Kagome topology. Such two-dimensional weaves have been found to be very strong. Each of the three topologies shown in Figure 1.46a–c is efficient at supporting structural loads, especially the shear loads encountered in panel bending.

Other lattice truss topologies have also been proposed based upon manufacturing considerations. Figure 1.46d–f shows examples that are easy to make from wires. The diamond textile structure is made from layers of a plain weave metal fabric that have been bonded to each other. A simple wire layup process can be used to create diamond and square truss structures shown in Figure 1.46e and f.

Lattice truss topology structures can be made by 3D printing, investment casting, perforated metal sheet forming, wire/hollow tube lay-up or by snap fitting laser cut out lattices. All but the 3D printing and the investment casting approach require assembly and bonding steps to create the cellular structure and for later attaching it to face sheets.

Truss core patterns with attached face sheets can be made from a volatile wax or polymer (e.g., polyurethane) by injection molding or 3D printing rapid prototyping methods. This pattern, together with a system of gating and risers, is coated with a ceramic casting slurry and dried. The wax or polymer is removed by melting or vaporization, and the empty mold is filled with liquid metal.[76]

The folding of a perforated or expanded metal sheet provides a simple means to make lattice trusses. A variety of die stamping, laser or water jet cutting methods can be used to cut patterns into metal sheets. For example, a tetrahedral lattice truss can be made by folding a hexagonally perforated sheet in such a way that alternate nodes are displaced in and out of the sheet plane.[75] By starting with a diamond perforation, a similar process can be used to make a pyramidal lattice. By using metal expansion techniques followed by folding provides a means for creating lattice truss topologies with little or no waste.[75]

1.5.2 3D Fabric Woven Cores

Three-directional (3D) spacer fabric construction is a newly developed core and skin combining concept. The fabric surfaces are strongly connected to each other by the vertical pile fibers which are interwoven with the skins. Therefore, the 3D spacer fabric can provide good skin-core debonding resistance, excellent durability, and superior integrity. In addition, the interstitial space of the construction can be filled up with foams to provide synergistic support with vertical piles. These products have broad application prospects in automobile, locomotive, aerospace, marine, windmills, building, and other industries.

As compared to materials such as foam and honeycomb, a 3D fabric core may provide many significant manufacturing advantages. As the material is provided as a fabric, it is initially very flexible and can be used easily as a core in nonconventional applications such as curved surfaces and sandwich pipes. Whereas plastic or metal cellular cores may need to be cut to accommodate smaller radii, a fabric can simply be rolled into place before the curing process. Furthermore, as the FRP skins and core are cured at the same time, they will ideally have an improved structural unity and this may eliminate the risk of delamination, which is a common issue in sandwich panel construction. Moreover, since the 3D fabric comes in a roll, it is a very easy to transport material and long lengths of composite beams can be produced without any seams or overlap in the core.[77]

Most usable 3D fiberglass fabric consists of two bi-directional woven fabric surfaces, which are mechanically connected with vertical woven piles. And two S-shaped piles combine to form a pillar, 8-shaped in the warp direction and 1-shaped in the weft direction. The 3D spacer fabric can be made of glass fiber, carbon fiber or basalt fiber as shown in Figure 1.47. Also their hybrid fabrics can be produced.[78]

The 3D space fabrics have the range of the pillar height of 3–50 mm and the range of the width up to 3,000 mm. The designs of structure parameters including the areal density, the height, and distribution density of the pillars are flexible. The mechanical and physical properties of the 3D fabrics and sandwich structures made with the 3D fabrics are listed in Table 1.30. As shown in the table, the performance of 3D fabric and truss-type core materials are fairly poor.

1.5.3 Core Mats

A number of thin, "fabric-like" materials are available, which can be used to slightly lower the density of a single-skin laminate. Materials consist of a nonwoven "felt-like" fabric impregnated with microspheres. They are usually below 10 mm in

FIGURE 1.47 The 3-D spacer fabric made of glass fiber, carbon fiber, or basalt fiber.[78]

TABLE 1.30
Typical Properties of One-Type 3D Fiberglass Spacer Fabrics and Sandwich Constructions[78]

Property		Unit	Typical Value						
Pillar height		mm	4	6	8	10	12	15	20
Warp density		root/10 cm	80	80	80	80	80	80	80
Weft density		root/10 cm	96	96	96	96	96	96	96
Face density	3D space fabrics	Kg/m²	0.96	1.01	1.12	1.24	1.37	1.52	1.72
	3D spacer fabrics and sandwich	Kg/m²	1.88	2.05	2.18	2.45	2.64	2.85	3.16
Flatwise tensile strength		MPa	7.5	7	5.1	4	3.2	2.1	0.9
Flatwise compressive strength		MPa	8.2	7.3	3.8	3.3	2.5	2	1.2
Flatwise compressive modulus		MPa	27.4	41.1	32.5	43.4	35.1	30.1	26.3
Shear strength	Warp	MPa	2.9	2.5	1.3	0.9	0.8	0.6	0.3
	Weft	MPa	6	4.1	2.3	1.5	1.3	1.1	0.9
Shear modulus	Warp	MPa	7.2	6.9	5.4	4.3	2.6	2.1	1.8
	Weft	MPa	9	8.7	8.5	7.8	4.7	4.2	3.1
Bending rigidity	Warp	N.m2	1.1	1.9	3.3	9.5	13.5	21.3	32
	Weft	N.m2	2.8	4.9	8.1	14.2	18.2	26.1	55.8

thickness and are used like another layer of reinforcement in the middle of a laminate, being designed to "wet out" with the laminating resin during construction. The microspheres displace resin and so the resultant "core" layer, although much heavier than cellular core materials, is lower in density than the equivalent thickness of FRP. The thin fabrics also conform easily to 3D curvature, and so are quick and easy to use. As these products are impregnated with resin in a similar fashion to fabrics, it is easy to maintain proper quality control. Mostly, the core mats are used for thin wall composite products and surface finishing material.

One type of core mats made with polyester nonwoven fabric with polymer spheres are stretchable and drapable in length and crosswise, in both dry and wet condition as shown in Figure 1.48. For this reason, it is well suited for making parts of complex component geometry. It is mainly used in the hand layup process but can also be

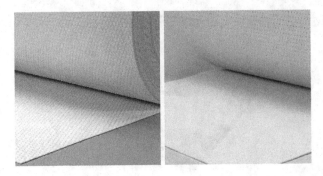

FIGURE 1.48 Core mat made by polyester nonwoven fabric and polymer sphere.[79]

used for surface infusion and wet pressing. As a cost-effective solution for open mold process, they can be core materials and/or print blockers compatible with all regular types of reinforcing fiber mats and resins, including polyester, vinyl ester, phenolic and epoxy. The physical and mechanical properties of one type of polyester nonwoven core mat are listed in Table 1.31.

Other products are made by combining high-modulus fibers (mainly glass fibers) with light polymer microspheres. Thus, it is possible to produce composite materials of both superior mechanical properties and low specific weight. One of these is based on stitched glass veil with thermoplastic hollow spheres as shown in Figure 1.49. It gets soft and shapeable when impregnated with resin and thus allows joint less overlapping and excellent bonding with the covering glass layers. It is also used to improve laminate surfaces as it eliminates print through of subsequent layers such as core layers. It is mainly used in hand layup process and panel production.

TABLE 1.31
Physical and Mechanical Properties of Core Mat Made by Polyester Nonwoven Fabric with Hollow Spheres*,[79]

Typical Property	Unit	Value			
Thickness	Mm	2.0	3.0	4.0	10.0
Dry weight	g/m^2	80	110	140	250
Resin uptake	kg/m^2	1.0	1.5	2.0	6.5
Density impregnated	kg/m^2	540	540	540	680
Property	**Test Standard**	**Unit**	**Value**		
Compression strength (10% Strain)	ISO 844	MPa	10		
Tensile strength (across Layers)	ASTM C297	MPa	4		
Shear strength	ASTM C273	MPa	3		
Shear modulus	ASTM C273	MPa	25		
Flexural strength	ASTM D790	MPa	8.5		
Flexural modulus	ASTM D790	MPa	1,250		

* Mechanical properties of honeycomb was made by nonwoven polyester fabric core mat impregnated with unsaturated Polyester.

Sandwich Structural Core Materials

FIGURE 1.49 Core mats made with fiber glass -woven fabric.[80]

1.6 SHEET FORMATS OF CORE MATERIALS

For use as the core of a sandwich composite, the core sheet or block must be processed into different formats or configurations based on the requirements of the application and laminating method. The different formats of core are shown in Table 1.32. Initially, the block or bun needs to be cut into a "plain" or rigid sheet with required thickness. Following that, different formats will be processed according to the specific processing method and contours of the mold or layup surface.

TABLE 1.32
Different Formats of Core Materials[81]

Surface Finishing	Figure	Remarks
Plain		After the sheets are cut, no extra surface treatment steps are needed. For a contoured surface, the sheet can be adapted by thermoforming the plain sheets.
Perforated		Cylindrical through holes are perforated in a regular pattern distributed over the sheet area. The diameter and distance between holes can be adjusted based on the application's requirements. By using this format core, the air under the sheet can be expelled out and the resin will flow to the other side through the holes.
Grooved		The sheet surface is grooved to provide resin flow channels. The grooves can be made on one side or both sides, in one direction or two crossing directions.

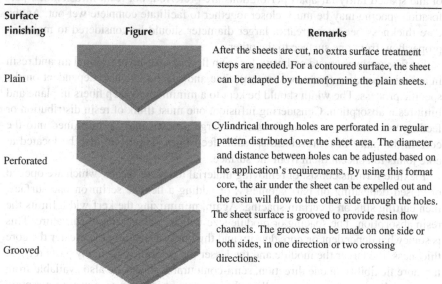

(Continued)

TABLE 1.32 (*Continued*)
Different Formats of Core Materials[81]

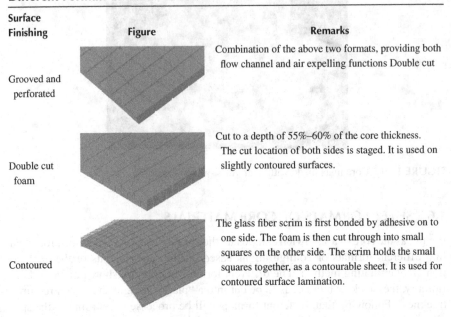

Surface Finishing	Figure	Remarks
Grooved and perforated		Combination of the above two formats, providing both flow channel and air expelling functions Double cut
Double cut foam		Cut to a depth of 55%–60% of the core thickness. The cut location of both sides is staged. It is used on slightly contoured surfaces.
Contoured		The glass fiber scrim is first bonded by adhesive on to one side. The foam is then cut through into small squares on the other side. The scrim holds the small squares together, as a contourable sheet. It is used for contoured surface lamination.

The diameter and spacing of perforations will depend on the lamination process. For open molding with vacuum bag curing and pre-preg molding, a limited number of and spaced fairly far apart perforations are preferred. For resin infusion, the perforation spacing may be much closer together to facilitate complete wet-out. As the core thickness becomes great, a larger diameter should be considered to maintain proper flow from one surface to the other.

Surface grooves create pathways for resin flow or take-up of residual air and resin in adhesive bonding (press or vacuum). Size and spacing are also dependent on the specific process. The width should be kept to a minimum to keep fibers in-plane and limit resin absorption. Considering infusion, one must think of resin distribution or feeding vs. laminate wet-out. Occasional large grooves may be machined into the core in the place of removable/disposable feed lines. These can also be located at core joints to conceal them from the surface.

Flexible formats consist of removing material to create "kerfs" which are opened or closed depending on part geometry, or adding a flexible scrim on one surface, then cutting slits on a certain spacing. Again, minimizing the kerf width limits the resin absorption. The spacing or module size will set the maximum curvature. This is somewhat dependent on core density and thickness. Generally the greater the core thickness, the larger the module and the lesser the flexibility. As many parts require far more flexibility in one direction, semi-contourable sheets are also available from most suppliers. These are generally only for vacuum infusion as a way to maintain flexibility and reduce resin uptake without thermoforming.

Sandwich Structural Core Materials

FIGURE 1.50 Core kits for a wind turbine blade.

When multi pieces of core are used for making a large composite construction, such as a boat hull and a wind turbine blade, they need to be cut into different thicknesses, different shapes, and even different 3D formats for meeting the requirements of the mechanical, geometric, and aesthetic designs of an end product. One set of these core sheets is often called a "kit". The core kit for one shell of a wind turbine blade is shown in Figure 1.50. Many specialized machines are designed and used for making core kits. There are a number of benefits to kitting core before installation:

- Design optimization
- Proper tapers and bevels on all edges
- Good fit between sheets
- Labels and instructions for piece placement
- Less labor
- Increased productivity
- Lower installed cost

REFERENCES

1. S. Black, "Getting To the Core of Composite Laminates." *Composites Technology*, October 1, 2003.
2. L. Lauri, S. Ang, J. Stigsson and R. Bressan, "Formulations for cellular, foamed-polymer products based on polyvinyl chloride, improved cellular, foamed-polymer products based on polyvinyl chloride and a process for producing said improved cellular foamed-polymer products," US Patent 8,168,293, May 1, 2012.
3. L. Lauri, "Neither toxic nor environmentally noxious foamed-polymer products," US Patent 5,352,710, Oct. 4, 1994.
4. "Strucell P Specification," http://www.strucell.com.
5. W. Ma and K. Feichtinger, "Polyester core materials and structural sandwich composites thereof," US Patent 7,951,449, May 31, 2011.
6. "The Fire Resistance Structural Foam," http://3accorematerials.com.
7. J. Sloan, "Structural polyurethanes: Bearing bigger loads." *Composites Technology*, March 10 2010.
8. "The production of rigid polyurethane foam," http://www.plastics.bayer.com.
9. "Air-cell polyester foam core," http://www.polyumac.com.

10. J. Sloan, "Core for composites: Winds of change", 6, 2010. https://www.composites-world.com/articles/core-for-composites-winds-of-change.
11. S. Ang, "Rigid Polymer Foams for Wind Blade Applications," *Polymer Foam Conf.*, New Jersey, Oct., 2012.
12. "Structural Core Materials," http://www.gurit.com/gurit-gorecell-t.aspx.
13. H. Seibert, "Applications for PMI foams in aerospace sandwich structures." *Reinforced Plastics*, vol. 50, no.1, pp. 44–48, 2006.
14. "ROHACELL® A Product Information," http://www.rohacell.com.
15. "Ultem foam for high-performance applications", https://www.sabic.com/en/products/specialties/ultem-resins/ultem-foam.
16. "High performance structure foam," http://3accorematerials.com.
17. "Divinycell F - A recyclable, prepreg-compatible sandwich core with excellent Fire, Smoke & Toxicity (FST) properties, suitable in commercial aircrafts interiors," https://www.diabgroup.com/en-GB/Products-and-services/Core-Material/Divinycell-F.
18. M. Islam and H.S. Kim, "Manufacture of syntactic foams: Pre-mold processing", *Materials and Manufacturing Processes*, vol. 22, pp. 28–36, 2007.
19. N. Gupita and E. Woldesenbet, "Characterization of flexural properties of syntactic foam core sandwich composites and effect of density variation," *Journal of Composite Materials*, vol. 39, no. 24, pp. 2197–2212, 2005.
20. "Gurit® Corecell™ S Structural Foam Core," https://www.gurit.com/Our-Business/Composite-Materials/Structural-Core-Materials/Corecell-S.
21. M. C. Wiemann, "Characteristics and Availability of Commercially Important Woods," In *Wood Handbook Wood as an Engineering Material*, Gen ed., Madison, WI: U.S. Dept. of Agriculture, Forest Service, Forest Products Laboratory, 2010, Chapter 2, p. 2.
22. Balsa Handbook, Gamble Bros. Inc., p. 5.
23. M. de Agricultura and G.A. Pesca, "Programa de Incentivos para la Reforestacion con Fines Comerciales" p. 32.
24. FPL report, C.A. Weipking and D.V. Doyle, "Strength & Related Properties of Balsa & Quipo Woods," Forest Products Laboratory, 1944.
25. M.C. Wiemann, "Characteristics and availability of commercially important woods," In *Wood Handbook: Wood as an Engineering Material*, Gen ed., Madison, WI: U.S. Dept. of Agriculture, Forest Service, Forest Products Laboratory, Chapter 2, p. 3, 2010.
26. Baltek Corporation Test Report, "Large Beam Flexural Testing of Selected Density vs. Alternating ("checkerboard") Density End-grain Balsa Core," 2007.
27. K. Feichtinger and W. Ma, "Durability of End-grain Balsa Core Material in the Marine Environment," *SAMPE Conference*, 2005.
28. Hexcel Corporation TSB 120 "Mechanical Properties of Hexcel Honeycomb Materials" p. 10, 1992.
29. Baltek Corporation Test Report, "BALTEK® SB Det Norske Veritas Type Approval Certification," 2010.
30. K. Feichtiner, "Test methods and performance of structural core materials -1. Static properties," *Journal of Reinforced Plastics and Composites*, vol. 8, pp. 334–357, July 1989.
31. K. Feichtinger, "Memo to Beacon Technology, High Strain-rate Compressive Properties of End-grain Balsa vs. Density, Energy Absorption," 1990.
32. K. Feichtinger and S. Martensen, "A Comparison of the Creep Properties of Aluminum and Nomex Honeycombs, PVCs and Balsa," In *3rd International Conference on Sandwich Construction*, 1995.
33. K. Feichtinger and W. Ma "Effect of Vacuum Infusion on Density & Mechanical Properties of End-grain Balsa Core Material," SAMPE 2006.
34. National Advisory Committee for Aeronautics, Technical Memorandum No. 984 & A.O. Desjarlais, "The Apparent Thermal Conductivity and Thermal Resistance of Four Specimens of End-Grain Balsa Insulation Materials," Dynatech report #BTX-1, 1985.

35. S.L. Zelinka and S.V. Glass "Moisture relations and physical properties of wood," In *Wood Handbook: Wood as an Engineering Material*, Gen ed., Madison, WI: U.S. Dept. of Agriculture, Forest Service, Forest Products Laboratory, Chapter 4, p. 41, 2010.
36. D. Spurr, *Heart of Glass*, 1st ed., pp. 121–122, 2004.
37. "Corecork® Technical data sheet," https://www.amorim.com/en/.
38. R. Elkin, "Corecork® evaluation of mechanical properties as a sandwich core material," Baltek Inc. Research paper, 2020.
39. "Honeycomb profile and core sheet after laminated with surface protecting film and veil," http://www.nidaplast.com.
40. R. Duchene, "Process for Producing of a Honeycomb Structure and Honeycomb Structure So Produced," US Patent 5,683,782, Nov. 4, 1997.
41. "Honeycomb Core Sheet Made from the Tubes," http://www.tubus-bauer.com.
42. "Honeycomb Core Made by PP and Tube Co-Extrusion/Block Heat Bonding," http://www.plascore.com.
43. I. Verroest and J. Pflug, "Half closed thermoplastic honeycomb, their production process and equipment to produce," EP 1,824,667 B1, 2010.
44. "Folded honeycomb production steps," http://www.econcore.com.
45. T. Bitzer, *Honeycomb Technology: Materials, Design, Manufacturing, Applications and Testing*, London, UK: Chapman & Hall, 1997.
46. H.N.G. Wadley, "Multifunctional Periodic Cellular Metals," *Philosophical Transactions of the Royal Society A*, vol. 364, pp. 31–68, 2006.
47. "Metallic honeycomb sheets," http://www.al-honeycomb-panels.com.
48. "Honeycombs Made by Stainless Steel Foils," http://www.plascore.com.
49. "Flexible metallic honeycomb," http://www.hexcel.com.
50. P. Lin, "High Shear Modulus Aramid Honeycomb," US Patent 5137768, Aug. 11, 1992.
51. "Expansion process for making composite honeycomb core," https://www.sme.org/honeycomb-heroes-making-composites-aerospace.
52. G.M. Naul, "Honeycomb Structure," US patent 4500583, Feb. 19, 1985.
53. W.T. Jackson, J.A. Stark and B.R. Garrett, "Reinforced plastic honeycomb method and apparatus," US Patent 3600249, Aug. 17, 1971.
54. "Fiberglass Composite Honeycomb," http://www.hexcel.com.
55. J. Pflug, I. Verpoest and D. Vandepitte, "Folded Honeycombs – Fast and continuous production of the core and a reliable core skin bond," In *International Conference on Composite Materials*, T. Massard, ed. Paris, 1999.
56. "Hexagon shape paperboard honeycomb made by expansion or corrugating process," https://bestem.com.pl/products.
57. "Paperboard Honeycomb with or without Impregnated Phenolic Resin," http://www.tricelcorp.com.
58. J. Banhart, "Manufacturing routes for metallic foams," *JOM*, vol. 52, no.12, pp. 22–27, 2000.
59. B.H. Smith, S. Szyniszewski, J.F. Hajjar, B.W. Schafer and S.R. Arwade, "Steel foam for structures: A review of applications, manufacturing and material properties," *Journal of Constructional Steel Research*, vol. 71, pp. 1–10, 2012.
60. L. Ma and Z. Song, "Cellular structure control of aluminum foams during foaming process of aluminum melt," *Scripta Materialia*, vol. 39, no.11, pp. 1523–1528, 1998.
61. F. Baumgärtner, I. Duarte and J. Banhart, "Industrialization of powder compact foaming process," *Advanced Engineering Materials*, vol. 2, p. 168, 2000.
62. "Different aluminum foams configurations," http://www.alumking.com.
63. "Aluminum Foam," http://www.alumking.com.
64. H.-W. Seeliger, "Application strategies for aluminium-foam-sandwich parts," In *Metal Foams and Porous Metal Structures*, J. Banhart, M.F. Ashby, and N.A. Fleck, eds., Bremen, Germany: MIT Verlag, 1999, p. 29.

65. U. Krupp, T. Hipke and S. Nesic, "Structural loading of cellular metals: damage mechanisms and standardization concepts," *Materials Science Forum*, vol. 933, pp. 220–225, October 2018.
66. J.-H. Eom, Y.-W. Kim and S. Raju, "Processing and properties of macro porous silicon carbide ceramics: A review," *Journal of Asian Ceramic Societies*, vol. 1, pp. 220–242, 2013.
67. K. Schwartzwalder and A.V. Somers, "Method of Making Porous Ceramic Articles," US Patent 3,090,094, May 21, 1963.
68. "Ceramic Foam," http://www.ultramet.com.
69. D.M. Spradling and R.A. Guth, "Carbon foam," *Advanced Materials & Processes*, vol. 161, pp. 29–31, Nov. 2003.
70. "Carbon Foams," https://www.cfoam.com/spec-sheets.
71. N. Zaidl, M. Rejab and N. Mohamel, "Sandwich Structure Based On Corrugated-Core: A Review," *MATEC Web of Conferences* 74, 00029, 2016.
72. "Triangle, trapezoidal and sinusoidal corrugated-cores," http://www.al-honeycomb-panels.com.
73. X. Yang, J. Ma, Y. Shi, Y. Sun and J. Yang, "Crashworthiness investigation of the bio-inspired bi-directionally corrugated core sandwich panel under quasi-static crushing load," *Materials and Design*, vol. 135, pp. 275–290, 2017.
74. T. Liu, Z.C. Deng and T.J. Lu, "Design optimization of truss-cored sandwiches with homogenization," *International Journal of Solids and Structures*, vol. 43, pp. 7891–7918, 2006.
75. H. Wadley, "Multifunctional periodic cellular metals," *Philosophical Transactions of the Royal Society A*, vol. 364, pp. 31–68, 2006.
76. H.N.G. Wadley, N.A. Fleck and A.G. Evans, "Fabrication and structural performance of periodic cellular metal sandwich structures," *Composites Science and Technology*, vol. 63, no. 16, pp. 2331–2343, 2003.
77. A. McCracken and P. Sadeghian, "Fiberglass Sandwich Beams With 3d Woven Fabric Cores," *Building Tomorrow's Society*, Fredericton, Canada, June 13–16, 2018.
78. "The 3-D spacer fabric made of glass fiber, carbon fiber or basalt fiber," http://www.ycglassfiber.com/3dcf/3dcf_01.php.
79. "Core mat made by polyester nonwoven fabric and polymer sphere," http://www.lantor.com.
80. "Core mats made with fiber glass woven fabric," http://spheretex.com.
81. "Foam core surface finishing." http://www.strucell.com.

2 Special Properties and Characterization Methods of Core Materials

Wenguang Ma

Russell Elkin

CONTENTS

2.1 Specialties of Sandwich Core Materials 74
2.2 Flatwise Compressive Properties 75
2.3 Flatwise Tensile Properties 76
2.4 Plate Shear Properties 78
2.5 Other Important Properties 79
References 84

An important structural element of the sandwich composite is the core that serves a far greater demand than simply separating the two facings. The prime requirements for a structural core sheet include high strengths and shear, compressive, and tensile moduli at densities much lower than the facings. Most sandwich constructions are large flat or contoured structures with a high ratio of length or width to thickness. The core in the sandwich structure suffers shear, compression, and tension stress during different times and in different locations in most applications. So the plane shear, flatwise tension, and compression properties of the core are important behaviors for all core materials. The water and moisture resistance, thermal conductivity, and temperature resistance are also important properties of the core materials. In this chapter, the testing principles, standard methods, fixtures, and calculation equations for characterizing the core materials will be introduced and discussed. The key parameters of each property are presented. This chapter gives the reader enough knowledge to evaluate and choose the core materials for designing the sandwich structure, product, and process.

2.1 SPECIALTIES OF SANDWICH CORE MATERIALS

As mentioned in the previous chapters, sandwich construction is a term used to describe a composite consisting of two relatively thin, high tensile, and compressive strength facings laminated around a lightweight, high shear strength core. The sandwich laminate is composed of a total of five structural elements. The two facings are generally materials exhibiting a high strength and modulus of elasticity, including reinforced plastics, single of multiple plies of wood, or metals such as aluminum or steel. As with the steel I-beam, when the panel is subjected to flexural loads, the loaded facing experiences compressive stresses while the opposite facing is in tension. When greater core thicknesses are utilized, the flexural rigidity is increased by much more than a linear relationship with only a minor increase in weight. The adhesion between each facing and the core constitutes the third and fourth structural elements of this assembly. This interface may be a film adhesive, as is the case for metal-faced laminations, or in the case of wet lamination, the unsaturated polyester or epoxy resin used as a matrix for the reinforcement would serve as the adhesive layer. Depending on the nature of the load imposed on the sandwich panel, the adhesive layer may be stressed in tension, shear, compression, or peel.[1]

The last and important structural element, the core, serves a far greater demand than simply separating the two facings. The prime requirements for structural core panels include high shear, compressive, and tensile strengths and moduli at densities much lower than the facings. Most sandwich constructions are large flat or contoured structures with a high ratio of length or width to thickness, such as a boat hull, truck sidewall, rail vehicle floor, and wind turbine blade.

The core is used as a flat sheet with large area but relatively small thickness. Most external forces, such as water and wind, uniformly load the construction, so that the flat sandwich structure usually reacts with a flexural response. When the sandwich structure experiences tensile stresses on one facing and compressive stresses in the other facing, the core must transfer the force from one facing to the other in shear, and that is the origin of the shear stress. When the sandwich panel serves as a floor, the core experiences the normal, flat-wise pressure. When the panel is used as a truck sidewall, a vertical force works on the top of the panel, so the facings experience a buckling force. If the facing is buckling in outward, the core with adjacent adhesive attempts to hold the facing and core together, and it experiences a flat-wise tensile stress as shown in Figure 2.1. In conclusion, the core in the sandwich structure suffers the shear, compression, and tension stress during different times and in different locations in most applications. The shear stress on the core is parallel to the face, but the compression and tension stress are perpendicular to the faces. In other words, the plane shear, flat-wise tension, and compression properties are important behaviors of all core materials.

The essential core static performance characteristics and test methods were identified and documented by the US Department of Defense in the form of MIL-STD-401B.[2] Consisting predominantly of American Society for Testing and Materials (ASTM) standard methods, MIL-STD-401B is used for specifying structural core materials. The International Organization for Standardization (ISO) also has similar test methods corresponding to the appropriate ASTM methods. The testing principles, fixtures, and important results for characterizing the core materials will be presented in the following sections.

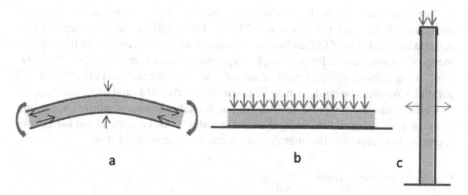

FIGURE 2.1 Sandwich panel in three different loading cases: (a) bending, (b) compression, and (c) vertical compression.

2.2 FLATWISE COMPRESSIVE PROPERTIES

As mentioned earlier, all sandwich core materials must have enough compression strength not to fail under load as well as suitable compression modulus to resist buckling. Thus, compressive properties through the thickness of the core are important performance characteristics for the applications of sandwich construction.

ASTM C365, D1621/B, ISO844, and GB/T8813 are standard test methods of the compressive strength and modulus of core materials, as tested in the plane of the core surface. The test setup is shown in Figure 2.2. The compressive loads between two flat anvils, one fixed and one gimbaled, are measured by means of a load cell. Strain is determined by precise measurements of the foreshortening of a gauge length, measured at the center (using a small hole drilled through) or on both sides of the sample. From the cross-sectional area and the load versus strain data, the core compressive strength and modulus may be determined. This improves on the older method where both indenters were fixed, and the extensometer was attached directly to the core sample.

FIGURE 2.2 Flatwise compressive test setup. Note LVDT through the sample on the left.

Test specimens should be of square or circular cross section having areas not exceeding 100 cm² but not less than 6.25 cm². The height of the specimen shall be equal to the thickness of the sandwich construction and thus changes with the requirements of the application. Proper sample preparation is important as well as "stabilization," using a thin coating of resin or adhesive to make the surface perfectly flat and parallel. Compression strength may be calculated at 2%, 10%, and maximum strain.[3]

Calculations for compressive strength and modulus are accomplished according to the standard procedures, using the load at failure and the slope of the load/strain response. For reference, the appropriate equations are provided below:

Core compressive strength: $\sigma = \dfrac{P}{A}$

where
σ = core compressive strength, MPa (psi);
P = ultimate load, N (lb); and
A = cross section area, mm² (in²);

Core compressive modulus: $E = \dfrac{St}{A}$

where
E = core compressive modulus, MPa (psi);
$S = \left(\dfrac{\Delta P}{\Delta \mu}\right)$ slope of initial linear portion of load-deflection curve, N/mm (lb/in);
μ = displacement with respect to the two loading anvils, mm (in); and
t = core thickness, mm (in).

High modulus and high strength are required for a good structural core material.

2.3 FLATWISE TENSILE PROPERTIES

Flatwise tensile strength is a valuable indication of the inherent bonding strength of the core material. While designers know to steer clear of applying forces to sandwich laminates in tension and peel, during certain times or localized loads, this is unavoidable. For evaluating the tensile properties of the foam core, both edgewise and flatwise tests are commonly used. Especially for a low-density and low-strength foam, the tensile grips used for conventional tensile "dog bone" specimens will crush the end of the foam sample. For that reason, the flatwise tensile test is used and accepted by more industries and researchers.

ASTM C297, ASTM D1623, and GB/T 1542 are used as standard test methods for evaluating the flatwise, or normal to facing, tensile properties of the foam core material. Two heavy steel or aluminum blocks, each having a through hole perpendicular to the axial pulling direction, are bonded to each facing surface of a core sample, as illustrated in Figure 2.3. Double gimbals are used for pinning the blocks to the load cell and base of the test machine, to ensure than only axial forces are imposed on the specimen. The test method consists of subjecting a cored sandwich specimen to a uniaxial tensile force normal to the plane of the foam sheet. The force is transmitted

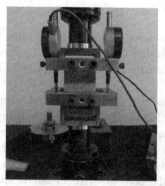 

FIGURE 2.3 Flatwise tension test setup.

to the foam through thick loading blocks, which are bonded to the core. Due to the small amount of strain before failure, it is critical to have enough resolution of the extensometers (or LVDTs) used. Note that this also applies to compression testing.[4] The ultimate failure load is normalized for the sample face area to obtain strength. Modulus is determined in the same manner as for compression but with forces acting in the opposite direction:

Core flatwise tensile strength: $\sigma = \dfrac{P}{A}$

where

σ = core tensile strength, MPa (psi);
P = ultimate load, N (lb); and
A = cross section area, mm² (in²);

Core tensile modulus: $E = \dfrac{St}{A}$

where

E = core tensile modulus, MPa (psi);
$S = \left(\dfrac{\Delta P}{\Delta \mu}\right)$ slope of initial linear portion of load-deflection curve, N/mm (lb/in);
μ = displacement with respect to the two loading blocks, mm (in); and
t = core thickness, mm (in).

Core tensile elongation at ultimate load: $\varepsilon = \left(\dfrac{\mu}{t}\right)100\ \%$

ε = elongation at ultimate load, %;
μ = displacement with respect to the two loading blocks, mm (in); and
t = core thickness, mm (in).

Not only exhibiting a high tensile modulus and a high strength, but an excellent core material should also possess a large ultimate elongation that is representative of the toughness of the core material.

2.4 PLATE SHEAR PROPERTIES

Shear is defined as the force that causes two contiguous parts of the same body to slide relative to each other in a direction parallel to their plane of contact. Shear strength is the stress required to yield or fracture the material in the plane of the core material.

When large sandwich structures, such as airplane fuselages and wings, boat hulls, truck sidewalls, and wind energy turbine blades, experience bending loads, the core materials function as a transition layer to transfer the force from one facing to the other and thus experience shear stresses more commonly than other stress. Shear stress also results when the sandwich structure simultaneously experiences compressive and tensile stresses, so when the core experiences any of these two stresses, it will transform them into shear stress within the core material. For these reasons, the shear properties of the core are the most important of all the mechanical properties.

Two prominent standard methods have been developed for evaluating the shear properties of sandwich core materials. The most commonly used is ISO1922 or ASTM C273 "Standard Test Method for Shear Properties of Sandwich Core Materials".[5] Samples are bonded between two steel plates by a structural adhesive and tested by forcing the plates to translate with respect to each other as shown in Figure 2.4. The plates can move in either tensile mode or compressive mode. From the load versus plate-to-plate displacement response, the shear modulus may be calculated. By normalizing the ultimate or breaking load for the sample dimensions, a core shear strength may be calculated. For reference, the appropriated equations are provided below.

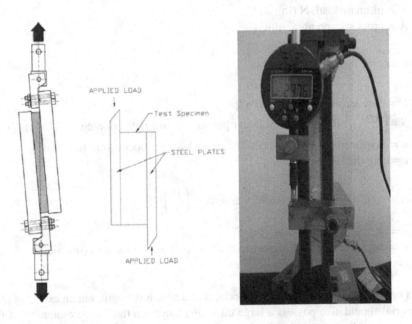

FIGURE 2.4 Plate shear test setup.

Core shear strength: $\tau = \dfrac{P}{Lb}$

where
- τ = core shear strength, MPa (psi);
- P = load on specimen, N (lb);
- L = length of specimen, mm (in); and
- b = width of specimen, mm (in).

Core shear modulus: $G = \dfrac{St}{Lb}$

where
- G = core shear modulus, MPa (psi);
- $S = \left(\dfrac{\Delta P}{\Delta \mu}\right)$ slope of initial linear portion of load-deflection curve, N/mm (lb/in);
- μ = displacement with respect to the two loading plates, mm (in); and
- t = thickness of core, mm (in).

Core shear ultimate elongation: $\gamma = \left(\dfrac{\mu}{t}\right)100\%$

where
- γ = core engineering shear elongation at specimen break, %;
- μ = displacement with respect to the two loading plates, mm (in); and
- t = thickness of core, mm (in).

What is important for shear testing is that the support block thickness must be proportional to the core thickness. This requires some large and heavy plates as the thickness approaches 50mm.

In addition to plate shear, the ASTM C393 "Core Shear Properties Testing of Sandwich Constructions by Beam Flexure" procedure can also be used as a test method for obtaining shear strength and modulus. For conducting the ASTM C393 test, the core must be laminated with facings that are sufficiently strong, such that failures are exclusively within the core, in shear. To accurately determine the properties of the core, the properties of the face sheets must be known before testing. Core and skin thickness must also be more precise than typically supplied. There is also a relationship between the performance of the skins and the measured modulus of the core.[6]

Another test method, ASTM C394 "Standard Test Method for Shear Fatigue of Sandwich Core Materials", can be used for evaluating the fatigue resistance of the foam core by repeatedly imposing shear stresses at levels below the core's static shear strength.[7]

2.5 OTHER IMPORTANT PROPERTIES

Even though the most common design criteria of structural core materials are their mechanical properties, in many cases, it is a combination of a core material's structural property and other characteristics that are of interest, for example, thermal insulation, sound dampening, dielectric properties, etc. The following section will give a brief description for each property.[8]

Thermal conductivity is the property or capability of a material to conduct heat, or the inverse value of thermal insulation capacity for a given material; e.g., foam cores have low thermal conductivity. Thermal conductivity is not a constant among various core materials and is affected by temperature, foam density, service history, and moisture content. A low-density material with air or a gas inside of the foam's cells generally provides a good insulating capacity, which explains the use of foam-cored sandwich structures as thermal insulation components where both insulation and load-bearing requirements are expected. Also, if the foam or material is prone to absorb water over time, the insulating capacity will get worse with age. This is a known problem for low density, polyisocyanurate foams.

ISO 8301 and ASTM C518 are two standard test methods for obtaining thermal conductivity of foam core materials by using the heat flow meter method to measure the steady state heat transfer through flat slab specimens and the calculation of the parameter. Table 2.1 lists thermal conductivity values of major foam core materials, polyethylene terephthalate (PET), polyvinyl chloride (PVC), and polyurethane foam cores at room temperature.

Maximum continuous operating temperature is the temperature at which the core, with or without facings, can be safely used without being reduced to below the minimum mechanical properties required by the particular application and imposed loads. Different core materials have special operating temperatures. Ceramic foam has the highest operating temperature, followed by metallic honeycomb, carbon foam, and fiberglass composite honeycomb. Maximum operating temperatures of plastic foam and honeycomb materials vary in a big range depending on the glass transition temperature (T_g) and heat distortion temperature (HDT) of each raw material. HDT is the temperature at which the foam or honeycomb deforms under a specified stress. The value of HDT can be determined by using a thermomechanical analyzer (TMA) or by testing the deformation of the sample at different temperatures in response to a specified stress. The HDT value varies with the density of the core, with the type of polymer used to produce the foam or honeycomb, and is a basic property for establishing the maximum processing and operating temperatures of a foam core material. Maximum process temperature is the temperature at which the core may be laminated with facing materials by thermal curing and consolidation processing without losing physical and mechanical properties. It is dependent on the time,

TABLE 2.1
Thermal Conductivity of Major Foam Core Materials at Room Temperature

		Density (kg/m³)			
	Material	80	100	130	200
Thermal Conductivity (W/m K)	PET Foam	0.030	0.034	0.037	0.045
	PUR Foam	0.030	0.034	0.035	0.042
	PVC Foam	0.033	0.035	0.039	0.048

pressure, and processing conditions, and is usually lower than the HDT but little high than the maximum operating temperature. The maximum operating temperatures of the most useful core materials are listed in Table 2.2 as reference.

Cellular foam will exhibit different behaviors than solid polymer with increasing temperature. The T_g for PVC foam is 82°C, but this does not prohibit the material from both service and lamination at temperatures well above this value. While the polymer may "soften", the gas pressure within each cell increases causing a stabilizing effect. Also, the T_g for PET foam can be as low as 61°C; however, this product will tolerate higher service and process temperatures than PVC or SAN because it is semi-crystallized polymer. It is important to know how the mechanical properties vary with temperature. Compression properties of a foam core at different temperatures represent its heat resistance. Figure 2.5 shows the compressive strength and modulus of PVC foam (top) and PEI foam at different temperatures. The modulus of PVC foam has a big reduction at 55°C, while that of PEI foam can keep with big change until 120°C.

Dielectric constant refers to the relative dielectric constant that is a ratio of capacitance for the material compared to capacitance for vacuum as the dielectric, when a high frequency potential is imposed upon it. It represents how readily electromagnetic signals (radio and radar waves) pass through the material. Usually foam core made by pure polymer has good dielectric properties that vary with the density of the foam, the polymer from which the foam was produced, and the frequency of the signal. A low value of dielectric constant means that electromagnetic signals travel through the material with only a small loss. Polymethacryl imide (PMI) foam, polyether imide (PEI) and polyether sulfone (PES) foam cores, fiberglass-reinforced polyimide honeycomb, and aramid honeycomb materials have good dielectric properties.

Fire, smoke, and toxicity (FST) behavior is a requirement in most public transportation and building construction applications, and the authority's specific requirements vary in different regions of the world. The FST behavior is defined by the aspects of flame spreading, dripping during burning, smoke density, heat release, toxic fumes, limiting oxygen index, etc. Metallic foam and honeycomb, ceramic foam, and carbon foam cores certainly have good FST properties. The honeycombs made by glass fiber fabric and aramid fiber paper saturated with phenolic resin all have excellent FST behavior. The plastic foam and honeycomb cores can be an inherent fire-retardant material or be modified into a fire-retardant grade by the use of an appropriate additive. Foam core materials made by crystalized PET, PMI, PEI, and PES have good FST properties also.

Core thickness can influence the fire performance. Some materials will not burn when the thickness is low; the skins act to snuff out the fire. As the core thickness becomes great, the distance between the face sheets becomes large enough to sustain combustion. Conditioning is also important. A number of products will exhibit greater flame spread if not conditioned before testing. It should also be noted that some fire-resistant resins do not perform as well when laminated in sandwich. The core acts as an insulator which prevents the energy from dissipating through the laminate, something to consider when designing laminates for fire.

TABLE 2.2
Maximum Operating Temperatures of the Most Useful Core Materials

Core Material	Silicon Carbide (SiC) Foam	Carbon Foam	Fiberglass-Reinforced Polyimide Honeycomb	NOMEX® Aramid Honeycomb	PEI Foam	PMI Foam	PET Foam	Crosslinked PVC Foam	PUR Foam	PP Honeycomb
Maximum Operating Temperature (°C)	1,650 in air 2,500 inert	350 3,000 inert	260	180	160	170	100	90	120	80

Characterization Methods of Core Materials

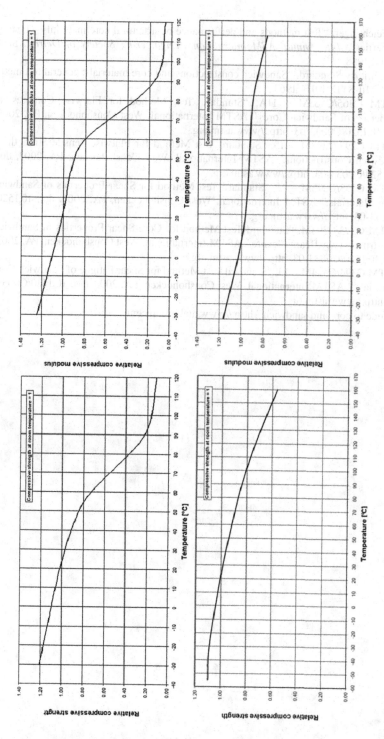

FIGURE 2.5 Compressive strength and modulus of PVC foam (top two charts, and PEI foam (bottom two charts).

REFERENCES

1. K. Feichtinger, "Test methods and performance of structural core materials – I. static properties". *44th Annual ASM International/Engineering Society of Detroit*, Sept. 13–15, 1988.
2. US Military Standard, "Sandwich constructions and core materials; general test methods." Mil-STD-401B, 1967.
3. ASTM C365/C365M – 11A, "Standard Test Method for Flatwise Compressive Properties of Sandwich Cores." ASTM International, West Conshohocken, PA, 2003. doi: 10.1520/C0033-03. http://www.astm.org.
4. ASTM C297/C297M – 15, "Standard Test Method for Flatwise Tensile Strength of Sandwich Constructions." ASTM International, West Conshohocken, PA, 2003, doi: 10.1520/C0033-03. http://www.astm.org.
5. ASTM C273/C273M – 11, "Standard Test Method for Shear Properties of Sandwich Core Materials." ASTM International, West Conshohocken, PA, 2003. doi: 10.1520/C0033-03. http://www.astm.org.
6. ASTM C393/C393M, "Standard Test Method for Core Shear Properties of Sandwich Constructions by Beam Flexure." ASTM International, West Conshohocken, PA, 2003, doi: 10.1520/C0033-03. http://www.astm.org.
7. ASTM C394/C394M – 13, "Standard Test Method for Shear Fatigue of Sandwich Core Materials." ASTM International, West Conshohocken, PA, 2003. doi: 10.1520/C0033-03. http://www.astm.org.
8. "Guide to core and sandwich." http://www.diabgroup.com.

3 Face Sheet Materials for Sandwich Composites

Trevor A. Gundberg

CONTENTS

3.1 Metal, Plastics, and Plywood Face Sheet Materials 86
    3.1.1 Metallic Face Sheet Materials 86
        3.1.1.1 Steel 87
        3.1.1.2 Aluminum 87
        3.1.1.3 Magnesium 90
        3.1.1.4 Titanium 90
        3.1.1.5 Miscellaneous Alloys 90
    3.1.2 Plastics 93
        3.1.2.1 Thermoplastics 93
        3.1.2.2 Commodity Plastics (PE, PP, PET, and PVC) 93
        3.1.2.3 Engineered Thermoplastics (PA6, PEI, PPS, etc.) 94
        3.1.2.4 High-Performance Thermoplastics (PAEK, PEEK, and PI) ... 94
    3.1.3 Wood 94
3.2 Reinforcement Fiber Materials for Composite Face Sheets 99
    3.2.1 Glass and Basalt Fibers 99
        3.2.1.1 E-Glass and ECR-Glass 99
        3.2.1.2 H-Glass and R-Glass 99
        3.2.1.3 Basalt 100
        3.2.1.4 S-Glass 100
        3.2.1.5 Quartz (Silica) 100
        3.2.1.6 A-Glass, C-Glass, and AR-Glass 100
    3.2.2 Carbon Fibers 101
        3.2.2.1 PAN-Based Carbon 101
        3.2.2.2 Pitch-Based Carbon 103
    3.2.3 Natural Fibers 103
        3.2.3.1 Bast 104
        3.2.3.2 Leaf 104
    3.2.4 Synthetic Fibers 104
        3.2.4.1 Para-Aramid 105
        3.2.4.2 High-Modulus Polypropylene 105
        3.2.4.3 UHMWPE, PBO, and LCP 105
    3.2.5 Boron and Ceramic Fibers 106
        3.2.5.1 Boron 106

DOI: 10.1201/9781003035374-3

		3.2.5.2	SiC and Alumina	106
	3.2.6	Architectural Forms of Reinforcement Materials		107

3.3 Plastic Matrix Materials for Composite Face Sheets 109
    3.3.1 Unsaturated Polyester and Vinyl Ester Resin 109
        3.3.1.1 Unsaturated Polyester (UPR) 110
        3.3.1.2 Vinyl Esters 110
    3.3.2 Epoxy Resin 111
    3.3.3 Phenolic Resin 111
    3.3.4 Polyurethane Resin 111
    3.3.5 High-Temperature Application Resins 112
    3.3.6 Thermoplastic Matrix 112
3.4 Composite Face Sheet Material Properties and Characterization 113
    3.4.1 Unidirectional Laminae 114
    3.4.2 Biaxial Laminates 116
    3.4.3 Quasi-Isotropic Laminates 116
    3.4.4 Test Methods for Laminated Composite Face Sheets 118
        3.4.4.1 Tensile Strength and Modulus 118
        3.4.4.2 Compressive Strength and Modulus 118
        3.4.4.3 In-Plane Shear Strength and Modulus 120
        3.4.4.4 Flexural Strength and Modulus in Sandwich ... 120
        3.4.4.5 Interlaminar Shear Strength 121
        3.4.4.6 Damage Tolerance and Impact Properties 121
References 122

Arguably, the most important components of a sandwich structure are the thin, stiff, and strong face sheets, facings, or "skins", which provide the in-plane tensile and compressive load resistance during bending or column compression. Due to the relative thinness of these elements, moderate to high density materials can be used with minimal impact on the overall sandwich laminate. This opens the door for many different structural materials to be utilized as face sheets including isotropic materials such as steel, aluminum, or plastic sheeting. Orthotropic materials such as wood veneers, plywood, or MDF also can be designed to be effective face sheet materials and are discussed in Section 3.1.3. A more detailed discussion is provided for polymer matrix composites (PMCs) given the large variety of options available for tailored face sheet properties.

3.1 METAL, PLASTICS, AND PLYWOOD FACE SHEET MATERIALS

3.1.1 METALLIC FACE SHEET MATERIALS

Due to the inherent stiffness, strength, and damage tolerance of metal materials, they make excellent facings for sandwich structures. While they typically are high in density relative to wood or polymer matrix composites, the sandwich construction with metal facings allows for lower weight and higher flexural stiffness and strength than monolithic metal sections. Several basic metal groups will be discussed in this section, following the group order detailed in the aerospace industry's MMPDS ("Metallic Materials Properties Development and Standardization") handbook.[1] Note that this is by no means a comprehensive list of metallic materials available for sandwich facings.

3.1.1.1 Steel

Generically, steel is a subgroup of iron-carbon alloy materials containing between 0.008% and 2.0% carbon but can also include a multitude of other components such as nickel, chromium, and molybdenum. For sandwich structures, only those steel alloys available in sheet or plate forms are of consequence, with several versions provided in Table 3.1. Steel materials provide high Young's modulus (180–210 Gpa), along with a range of strength, hardness, corrosion and temperature resistance, and ductility values depending on the alloy and any heat treatments or working done to the material. Drawbacks in use for sandwich constructions include high density, difficulty in forming complex shapes, and the need to adhesively bond (and properly treat the bonding surface) to core materials.

Another available format for steel is fibers. Various twisted strand wires are bonded to a scrim. While this greatly reduces the effective modulus, these tapes can be laminated like FRP. This creates opportunity for steel fibers to be braided or woven into fabrics, allowing the designer to tailor special or specific performance into the laminate or part.

Steel fiber and other metal fiber can be used as reinforcement fiber for composite materials, improving breaking behavior upon impact and electrical conductivity. Traditional carbon or glass fiber reinforcement fibers have very limited elongation possibilities, which results in a brittle and explosive breaking behavior. Metal fibers act perfectly complementary to this and can absorb much more energy before breaking. Processing is no different from any other reinforcement fiber for composite materials. It is even possible to combine metal fibers with other fibers into a "hybrid" composite structure, which combines all the benefits of carbon, glass, and steel.

If a material can be drawn into fibrous form (continuous or discontinuous), anything can become a composite laminate. Sandwich creates opportunities for lower performing, but low-cost materials to still be used to fabricate light and stiff parts.

3.1.1.2 Aluminum

The term "aluminum" in the case of structural metallic materials encompasses a variety of lightweight wrought or cast alloys (mainly wrought, for the sake of thin sheet material) with the aluminum element as its main ingredient. While elemental aluminum is rather soft, alloying it with other materials such as copper, magnesium, or zinc can increase its strength dramatically. As with steel, aluminum alloy properties can also be manipulated via heat and working treatments. Typical aluminum alloys have much lower Young's moduli than steel (69–79 Gpa) but are also much lower in density (2.77 g/cm^3 vs 7.83 g/cm^3), providing similar specific stiffnesses. Issues seen in use for sandwich structures again include adhesive bonding between skins and core and forming into complex shapes, but the lower density, corrosion resistance, and relative softness of the materials can make them much more attractive than steel materials (Table 3.2).

TABLE 3.1
Design Mechanical and Physical Property Ranges of Various Steels[1]

Steel Grade and Specification		Tensile Yield Strength F_{ty} (Mpa)	Compression Yield Strength F_{cy} (Mpa)	Shear Ultimate Strength F_{su} (Mpa)	Young's Modulus E (Gpa)	Shear Modulus G (Gpa)	Poisson's Ratio ν	Density ρ (g/cm³)
Carbon Steels	AISI 1025 (AMS 5046)	248	248	241	200	76	0.32	7.861
Alloy Steels (Low to High)	AISI 4130 & AISI 8630	483–517	483–517	372–393	200	76	0.32	7.833
	4335V	1,310	1,366	910	200	76	0.32	7.833
	9Ni-4Co-0.20C (AMS 6523)	1,193	1,290	786	199	76	0.3	7.833
	250 Maraging (AMS 6520)	1,641–1,690	1,524–1,759	1,020–1,069	183	N/A	0.31	7.916
Stainless Steels	AM-350 (AMS 5548)	1,014–1,034	1,124	834	200	76	0.32	7.806
	15-5PH (AMS 5862)	986–1,000	1,007–1,048	662–669	197	77	0.27	7.833
	AISI 301 ½ Hard (AMS 5518)	634–759	421–800	531–566	179–193	72	0.27	7.916

TABLE 3.2
Design Mechanical and Physical Property Ranges of Various Wrought Aluminum Alloys[1]

Alloy Grade and Specification	Tensile Yield Strength F_{ty} (Mpa)	Compression Yield Strength F_{cy} (Mpa)	Shear Ultimate Strength F_{su} (Mpa)	Young's Modulus E (Gpa)	Shear Modulus G (Gpa)	Poisson's Ratio ν	Density ρ (g/cm^3)
2000 Series							
2024-T3 (AMS-QQ-A-250/4)	290–331	269–317	269–283	72	28	0.33	2.768
2090-T83 (AMS 4251)	386–483	400–490	255	76–79	30	0.34	2.602
5000 Series							
5052-H34 (AMS 4017)	172–179	172–179	138	69	27	0.33	2.685
5083-O (AMS 4056)	124–131	124–131	172–179	70	27	0.33	2.657
6000 Series							
6013-T6 (AMS 4347)	317–331	331–337	220	68	2.6	0.33	2.713
6061-T6 (AMS 4025)	241–262	241–262	186–193	68	2.6	0.33	2.713
7000 Series							
7075-T6 (AMS 4045)	434–503	469–517	317–331	71	2.7	0.33	2.796
7475-T61 (AMS 4084)	441–455	441–468	310	69	2.6	0.33	2.796

TABLE 3.3
Design Mechanical and Physical Property Ranges of Wrought Magnesium Alloy[1]

Alloy Grade and Specification	Tensile Yield Strength F_{ty} (Mpa)	Compression Yield Strength F_{cy} (Mpa)	Shear Ultimate Strength F_{su} (Mpa)	Young's Modulus E (Gpa)	Shear Modulus G (Gpa)	Poisson's Ratio ν	Density ρ (g/cm^3)
AZ31B (AMS 4377)	200–221	166	124	45	17	0.35	1.772

3.1.1.3 Magnesium

As with aluminum, magnesium requires alloying with other ingredients such as aluminum and zinc to produce materials suitable for structural use. Again, wrought versions are available in thin sheet forms suitable for sandwich facing applications. Magnesium alloys are attractive materials as they are much lower in density (1.77 vs 2.77 g/cm^3) than aluminum alloys despite the lower Young's modulus. Flammability concerns have limited some use of magnesium alloys; however, work on better performing options has been in development (Table 3.3).

3.1.1.4 Titanium

This metallic element and its alloys can provide low density, high specific strength and stiffness, and corrosion-resistant face sheets when compared to other isotropic metals such as steel. Typical alloying elements for titanium include aluminum and vanadium, which provide various benefits to the base metal as do various heat treatments. Unlike aluminum and magnesium, titanium can also be used for structural applications in its "pure" or unalloyed form (Table 3.4).

3.1.1.5 Miscellaneous Alloys

There are a multitude of other metallic materials and alloys which can be suitable for sandwich facings which require specialized properties. Heat-resistant alloys based on iron, nickel, or cobalt can be used for sandwich structures requiring +540 °C service temperatures. Beryllium has excellent stiffness and low density but can be a health hazard due to its high toxicity. It is also highly anisotropic, with relatively low properties transverse to the grain. Copper and its alloys are excellent materials for thermal and electrical management and provide good galling and corrosion resistance.[1] Aluminum alloy sheets are laminates of thin aluminum alloy and fiber reinforced thermoset resin or adhesive. A common alloy sheet material uses 2024-T3 aluminum alloy with alternating aramid fiber-reinforced composite (Table 3.5).

TABLE 3.4
Design Mechanical and Physical Property Ranges of Titanium and Its Alloys[1]

Alloy Grade and Specification		Tensile Yield Strength F_{ty} (Mpa)	Compression Yield Strength F_{cy} (Mpa)	Shear Ultimate Strength F_{su} (Mpa)	Young's Modulus E (Gpa)	Shear Modulus G (Gpa)	Poisson's Ratio ν	Density ρ (g/cm^3)
Pure Titanium	AMS 4901	482	482	284	107	45	N/A	4.512
Alpha and Near Alpha Alloys	AMS 4910	759–834	793–869	507–554	107	N/A	N/A	4.484
	Ti-8Al-1Mo-1V (AMS 4915)	931	993–1,027	628	121	46	0.32	4.373
	Ti-8Al-1Mo-1V (AMS 4916)	828	869	568	121	46	0.32	4.373
	Ti-6Al-2Sn-4Zr-2Mo (AMS 4919)	862–938	910–979	N/A	114	43	0.32	4.540
Alpha-Beta Alloys	Ti-6Al-4V (AMS 4911)	869–903	917–972	588–608	110	43	0.31	4.429
	Ti-6Al-6V-2Sn (AMS 4918)	966–1,062	959–1,041	615	110	43	0.31	4.540
	Ti-4.5Al-3V-2Fe-2Mo (AMS 4899)	869–945	883–986	608–669	110	N/A	N/A	4.540
Beta, Near-Beta, and Metastable-Beta Alloys	Ti-13V-11Cr-3Al (AMS-T-9046)	828–869	828	622	100	N/A	N/A	4.816
	Ti-15V-3Cr-3Sn-3Al (AMS 4914)	966	959–993	622	105–108	N/A	N/A	4.761

TABLE 3.5
Design Mechanical and Physical Property Ranges of Various Other Alloys[1]

Alloy Grade and Specification	Tensile Yield Strength F_{ty} (Mpa)	Compression Yield Strength F_{cy} (Mpa)	Shear Ultimate Strength F_{su} (Mpa)	Young's Modulus E (Gpa)	Shear Modulus G (Gpa)	Poisson's Ratio ν	Density ρ (g/cm^3)
Iron-Chromium-Ni-Base A-286 (AMS 5525)	655	655	628	200	77	0.31	7.944
Nickel-Base Inconel 600 (AMS 5540)	241	241	352	207	76	0.29	8.415
Inconel 625 (AMS 5599)	366–428	386–455	545–579	206	81	0.28	8.442
Inconel 706 (AMS 5605)	966–1,000	1,007–1,048	731–752	210	76	0.38	8.083
Cobalt-Base L-605 (AMS 5537)	379–428	283–421	628–655	225	87	0.29	9.134
HS 188 (AMS 5608)	379–393	379	766	237	88	0.31	8.968
Beryllium AMS 7902	207–441	N/A	N/A	293	138	0.1	1.855
Al Alloy/Aramid Sheet AMS 4254 (3/2 layup scheme)	207–338	207–241	103[a]	52–68	16–17	0.27–0.34	2.325

[a] Shear yield strength as ultimate is not available.

3.1.2 Plastics[2-8]

Plastic materials work well as sandwich facings for applications requiring corrosion resistance, low weight, and overall durability. Plastic is the colloquial term for "polymers" which are hydrocarbon-based materials based on many "mers" or repeating chains of molecules. While unreinforced plastics are relatively compliant, they can still provide high stiffness when coupled with a suitable core material. Plastics are available in two broad categories: thermoplastics and thermosets. The former includes the more commonly known plastics (polypropylene, polyethylene, polystyrene, etc.) used for beverage containers, packaging, etc. and are ubiquitous in everyday life. They can be melted and reformed many times over and can be welded together to form efficient joints. The thermoset plastics (such as epoxies, polyurethanes, and vinyl esters) are more typically used as adhesives and resins for reinforced continuous fiber composites, even though thermoplastic composites have been recently gaining ground in the latter application. These plastics cannot be melted but will degrade and eventually combust at very high temperatures. Due to this, thermosets typically have higher operational temperatures and properties at these temperatures than thermoplastics. Note that some polymer chemistries can have both thermoplastic and thermoset varieties (polyurethanes and polyesters for two examples), with the main molecular differences being the polymer chain lengths and degree of atomic bonding.

3.1.2.1 Thermoplastics

This large group of plastic materials are categorized in a couple of different ways. The two broadest categories focus on the material molecular structure: semi-crystalline and amorphous. The former category indicates that these plastics have an ordered crystalline structure like most metals for at least a portion of its makeup. This generally provides better solvent resistance and a more well-defined melt temperature but can also produce more shrinkage during cooling from the liquid state compared to amorphous plastics. The amorphous materials contain no ordered structure, which can impart some unique attributes such as transparency (polycarbonate and PMMA) and high melt viscosity (desirable for certain fabrication processes such as blow molding).

A second grouping of thermoplastics is based on the mechanical properties and melt temperatures: commodity, engineered, and high-performance thermoplastics as shown in Figure 3.1.

3.1.2.2 Commodity Plastics (PE, PP, PET, and PVC)

Commodity plastics encompass materials mostly used in consumer products such as packaging and food or beverage containers, are generally the weakest and most compliant, and have the lowest maximum operating temperatures of all the thermoplastics. Partly due to their widespread use, commodity plastics are relatively cost-effective, and their combination of high chemical resistance, formability, and low density makes them an important material group. To increase the strength, stiffness, and other attributes for structural uses, commodity plastics are typically used with fiber reinforcements (see Section 3.2) in various length scales from milled up to continuous lengths.

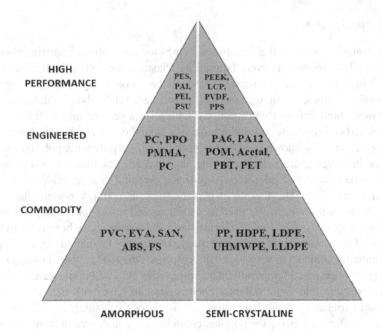

FIGURE 3.1 Different types of plastics can be used as facing materials of the sandwich composites.

3.1.2.3 Engineered Thermoplastics (PA6, PEI, PPS, etc.)

This grouping of thermoplastics generally provides higher mechanical properties and operating/melt temperatures than the commodity materials. As such, producing and processing these plastics can be more expensive, which can limit their use to more structural applications. As with the commodity plastics, these materials are also used with discontinuous and continuous reinforcement fibers to increase their properties, but neat or alloyed versions can still be strong enough for a multitude of applications including automotive and industrial structures.

3.1.2.4 High-Performance Thermoplastics (PAEK, PEEK, and PI)

This group provides the highest performance and operating temperature levels of all thermoplastic materials. As with the engineered plastics, the higher required processing temperatures and more onerous production of these materials also make them the most expensive and difficult to form plastics. With their high operating temperatures, these plastics are typically used in demanding aerospace applications and can often be used as matrices for carbon fiber reinforcements (Tables 3.6 and 3.7).

3.1.3 Wood

Wood, one of the earliest used materials used by man (including in sandwich structures), is still a very viable material for facings due to its relatively light weight and orthotropic nature as shown in Figure 3.2. The mechanical properties of wood depend not only on the species of tree from which it is harvested but also on the

TABLE 3.6
Nominal Mechanical and Physical Properties of Various Semi-Crystalline Thermoplastic Materials at Room Temperature[3]

Nominal Thermoplastic Grade		Tensile Yield Strength F_{ty} (Mpa)	Young's Modulus E (Gpa)	Flexural Modulus E_{flex} (Gpa)	Poisson's Ratio ν	Melting Point MP (°C)	Density ρ (g/cm³)
Commodity	HD Polyethylene	22.8–29.7	0.90–1.55	0.97–1.38	0.46	126–135	0.94–0.97
	Polypropylene	31.0–44.8	1.95	1.35–1.80	0.43	160–163	0.91
Engineered	Polyamide 6[4]	82.8	2.85	1.80	0.42	222	1.13
	PBT	57.9–60.0	1.60–2.70	1.60–2.40	0.42	220–225	1.30
	PET	60.0–85.5	2.80–3.17	3.39	0.43	255	1.36–1.41
	PPS	33.1–89.7[a]	3.81–4.21	3.75–3.90	0.43	280	1.36
High Performance	PAEK[5]	89.0	3.70	3.70	0.43	340	1.33
	PEEK	100.0–115.2	3.73	40.7–4.28	0.39	343–387	1.33

[a] Ultimate strength range.

growth conditions (inclusion of knots/defects, weather/climate conditions), moisture content, and specific gravity. As mentioned, wood is highly orthotropic in nature, with the highest stiffness and strength parallel to the grain direction, and varyingly lower properties in the tangential and radial grain directions.[9] Wood is categorized into two board tree groups: hardwoods (angiosperms) and softwoods (gymnosperms or conifers), names which are not necessarily representative of either group's hardness properties.

Tables 3.8 and 3.9 cover select hardwoods and softwoods, respectively, and note the rather "wood-centric" property listings. While several properties are measurements of the same variables as with metals and plastics, others such as modulus of rupture and directional-to-grain strengths are unique to wood materials but are analogous to other more universal property designations.

Plywood, as the name suggests, is a collection of oriented wood veneer plies into layers (consisting of single or multiple plies) to minimize the orthotropy and make the overall panel quasi-isotropic. The plies are adhered together with a moisture-resistant adhesive to effectively transfer loads between them. Plywood panels are usually made up of an odd number of layers (typically 3 or 5) and are oriented according to the longitudinal grain direction in a cross-ply configuration (adjacent layers are oriented 90° to the previous layer). This provides the overall panel with the maximum strength and stiffness possible in the major panel axes and the minimum thermal and moisture expansion coefficients[9]. Due to plywood's good mix of properties, relatively low density (usually between 0.61 and 0.69 g/cm³), and low cost, it has not only been used as sandwich facings but also for high compression-resistant core materials for such applications as outboard transom motor mounts in fishing and pleasure boats.

TABLE 3.7
Nominal Mechanical and Physical Properties of Various Amorphous Thermoplastic Materials at Room Temperature[3]

Nominal Thermoplastic	Grade	Tensile Yield Strength F_{ty} (Mpa)	Young's Modulus E (Gpa)	Flexural Modulus E_{flex} (Gpa)	Poisson's Ratio ν	Glass Transition Temperature Tg (°C)	Processing Temperature PT (°C)	Density ρ (g/cm³)
Commodity	RPVC	38–55	2.46–4.70	2.50–4.50	0.40	60–100	199–204	1.30–1.48
	PC	55–67	2.35–2.40	2.23–2.50	0.41	160–200	310–330	1.19–1.22
	GPPS	44–52	3.10–3.17	2.90–3.48	0.35	90	180–260	1.05
	PMMA*	80	3.40	3.50	0.36	90–110	210–280	1.19
Engineered	PEI[5]	110	3.58	3.51	0.36	215	>218	1.27
High Performance	PAI[6]	152–192	4.48	5.03	0.45	275	>278	1.41
	PI[7]	121	3.72	3.79	N/A	250	304	1.36

* Albis Plastics ALCOM® LDDC PMMA 1000 UV 18123 BK1016-11 PMMA, UV stabilized

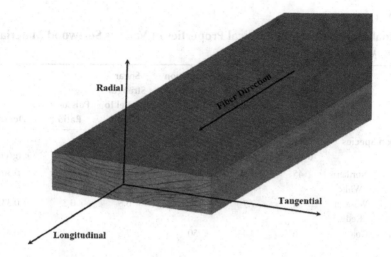

FIGURE 3.2 Fiber directions of wood plank.

TABLE 3.8
Nominal Mechanical and Physical Properties of Various Hardwood Materials at 12% Moisture Content[9]

Common Species Grade		Modulus of Rupture[a] MOR (Mpa)	Young's Modulus E (Mpa)	Compression Strength Parallel to Grain F_{cu} (Mpa)	Shear Strength Parallel to Grain F_{su} (Mpa)	Poisson's Ratio ν_{LT}	Density ρ (g/cm³)
Ash	White	103	12.0	51	13.2	0.44	0.609
Basswood	American	60	10.1	33	6.8	0.41	0.360
Birch	Yellow	114	13.9	56	13.0	0.45	0.609
Maple	Red	92	11.3	45	12.8	0.51	0.554
Oak	Northern Red	99	12.6	47	12.3	0.45	0.637
Oak	White	105	12.3	51	13.8	0.43	0.692
Walnut	Black	101	11.6	52	9.4	0.63	0.554

[a] Modulus of rupture is analogous to ultimate flexural strength assuming no yielding/plasticity.

Oriented strand board, or OSB, is a similar concept except that compressed and oriented wood strands are adhered together to form the final panel. OSB is typically cheaper than standard plywood but also is not as structural a material and not suitable for as many applications. Table 3.10 provides nominal engineering data for plywood and OSB panels, grouping by "span rating" (see reference 10), and note the plywood/OSB panel-specific properties when comparing to other structural facing materials.

TABLE 3.9
Nominal Mechanical and Physical Properties of Various Softwood Materials at 12% Moisture Content[9]

Common Species	Grade	Modulus of Rupture[a] MOR (Mpa)	Young's Modulus E (Mpa)	Compression Strength Parallel to Grain F_{cu} (Mpa)	Shear Strength Parallel to Grain F_{su} (Mpa)	Poisson's Ratio ν_{LT}	Density ρ (g/cm^3)
Cedar	Northern White	45	5.5	27	5.9	0.34	0.304
Cedar	Western Red	52	7.7	31	6.8	0.3	0.332
Douglas-Fir	Coast	86	13.4	50	7.8	0.45	0.471
Hemlock	Western	78	11.2	50	8.9	0.42	0.443
Pine	Red	76	11.2	42	8.3	0.32	0.471
Redwood	Young-Growth	54	7.6	36	7.7	0.35	0.360
Spruce	Sitka	70	10.8	39	7.9	0.47	0.388

[a] Modulus of rupture is analogous to ultimate flexural strength assuming no yielding/plasticity.

TABLE 3.10
Nominal Mechanical Properties of Various Plywood Panels at 12%–15% Moisture Content[10]

Span Rating and Ply Number		Axial Tensile Strength F_tA (kg/m width) × 10^3	Axial Compression Strength F_cA (kg/m width) × 10^3	Axial Stiffness EA (kg/m width) × 10^6	Bending Strength F_bS (N-m/m width)	Bending Stiffness EI (kg-m^2/m width)	Shear in the Plane F_s [lb/Q] (KN/m width)
24/0	3-ply	3.52	4.24	4.99	93	63	2.26
32/16	3-ply	4.17	5.28	6.18	137	120	2.92
32/16	OSB	4.17	5.25	6.18	165	110	2.41
20 oc	5-ply	5.58	9.38	7.44	213	221	3.87
20 oc	OSB	4.32	6.25	7.44	213	202	2.99
24 oc	5-ply	6.47	11.62	8.71	389	317	4.74
48 oc	5-ply	10.86	18.28	12.20	704	1,214	7.30

3.2 REINFORCEMENT FIBER MATERIALS FOR COMPOSITE FACE SHEETS

Most, if not all, of the plastic materials detailed in Sections 3.1.2 and 3.3 (and in some cases the metallics in Section 3.1.1) above can be reinforced via stiffer/stronger materials to provide a synergistic effect on the mechanical properties of the union, creating what is known as a composite material. Composite materials can take many forms, but the variants discussed here will focus on fiber-based composites (Section 3.4), which are the most widely used and efficient forms for sandwich structure face sheets. The current section will detail the many forms of fibers used in composite materials, all of which can be used in short, long, or continuous fiber forms depending on the end use requirements.

3.2.1 GLASS AND BASALT FIBERS[11]

Glass-based fiber, also colloquially known as "fiberglass", is the most widely used reinforcement fiber in polymer-based composite materials. Within the glass fiber category are multiple different types of glass fibers which provide certain characteristics that differentiate themselves from the other variants while having several properties in common. These commonalities include being electrically and thermally isolative, nonmagnetic, and highly resistant to corrosion from a variety of chemicals. Glass fiber is available is several formats including yarns (small, twisted assemblies of filaments) and rovings (larger nontwisted assemblies of filaments). These yarns and rovings are then further processed either into the composite itself, as in many filament winding and pultrusion processes, or into intermediate forms such as mats or fabrics, which is common for most continuous reinforcements. The fiber yarns/rovings can also be chopped or milled into long and short fibers for use in thermoplastic-based composites or other random materials such as SMC, BMC, or GMT (see Section 3.3 for more details).

3.2.1.1 E-Glass and ECR-Glass

E-glass, or "electrical grade" glass fiber (alumina-calcium-borosilicate), has been the most used fiber reinforcement in composite materials mainly due to its combination of moderate specific strength and stiffness (modulus) and low cost. In an effort to reduce environmental issues with some fiber constituents, namely boron oxide, and increase acid resistance, many glass fiber producers have transitioned over from producing E-glass to ECR (electrical corrosion resistant) glass, also known as "boron-free" E-glass. This transition also provided some slight increases in fiber mechanical performance in addition to the higher acidic resistance. Due to E/ECR-glass composites' overall low modulus compared to carbon fiber composites or metals, they are used readily in sandwich structures where their flexural stiffness can be amplified via the separation provided by the core material.

3.2.1.2 H-Glass and R-Glass

As specific large-scale applications (such as wind turbine blades) began to grow, new glass fibers were developed to fit the niche between E/ECR-glass and S-glass in both mechanical properties and cost. Both H-glass and R-glass classifications (calcium-alum inosilicate) provide higher tensile strength and tensile modulus properties than the E grade fibers, with R-glass grades typically providing higher values compared to H-glass. Consequently, R-glass fibers are typically more expensive than H-glass. Both glass variants are "boron-free", providing similar corrosion resistance to ECR-glass fibers.

3.2.1.3 Basalt

Basalt fiber is, as the name implies, produced from crushed and filamented basalt volcanic rock. In terms of mechanical properties, basalt fiber is akin to an R-glass, higher strength, and modulus compared to E/ECR/H-glass fibers, and just short of those possible with the S-glass fibers. Unlike glass fibers, basalt does not become semi or fully transparent when impregnated with matrix resin, due to the iron oxide present within the parent rock material. High-quality basalt fiber is now widely commercially available as problems with quality plagued initial efforts due to raw material inconsistencies from different quarry locations.

3.2.1.4 S-Glass

S-glass fibers (magnesium aluminosilicate) were initially developed for high-performance military applications where E-glass was found to be unsuitable. S-glass and its variants provide the highest tensile strength and modulus properties of all commercially available glass fibers, while still providing high corrosion resistance and high temperature capability. Due to the processing and materials required to make these high-performance fibers, the cost is higher compared to almost all of the other glass fiber variants. As such, several S-glass variants (S-1™, S-2™, and S-3™) have been developed to target specific applications where the cost can be justified for the performance required.

3.2.1.5 Quartz (Silica)

Quartz fibers, which are 99.99% pure silica (SiO_2), serve niche markets requiring high temperature performance, very low coefficient of thermal expansion (CTE), and superior electromagnetic properties. One major market is for radio frequency transparency use on radomes of aircraft and land-based communications and radar systems. Due to the high silica purity and low volume production, quartz fiber can be extremely expensive compared to other glass fibers, reaching into triple digits of US dollars per pound.

3.2.1.6 A-Glass, C-Glass, and AR-Glass

Specialty glass fibers suited for specific applications requiring high alkaline or acidic resistance has also been developed, with less of an emphasis on high mechanical properties. A-glass fibers (soda lime silicate) are typically made from recycled light bulbs and have strengths and moduli well below standard E-glass while still providing high corrosion resistance. C-glass or "corrosion" glass (calcium borosilicate) has typically been used in composite corrosion tank/pipe applications as a high resin content surface barrier in contact with the retained corrosive liquid. Both A-glass and C-glass fibers are typically available in nonwoven veil formats to produce these types of barrier layers. AR-glass (alkali zirconium) or "alkali-resistant" glass is a highly alkali-resistant fiber used for cement and concrete reinforcement, usually in short fiber format and mixed into the concrete itself (Table 3.11).

TABLE 3.11
Nominal Mechanical and Physical Properties of Various Glass and Basalt Fibers at Room Temperature[11]

Nominal Fiber Designation	Tensile Strength (Impregnated Strand) F_{ty} (Gpa)	Young's Modulus (Impregnated Strand) E (Gpa)	Coeff. of Thermal Expansion CTE (10^{-6} mm/mm-°C)	Density ρ (g/cm^3)
A-Glass	2.21*	68.9	9.00	2.436
C-Glass	2.21*	68.9	6.30	2.519
AR-Glass	2.16	73.1	6.50	2.685
E-Glass	2.00–2.50	77.9–80.0	5.40	2.547
ECR-Glass	2.20–2.60	81.1	5.99	2.630
H-Glass	2.40–2.90	90.1	5.29	2.602
R-Glass	3.06–3.40	90.1	4.10	2.547
Basalt	2.90–3.30	84.1–94.1	5.49	2.630
S-Glass	3.40–3.83	91.4	3.40	2.464
Quartz	2.74*	69.0	0.54	2.159

[a] Estimated strengths from pristine properties.

3.2.2 CARBON FIBERS

Carbon fiber is a broad classification of composite reinforcement fibers that are generally greater than 90% carbon atoms arranged in long molecules oriented along the fiber longitudinal axis. This molecular orientation produces very high strength and modulus while also having very low density (around 30% less than E-glass) making it an excellent high-performance fiber for composite applications. Carbon fibers are classed mainly in terms of raw material source, often called precursor, and then subclassed by fiber modulus range. The first type of carbon fiber developed in the 1960s was based on rayon (cellulose), which had found applications in ablative surfaces but did not have the high strength and modulus seen in today's more commoditized carbon fiber based on polyacrylonitrile (PAN) precursor fibers. Properties shared by most carbon fibers are high fatigue resistance, good thermal and electrical conductivity, and slightly negative longitudinal coefficients of thermal expansion (LCTE). Due to the relatively high amounts of energy needed to convert modern precursor material into carbon fiber, the relative cost is much higher than most glass fiber options, but economies of scale, alternate precursor sources, and conversion methods are being used and developed to help bring this cost down.

3.2.2.1 PAN-Based Carbon

The most widely produced and used carbon fiber, PAN-based carbon, is produced by effectively stretching and heating (oxidizing, carbonizing, and sometimes graphitizing) this acrylic-based precursor fiber to achieve 95%–99% aligned carbon material. As the carbon fiber is processed, more crystalline basal planes form on the outer filament surface, providing higher modulus as these layers grow. Consequently, as the crystallization increases (typically through the "graphitization" process), the random molecular structure, which contributes more to the strength of the fibers, in the center of the filament, decreases and typically produces lower strength. Therefore, higher-modulus carbon fiber has traditionally been weaker than the lower-modulus counterparts. However, developments over the last couple of decades have produced higher strength versions of the high modulus carbon fibers.

Also referred to as "high strength", the standard modulus version (with a Young's modulus range of 228–255 Gpa) of PAN-based carbon fiber is the most widely used due to its relatively low cost and board range of tow sizes. A carbon fiber "tow" is a collection of carbon filaments and is denoted by how many thousands of filaments are present in any given tow (i.e., "12K" means 12,000 filaments per tow). Standard modulus carbon fiber generally is available in tow sizes from 1K to 60K, with some lower cost versions having well over 300K filaments per tow. Popular tow sizes are 3K, 12K, 24K, and 48K/50K, with the smaller tows typically used for light areal weight fabrics or unidirectional prepregs, and the larger tows in filament winding, pultrusion, heavier areal weight fabrics, and prepregs. Due to this more ubiquitous nature, most industrial and commercial applications have been developed around this fiber type.

Intermediate modulus (typically 276–324 Gpa Young's modulus) carbon fiber typically produces higher tensile modulus and strength compared to the standard modulus versions. Filament diameters are generally smaller (5 µm versus 7–8 µm) than standard modulus fiber, which allows for less defects within the fibers, thereby producing higher properties. Due to the smaller filament scale of these fibers and the increase in energy consumption required to produce them, intermediate modulus carbon fiber is more expensive (2–4 times) than standard modulus. Typical tow sizes are 6K to 24K, but some new development versions are using larger bundles to help reduce costs. Due to the relatively high costs, these fibers are typically used in applications where high properties and low weight are critical, such as in aerospace and defense primary and secondary structures.

High modulus carbon fiber encompasses a group of PAN-based fibers with a large range on Young's modulus values (310–600 Gpa) that are "graphitized" after the subsequent oxidation and carbonization steps during fiber manufacture. As such, these fibers are also known as graphite fiber and contain >99% carbon composition compared to the standard and intermediate versions that have typically >95% carbon. The addition of this graphitization stage requires even higher energy consumption, and as such, high modulus carbon is the most expensive PAN-based carbon and generally not produced in as high volume as the other versions. Due to the higher crystallinity of these fibers, the tensile strengths typically degrade as the modulus increases, while linear coefficients of thermal expansion become even more negative, and thermal and electrical conductivities greatly increase. With these mixes of properties, high modulus carbon makes ideal reinforcements for space vehicles where thermal stability, low weight, and high stiffness are key. Their high stiffness also makes them useful in some hand-held sporting goods applications such as high-end golf shafts and tennis rackets. Tow sizes typically range from 3K to 12K, and filament diameters are more in line with intermediate modulus carbon (~5 µm).

3.2.2.2 Pitch-Based Carbon

This type of carbon fiber is produced from coal and tar processing by-product called pitch. Depending on how the pitch is processed, two versions can be produced: isotropic and mesophase. Isotropic pitch-based carbon fiber is generally low modulus and not used for many composite or sandwich composite applications but mainly for thermal management. Conversely, mesophase pitch-based carbon can have very high Young's modulus (517–938 Gpa) and are generally known as "ultra-high modulus" carbon fiber. The ultra-high modulus carbon has similar physical properties, compromises, and uses as high modulus PAN-based carbon. Typical tow sizes range from 1K to 16K, and due to precursor costs and energy consumption, this type of carbon can be as expensive as or costlier than any of the other carbon fiber grades (Table 3.12).

3.2.3 Natural Fibers[12]

The phrase "natural fibers" can encompass many different forms of fiber material, including lignocellulosic, animal, and mineral fibers. For this description, only lignocellulosic (plant or cellulosic based) fibers will be detailed and particularly, nonwood fibers (see Section 3.1.3 for wood materials). Nonwood fiber can be further separated into seed, leaf, bast, fruit, and stalk fibers; however, the leaf and bast groupings are of the most interest for composite reinforcing. These types of natural fibers provide low densities, low energy consumption, low cost, and good machinability and can be recycled and/or are biodegradable. Downsides for use are high variations in strength/stiffness, uncertain fatigue life/durability, and susceptibility to moisture and fire degradation. Also, there are some challenges in obtaining sufficient bonding between untreated natural fibers and polymeric resins, and so much research has been conducted in sizing and surface treatments.[12] Generally, natural fibers are produced as short and long fibers and random fiber mats.

TABLE 3.12
Nominal Mechanical and Physical Properties of Various Carbon Fibers at Room Temperature

Nominal Precursor and Fiber Designation		Tensile Strength (Impregnated Strand) F_{ty} (Gpa)	Young's Modulus (Impregnated Strand) E (Gpa)	Coeff. of Thermal Expansion CTE (10^{-6} mm/mm°C)	Density ρ (g/cm^3)
PAN	Standard Modulus	3.45–5.07	228–255	−0.38→−0.45	1.77–1.80
	Intermediate Modulus	4.14–7.01	276–324	−0.56→−0.59	1.77–1.80
	High Modulus	1.70–4.83	345–600	−0.74→−1.10	1.74–1.94
Mesophase Pitch	Ultra-High Modulus	2.59–3.79	517–938	−1.40→−1.49	2.10–2.21

3.2.3.1 Bast

Bast fibers are found in the phloem (tissue used for transporting nutrients throughout the plant) around the stem. They support the plant, providing strength and stiffness and allowing it to be freestanding and withstand wind, animal contact, etc. Bast fibers generally produce the highest tensile strengths and stiffnesses of all nonwood natural fibers. Popular bast fibers for reinforcing composites are flax, jute, hemp, and kenaf (see Table 3.13).

3.2.3.2 Leaf

As the name implies, leaf fibers are sourced from the leaves of nonwood lignocellulosic plants. While tensile moduli for these fibers are generally lower than bast fibers, the strengths can be on par or higher in some cases. Typical leaf fibers used for composite reinforcements include sisal, abaca, and pineapple (PALF).

3.2.4 SYNTHETIC FIBERS

This group of man-made reinforcement fibers are mostly polymer or ceramic-based and can have several uses outside of polymer composites including soft/hard armor, protective clothing, sporting goods, and high temperature-resistant metal/ceramic matrix composites. The polymer-based fibers are generally low in density, have moderate modulus (due to molecular structure and orientation), moderate to high strength, and negative axial CTE. These fibers can have issues with UV degradation and moisture absorption, and those detailed in Section 3.2.4.3 require special fiber surface modifications such as plasma treatment to properly bond to composite matrix resins. The ceramic and boron fibers provide good mechanical properties and elevated temperatures, but can have limited uses due to cost, availability, their brittle nature, and noncompliant behavior.

TABLE 3.13
Nominal Mechanical and Physical Properties of Various Natural Fibers at Room Temperature[12]

Nominal Fiber Designation		Tensile Strength (Impregnated Strand) F_{ty} (Mpa)	Young's Modulus (Impregnated Strand) E (Gpa)	Density ρ (g/cm^3)
Bast	Flax	345–2,000	27.6–102.8	1.41–1.49
	Jute	317–800	8.3–77.9	1.30–1.49
	Hemp	269–897	23.4–90.3	1.41–1.49
	Kenaf	221–931	14.5–53.1	1.41
Leaf	Sisal	366–703	9.0–37.9	1.33–1.49
	Abaca	400–979	6.2–20.0	1.49

Face Sheet Materials

3.2.4.1 Para-Aramid

These aramid (aromatic polyamide)-based fibers are usually known by the legacy trade names Kevlar® (Dupont) and Twaron® (Teijin). Originally developed for replacing steel chord in radial tires, higher modulus versions of these para-aramid fibers have been used extensively in polymer composite reinforcements over the last several decades. These fibers exhibit low density and high strength with high impact and fatigue resistance when used as composite reinforcements. They also produce a "soft" failure mode in that the composite does not shatter upon impact or flexing. However, these fibers produce low composite compression strength and are susceptible to UV degradation and high moisture absorption issues, which must be considered. Fabrics require special scissors to cut and will look dry when saturated with resin, and the cured composite needs special techniques to cut and machine. Other "aramid" fibers are available, most notably meta-aramid grades, which are not quite suitable for composite skin/face sheet reinforcement due to their lower modulus. However, these meta-aramid fibers are used extensively in sandwich composites as reinforcements for honeycomb core materials.

3.2.4.2 High-Modulus Polypropylene

Relatively new to the composite reinforcement market, HMPP fibers provide similar impact, ductile failure modes, and dynamic properties as para-aramid and UHMWPE fibers and at a much lower density (0.83 g/cm^3). Used on its own, HMPP fibers do produce lower composite mechanical properties, which makes hybridizing with high modulus fibers (such as carbon) an attractive option. Cost for HMPP fibers is typically much lower than para-aramid, UHMWPE, and standard modulus carbon fibers, so hybridization also can come with an added cost benefit. As with most olefin-based materials (polypropylene and polyethylene), sufficient bonding to resin matrices can be difficult, so compatible matrix materials must be vetted thoroughly for proper use. Processing temperature is also limited to about 300°F due to the nature of the polypropylene base material.

3.2.4.3 UHMWPE, PBO, and LCP

UHMWPE polyolefin fiber is made of extremely long polymer chains of polyethylene that produces a low-density reinforcement with high resistance to chemicals, water, and ultraviolet light. This structure can produce fibers 40% stronger than para-aramids and are capable of withstanding high-load and strain rates. Due to its morphology, UHMWPE fibers can have issues with creep under sustained loads and has a low maximum processing temperature of approximately 300°F, similar to HMPP.

PBO (poly-p-phenylene benzobisoxazole) fiber has moderately higher nominal tensile properties compared to UHMPE fibers but with higher heat resistance. Liquid crystal polymer, or LCP, fibers provide excellent creep and abrasion resistance, with minimal moisture absorption. LCPs also exhibit good chemical resistance, negative CTE, high dielectric strength, and high impact strength (Table 3.14).

TABLE 3.14
Nominal Mechanical and Physical Properties of Various Polymer Fibers at Room Temperature

Nominal Fiber Designation and Trade Name		Tensile Strength F_{ty} (Gpa)	Young's Modulus E (Gpa)	Coeff. of Thermal Expansion CTE (10^{-6} mm/mm°C)	Density ρ (g/cm³)
Para-Aramid	Kevlar® 49	3.62	124	−4.9	1.44
	Kevlar® 149	3.46	179	N/A	1.47
HMPP	Innegra™ S	0.67	15	−8.0	0.83
UHMWPE	Spectra® 900	2.18–2.60[a]	73–79[a]	−12.0	0.97
	Spectra® 2000	3.26–3.34[a]	113–124[a]	−12.0	0.97
	Dyneema SK60	3.60–2.80[a]	83–88[a]	−12.0	0.97
	Dyneema SK76	3.20–3.50[a]	109–115[a]	−12.0	0.97
PBO	Zylon® AS	5.81[a]	180[a]	N/A	1.55
	Zylon® HM	5.81[a]	270[a]	−6.0	1.55
LCP	Vectran™ HT	3.20[a]	75[a]	−4.8	1.41
	Vectran™ UM	3.00[a]	103[a]	N/A	1.41

[a] Dry fiber properties, not impregnated strand.

3.2.5 BORON AND CERAMIC FIBERS[13]

These high-performance fibers were developed to provide high strength and stiffness at high operating temperatures, including use in metal and ceramic matrix composites (MMC and CMC). Use of these types of reinforcement fibers are currently limited by high cost and low volume of production; however, their unique properties can provide specific niche solutions when paired with high-temperature polymers (see Section 3.3.5).

3.2.5.1 Boron

Boron fibers are produced by a chemical vapor deposition (CVD) process in which elemental boron is added onto fine tungsten wire. This produces relatively large diameter monofilament fibers (typically 0.10 and 0.14 mm) that cannot be woven into fabric and typically are used only in unidirectional prepreg or fabric with a polyester fiber weft. While boron fibers have good tensile strength and even better modulus, it is their high compression properties and temperature resistance that separate them from other reinforcement fiber materials. Unfortunately, the high cost and brittle nature of these fibers have limited their use to high-end military applications.

3.2.5.2 SiC and Alumina

Silicon carbide, or SiC, fibers were developed for use at very high operating temperatures up to 2200°F with little to no loss in properties. While capable of being used in

TABLE 3.15
Nominal Mechanical and Physical Properties of Various Boron/Ceramic Fibers [13]

Nominal Fiber Designation and Trade Name		Tensile Strength F_{ty} (Gpa)	Young's Modulus E (Gpa)	Coeff. of Thermal Expansion CTE (10^{-6} mm/mm°C)	Density ρ (g/cm^3)
Boron	4 mil	3.6	400	4.50	2.60
	5.6 mil	4.0	400	4.50	2.46
SiC	SCS-6	3.9	386	4.14	3.04
	SCS-Ultra	5.9	414	4.14	3.04
Alumina/aluminosilica	Nextel™ 610	2.8	372	7.92	3.90
	Nextel™ 720	1.9	248	5.94	3.40

polymer matrices, SiC fibers are more ideally used to reinforce metals or ceramics, as it is chemically stable within these matrices during processing.

Alumina- and aluminosilica-based fibers have similar uses and max operating temperatures as the SiC fibers, albeit with some lower tensile properties and higher density. The structural versions of alumina fibers are crystalline (based on α-Al_2O_3) in nature with the higher temperature versions containing some SiO_2 (hence aluminosilica) to inhibit grain boundary movement during elevated temperature service (Table 3.15).

3.2.6 Architectural Forms of Reinforcement Materials

Fiber materials have different formats for various processing methods and applications. The basic format of continuous fibers is roving as shown in Figure 3.3, which can be used directly for filament winding and pultrusion processes and can be chopped into short length fibers for spray-lamination and for producing sheet molding composite prepreg, chop strand mats, and surface finishing veil as illustrated in Figure 3.4. Roving is the basic starting material for making a variety of different fabric formats as for wet laminations. Figure 3.5 shows different formats of fabrics that are widely used in composite lamination.

One widely used fabric is woven roving that is produced by interlacing continuous fiber roving into relatively heavy-weight fabrics. Woven roving is generally compatible with most resin systems and used primarily to increase the flexural and impact strength of facing laminates. It is ideal for multilayer hand lay-up applications where increased material strength is required, along with good drapeability, ease of wet-out, and cost efficiency. Woven roving is a "plain weave" style, available in a variety of weights, widths, and finishes to suit a wide range of applications.

Multiaxial nonwoven stitched fabrics are ideally suited for selective reinforcement. Fibers are bundled and stitched together, rather than woven, so there is no

FIGURE 3.3 Roving of different fiber materials.

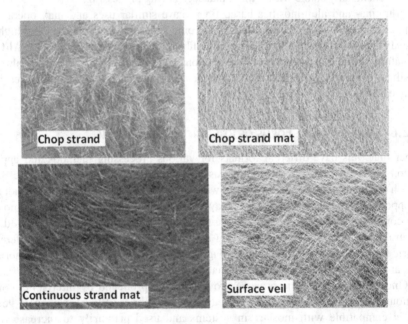

FIGURE 3.4 Different fiber mats.

crimp in the rovings. As a result, these reinforcements offer optimized directional strength for their finished composite parts. For this reason, they are popular in the wind energy industry. These fabrics are ideal for wet lay-up, pultrusion, and vacuum infusion processing, to create strong, stiff composite parts, and are compatible with polyester, vinyl ester, and epoxy resins.

Face Sheet Materials

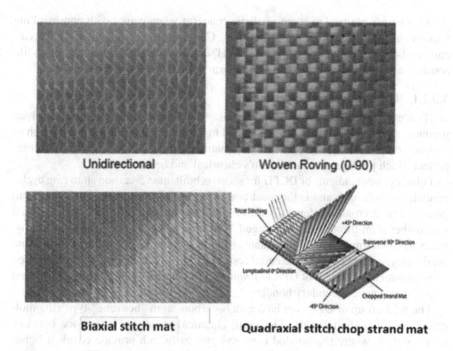

FIGURE 3.5 Different fibrous fabrics.

3.3 PLASTIC MATRIX MATERIALS FOR COMPOSITE FACE SHEETS[14]

As discussed in Section 3.1.2 above, thermoset plastics are very rarely used unreinforced and as such are not suitable as face sheet materials on their own, due to high brittleness and low strength. However, as a reinforced matrix, thermoset plastics provide several benefits over thermoplastic options such as high temperature resistance, specific chemical resistance, and most importantly, low initial viscosity for suitable processing at ambient temperatures.

3.3.1 UNSATURATED POLYESTER AND VINYL ESTER RESIN

Unsaturated polyester resins (UPRs) are short-chain molecule versions of thermoplastic polyesters with unsaturations (bonding sites) occurring along the length and at the ends of these chains. Vinyl esters are similar in construction except that their unsaturations are only located at the molecular chain ends. Both resins are supplied in liquid form with a "reactive diluent" which is typically styrene or a similar solvent. Hence, these materials are often referred to as "styrenated" resins and carry a rather distinct smell. Upon cure initiation (generally from addition of a metallic accelerant and free-radical generating peroxide), the reactive diluent covalently bonds to the polymer chains, cross linking them together to eventually form a solid material. This curing reaction is exothermic and densifies the resin (i.e., volumetrically shrinks) which can lead to some aesthetic concerns when paired with nonshrinking reinforcement creating fiber "print-through". Certain fillers or extenders such as calcium carbonate and alumina trihydrate (ATH) can be added to the resin to impart fire resistance and lower shrinkage/peak exotherm or just to reduce cost.

3.3.1.1 Unsaturated Polyester (UPR)

UPRs are generically a chemical combination of dibasic acids and glycols which produce polymer chains that are then added to a mixture of reactive diluents such as styrene or alpha methyl styrene. UPRs are named mainly for the aromatic acid component which provides the cured resin's chemical and heat resistance.

Polydicyclopentadiene, or DCPD for short, is built upon 5-carbon atom ring cyclopentadiene molecules and is the most cost-effective UPR available. DCPDs tend to cure to a high crosslink density with minimal volumetric shrinkage and are used where fiber print-through is discouraged. This high degree of cure also has its drawbacks as it does not allow for adequate secondary bonding, due to low bonding site availability, and it can make the cured resin rather brittle. DCPDs are generally used blended with other grades of UPR to lower overall cost, add some shrink control, and allow for adequate secondary bonding.

The next group of UPRs are based on 6-carbon-atom "benzene"-type ring molecules called phthalic acids. The main chemical structural difference between these acids is where the bonded carboxyl groups (which provide covalent bonding sites) appear along the main benzene ring of the molecule. The most prevalent UPR of this group is the orthophthalic resin, often called "general purpose" or just "ortho resin". Ortho resins are very cost-effective and produce balanced mechanical and physical properties. Isophthalic-based resins (often called just "iso resins") provide higher properties and temperature/chemical resistance than ortho resins but are typically more expensive. Iso resins are typically blended with DCPD to reduce cost and overall volumetric shrinkage. Terephthalic-based resin, while not as widely used, is similar to iso resin in terms of corrosion resistance and properties.

Other UPR versions include chlorendics and bisphenol-A fumarates but are not as widely used as the resins listed above. This is due to regulations of some of the more toxic component materials and the overall performance of alternate systems such as vinyl esters eclipsing those of these older resins.

3.3.1.2 Vinyl Esters

Vinyl esters were originally derived as a compromise between UPRs and epoxy resins. Epoxy molecular backbones were chemically altered to contain UPR-style bonding sites only at the ends. These polymers are then diluted in styrene or other suitable monomer and cured in the same fashion as UPR. The epoxy-based molecules provide higher mechanical and fatigue properties, increased corrosion resistance, and better moisture/solvent resistance than any of the typical UPR grades, while the addition of the monomer reactive diluents decreases the viscosity and more importantly the cost compared to standard epoxies. As with other styrenated resins, vinyl esters are blended with other UPR systems to provide a compromise of properties and cost.

3.3.2 Epoxy Resin[15]

Epoxy resin is a large category of thermosetting plastics that include adhesives, liquid resins, and B-staged matrices used in prepreg (preimpregnated fiber/fabric) materials. Generically, an epoxy is any molecule containing the epoxy group (C-O-C triangular ring), but when discussing epoxies for structural composite applications, the base molecules are generally diglycidyl ether of bisphenol A (DGEBA), diglycidyl ether of bisphenol F (DGEBF), Novolac, or several other long chains with epoxy end groups. Epoxy resins are typically cured using stoichiometrically ratioed hardeners based on amine or anhydride chemistry, which react with the epoxy groups and crosslink the base molecules together into a solid. Some aliphatic amine cured epoxies can be processed at room temperature with a final elevated post cure, while more aromatic-based amine and anhydride-cured systems require elevated temperature for both initial processing and post cure. Liquid epoxy resins are typically much higher in viscosity compared to UPRs and vinyl esters (which use very low viscosity monomers as reactive diluents), so the elevated temperatures are required to provide better molecular movement and crosslinking. Since epoxies do not produce any volatiles during cure, they do not shrink as much as UPRs and they also produce higher composite mechanical properties due to their higher strength and greater ability to adhere to the reinforcement fiber surface. Fillers or additives can be readily incorporated into epoxies for greater toughness, flexibility, and lower cost.

3.3.3 Phenolic Resin[16]

Phenolic resins are produced from reacting phenol and formaldehyde and are well known for their excellent FST (fire, smoke, and toxicity) and high-temperature properties without the addition of fillers or additives. When phenolic resins are cured, water is a byproduct, producing a more porous material than traditional UPRs, vinyl esters, or epoxies. This produces composites with typically lower mechanical properties, so phenolic composites are generally used in areas requiring good FST resistance, but that are not as structurally demanding. Despite some of their drawbacks, phenolic resins have been successfully used in processes such as pultrusion and vacuum infusion.

3.3.4 Polyurethane Resin

Polyurethanes encompass a wide variety of plastic materials, including thermoplastic (TPUs), but in general, a urethane is defined as the product of a reaction between an isocyanate and a polyol. Polyurethanes are known for their high flexibility and toughness, making them not only excellent composite matrices but also adhesives often on par with epoxies. The toxicity and affinity for water of the isocyanate component has limited the use of most polyurethanes to closed mold processes, and the fast reaction rates have further limited use to such processes as reaction injection molding (RIM/SRIM) and die-injected pultrusion.

To widen the use of polyurethanes into more composite manufacturing processes, several hybrid versions have recently been developed. Combining urethane chemistry with UPR-based materials have created urethane ester and urethane acrylate hybrids, providing the processing ease of UPR's with the benefit of the toughness and structural properties of polyurethanes.

A variation of polyurethane chemistry, called polyurea, is the product of an isocyanate with a multifunctional amine instead of polyol. Polyureas have similar properties and applications as urethanes and are known for their very high reaction rates. Due to this high reactivity, polyurea has seldom been used as a composite material matrix, despite its obvious positive attributes.

3.3.5 High-Temperature Application Resins[15]

One aspect of traditional polymer matrix composites that often falls short in high-performance applications is retaining adequate mechanical properties at elevated temperatures (+120 °C). This shortcoming is due to the polymer chemistry itself, and in response, multiple chemistries have been developed to push the use of PMCs into more thermally demanding applications such as aerospace engine components and space vehicle structures.

Cyanate esters are matrices used for critical spacecraft applications where low outgassing and high-temperature performance are required. Typical operating temperatures peak around 205 °C, and cyanate esters generally perform better than epoxies in fire-related testing. Extreme care must be taken when processing cyanate esters, as excess moisture present during cure can have a deleterious effect on the end composite.

Bismaleimides (BMIs) are utilized at operating temperatures generally up to 500°F while still maintaining high mechanical properties, low moisture uptake, and chemical resistance. Generally used as a prepreg, BMI composites generally require relatively long processing times at elevated temperature (177 °C) for proper cure compared to epoxy. Like epoxies, BMIs can incorporate additives to increase toughness characteristics.

Polyimides (PIs) are an older thermosetting technology and has been one of the more widely used high-temperature resin group used in aerospace engine applications. Operating temperature range is 260–315 °C, depending on type of PI cure process and raw materials, and as with BMIs, long process times at elevated temperatures are required for complete cure.

Other high-temperature resins such as benzoxazines, phthalonitriles, and pDCPDs are constantly being developed to push the operating envelope of PMCs, making them competitive with ceramic and metallic materials.

3.3.6 Thermoplastic Matrix[17]

Thermoplastics can be used as matrix materials with fiber reinforcements forming the facing sheets of the sandwich structures. Thermoplastic materials are characterized by molecule chains associated reversibly through intermolecular forces (such as Van der Waals' forces), which allow disconnection and thereby softening (melting) of the material by the input of heat or mechanical forces. Material softening offers the potential to be thermoformed or fusion-joined. Cooling returns the polymer to a solid. The fact that the polymer is thermoplastic means that mere physical processes are required to fabricate the structure.

TABLE 3.16
Typical Thermoplastics for Facing Sheet of the Sandwich Composites [17]

Matrix Polymer	Reinforcement	Fiber Content	Product Format
Polypropylene (PP)	Glass fiber unidirectional	60–75 wt%	Unidirectional tap or 0/90 multiplayer sheet
Polypropylene (PP)	Glass fiber fabric	60–75 wt%	Commingling of glass fiber and PET fiber woven roving
Polypropylene terephthalate (PET)	Glass fiber unidirectional	65 wt%	Unidirectional tap or 0/90 multiplayer sheet
Polypropylene terephthalate (PET)	Glass fiber fabric	65 wt%	Commingling of glass fiber and PET fiber woven roving
Polyamide (PA)	Glass, Carbon or Aramid fiber fabrics	40–60 wt%	Dry prepreg fabrics
Polyetherimide (PEI)	Glass, Carbon or Aramid fiber fabrics	60–70 wt%	Dry prepreg fabrics
Polyether ether ketone (PEEK)	Glass fiber unidirectional	60–70 wt%	Impregnated tape or 0/90 sheet

Skins for the sandwiches consist of fiber-reinforced laminates with a thermoplastic matrix in which either short or continuous fibers are used. Thermoplastic composite skins are available as fully consolidated laminates, semi-prepregs, prepregs, or fiber commingled fabrics. The thermoplastic matrix polymers in general range from commodity plastics such as PP or engineering polymers such as polyamide (PA) to high-performance polymers such as PEI or polyetheretherketone (PEEK). Table 3.16 gives an overview of thermoplastic composite sandwich skin materials. Commodity plastics in combination with reinforcing glass fibers are mostly employed as skin materials for the sandwich structures. Especially, PP-based composites are applied. This can be explained by the low price of the material and the ease of processability of PP since many commercial products are available in the market, and most investigation has been performed for many applications where manufacturing times need to be reduced drastically. Another most widely applied high-performance polymer in the sandwich skins of interest to the aviation industry is PEI, often combined with reinforcing glass fibers. Glass fibers are mostly used reinforcement materials, but for some high-performance applications, such as airplane and airspace, and military application, carbon fiber and aramid fiber also are used. More information of thermoplastics used as the facing matrix can be found in Chapter 5.

3.4 COMPOSITE FACE SHEET MATERIAL PROPERTIES AND CHARACTERIZATION[18]

While the detailed information given in Sections 3.2 and 3.3 (and 3.1.2.1) is very important in selection of face sheet materials, the properties of fibers and polymers/resins by themselves are not wholly adequate. The composite of these materials, a combination of fiber and polymers each at a specific ratio, number of layers, layer thicknesses, and layer orientations, is key to optimizing the final face sheet design. In this section, the basic building blocks will be provided, which can then be utilized in a classical lamination theory (CLT)-based methodology to determine adequacy for the final sandwich structural design. To verify the design properties of the face sheet materials and sandwich structure, several standardized test methods from ASTM methods will also be discussed along with direction to other methods that may be required for final sandwich design.

3.4.1 UNIDIRECTIONAL LAMINAE

The main building block for all laminated composites, the unidirectional lamina (lamina = 1 unidirectional ply, laminate = more than 1 ply), is composed of continuous fibers all oriented in one direction ("1" axis in the picture below) embedded in the matrix material. Unidirectional lamina properties are based on several factors: the constituent fiber and matrix mechanical and physical properties, the ratio of fiber to matrix (either by weight or by volume), and the overall amount of fiber/matrix. The fiber material provides the fiber-direction stiffness and strength of the composite, while the matrix and fiber-matrix interface provide the stiffness and strength in the transverse (90° to the fiber orientation or the "2" axis as shown in Figure 3.6) direction. Typically, the more fiber that is present in the composite, the higher the strength and stiffness is in the fiber direction. This is true up to a point where there is not enough matrix material to properly transfer stresses between fibers, and then these properties tend to fall precipitously. The amount of fiber and matrix present in a single ply dictate the ply thickness. Generally, composite laminae are relatively thin (less than "1.27 mm") and allow for analysis using "plane stress" assumptions in the CLT methodology. This basically assumes the "3" axis (through the thickness direction) of the face sheets do not carry any loads or provide any stiffness. Any shear deflection due to out of plane loading is assumed to be done by the core material only.

Table 3.17 below provides some basic composite properties of various fiber reinforcements preimpregnated (prepreg) in an epoxy resin matrix. Prepreg materials

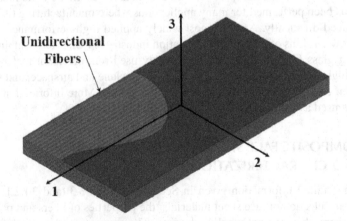

FIGURE 3.6 Three directions of unidirectional laminae.

TABLE 3.17
Nominal Mechanical and Physical Properties of Various Epoxy Prepreg Unidirectional Composites at 60 Volume Percentage Fiber Content[19]

Nominal Matrix and Fiber Designations		Tensile Strength		Compression Strength		Shear Strength	Young's Modulus		Shear Modulus	Poisson's Ratio	Density[a]
		F_{1t} (Gpa)	F_{2t} (Mpa)	F_{1c} (Gpa)	F_{2c} (Mpa)	F_{12} (Mpa)	E_1 (Gpa)	E_2 (Gpa)	G_{12} (Gpa)	ν_{12}	ρ (g/cm³)
		0°	90°	0°	90°	0°/90°	0°	90°	0°/90°	0°/90°	
Epoxy	E-Glass	1.07	31.0	0.61	117	72	39	9.7	6.21	0.26	1.97
	S-Glass	1.59	60.7	0.86	234	90	52	11.7	7.59	0.28	1.91
	Para-Aramid	1.38	29.7	0.28	138	43	76	5.5	2.07	0.34	1.30
	SM Carbon	1.55–2.14	40.0–51.7	1.45–1.59	207–248	117–131	138	9.0–10.3	6.90	0.29–0.30	1.52
	IM Carbon	2.55–2.76	51.0–55.9	1.54–1.90	69–152	108–124	166–172	8.3–11.0	6.21–8.28	0.18–0.30	1.52
	HM Carbon	0.76	26.2	0.70	34	27	325	6.2	4.14	0.26	1.61
	UHMWPE	1.10	8.3	0.08	48	24	31	3.4	1.38	0.32	1.02
	Boron	1.59	60.7	2.93	200	108	207	18.6	5.52	0.23	1.99

[a] Composite density assuming a cured epoxy density of 1.08 g/cm³.

(processed typically in autoclaves or vacuum-oven equipment) generally produce the highest composite mechanical properties of all precursor composite material systems due to the high-quality controls implemented in their production (i.e., low void content, high fiber alignment, tight control of fiber content and high fiber volume fractions, and tightly controlled curing processes). On the opposite end of the spectrum are unidirectional laminae produced via hand lamination, which is simply applying liquid resin to dry fiber fabric and manually compacting and impregnating via FRP rollers/squeegees/paint brushes. This type of process generally produces lower properties due to a higher degree of fiber misalignments, lower fiber content and higher variation, and greater void content. There is also a lack of off-axis property databases due to lower need for such information as structures made by this type of process are generally not as highly engineered. Most hand lamination and other nonprepreg systems are used in multiaxial laminates, and much more data are available for these types of constructions.

3.4.2 Biaxial Laminates

Biaxial laminates use, at minimum, two individual laminae oriented at 90° to one another. This includes 0°/90°, +45°/−45°, and other such biased ply stacking sequences and woven fabrics (with the 0° fiber as the "warp" and the 90° fiber as the "weft" or "fill"). These types of laminates can also be considered orthotropic and even quasi-isotropic (see Section 3.4.3 below), as they typically provide similar properties in both fiber orientation directions, assuming a balanced stacking sequence or weave. Outside of the same processing differences referenced in the section above, other factors affecting properties in this group on materials are stacking sequences for unidirectional plies/NCF and fiber crimping in woven fabrics. Best performance is generally seen with balanced and symmetric laminates and woven fabrics using high harness (low fiber crimp) weaves.

Tables 3.18 and 3.19 below provide nominal properties for biaxial laminates made via prepreg/autoclave and vacuum infusion processes, respectively. Current biaxial prepreg materials with available property databases (such as through NCAMP) are generally made with woven fabric reinforcements, while most vacuum-infused laminates will tend to use noncrimped fabrics (NCF). Most of the property differences seen below in similar composite materials but made via different processes can be attributed to several aspects including fiber crimp of woven fabric, and fiber alignment and void content differences.

3.4.3 Quasi-Isotropic Laminates

One of the strengths of laminated composites is the ability to customize strength and stiffness is various directions; however, there are times (unknown or vaguely defined loading conditions) where balanced properties are preferred. Laminated composites can mimic isotropic material properties through their stacking sequences. By orienting plies in a manner resulting in fibers running in the 0°, 90°, −45°, and +45° directions (or even 0°, −60°, and +60°), the in-plane mechanical properties become less sensitive to orientation changes.

Face Sheet Materials

TABLE 3.18
Nominal Mechanical and Physical Properties of Various Prepreg Biaxial Composites[20-24]

Nominal Matrix/Fiber and Fabric Designations	Fiber Volume Fraction (%)	Tensile Strength F_{1t} (Mpa) 0°	F_{2t} (Mpa) 90°	Compression Strength F_{1c} (Mpa) 0°	F_{2c} (Mpa) 90°	Shear Strength F_{12} (Mpa) 0°/90°	Young's Modulus E_1 (Gpa) 0°	E_2 (Gpa) 90°	Shear Modulus G_{12} (Gpa) 0°/90°	Poisson's Ratio ν_{12} 0°/90°	Density ρ (g/cm³)
Epoxy/E-Glass 8HS Satin	50	483	483	476	476	N/A	27	27	N/A	N/A	2.21
Epoxy/S-Glass 8HS Satin	47	559	559	572	476	63	29	28	3.8	0.14	1.80
Epoxy/P-Aramid 4HS Satin	48	524	524	400	400	N/A	30	30	N/A	N/A	1.41
Epoxy/SM Carbon Plain Weave	49	897	869	697	634	68	61	61	3.6	0.04	1.22
Epoxy/IM Carbon Plain Weave	58	841	821	710	676	101	67	67	5.1	0.05	1.55
Epoxy/HM Carbon 5HS Satin	40	710	710	441	441	N/A	115	115	N/A	N/A	N/A

TABLE 3.19
Nominal Mechanical and Physical Properties of Various Vacuum-Infused Biaxial Composites[25]

Nominal Matrix/Fiber and Fabric Designations	Fiber Volume Fraction (%)	Tensile Strength F_{1t} (Mpa) 0°	F_{2t} (Mpa) 90°	Compression Strength F_{1c} (Mpa) 0°	F_{2c} (Mpa) 90°	Shear Strength F_{12} (Mpa) 0°/90°	Young's Modulus E_1 (Gpa) 0°	E_2 (Gpa) 90°	Shear Modulus G_{12} (Gpa) 0°/90°	Poisson's Ratio ν_{12} 0°/90°	Density ρ (g/cm³)
UPR/E-Glass NCF	52	510	510	510	510	90	26.9	26.9	4.55	0.13	1.91
UPR/S-Glass NCF	53	614	614	614	614	94	31.0	31.0	4.76	0.12	1.85
UPR/P-Aramid NCF	53	566	566	83	83	68	35.9	35.9	3.38	0.04	1.33
VE/SM Carbon NCF	54	683	683	483	483	82	62.1	62.1	4.07	0.02	1.52

Prepreg-style laminates generally use oriented unidirectional plies and woven biaxial to produce these types of laminates, while NCFs colloquially known as "quads" or quadaxials provide multidirectional reinforcement within a single fabric. Table 3.20 below provides nominal properties of the latter produced via vacuum infusion.

3.4.4 Test Methods for Laminated Composite Face Sheets[26]

All the nominal properties listed in Tables 3.1–3.19 above were obtained, measured, and verified using some sort of test methodology. While public test databases like MMPDS exist for isotropic materials, similar databases are harder to develop for laminated composites. Given the huge variety of matrix resins, reinforcing fibers, fiber orientations, stacking sequences, manufacturing processes, and matrix/fiber ratios available, an almost infinite number of combinations are possible, and any comprehensive database is exceedingly expensive to develop. Due to this, it is recommended that material property test validation occur once the matrix, fiber/fabric, and manufacturing process have been selected to determine design values for the overall structure. This section will detail the current ASTM test methods used to measure key mechanical properties of composite laminates, with a list of these methods as well as alternates available in the publication ASTM D4762 "Standard Guide for Testing Polymer Matrix Composite Materials".[26]

3.4.4.1 Tensile Strength and Modulus

Tensile strength and modulus are usually the most sought-after properties of composite laminates, especially when they are to be used as face sheets for sandwich constructions. The current industry standard for these properties is ASTM D3039 "Standard Test Method for Tensile Properties of Polymer Matrix Composite Materials". This method uses rectangular specimens with straight sides as opposed to the ASTM D638 method for plastics which requires an hourglass or "dog-bone"-shaped specimen, which is more appropriate for chopped fiber-reinforced composite materials. End tabs may be required to reduce specimen splitting due to load introduction, depending on the laminate being tested. ASTM D5083 is a similar method which does not require end tabbing, as is more directly a straight-sided offset to the ASTM 638 method. Poisson's ratio measurements are also typically taken from this test via strain gauges or transducers.

3.4.4.2 Compressive Strength and Modulus

There are several compression property test methods in current use, and all vary based on how load is introduced into the specimen. One of the older methods is a modified ASTM D695 method originally published by SACMA (the now defunct Suppliers of Advanced Composite Materials Association) as the SRM 1R-94 "Test Method for Compressive Properties of Oriented Fiber-Resin Composites" method. The SRM 1R-94 test introduces load on the specimen ends with end tabbing required to reduce/eliminate premature end brooming/crushing failures. Another older standard method is ASTM D3410 "Standard Test Method for Compressive Properties of Polymer Matrix Composite Materials with Unsupported Gage Section by Shear Loading". This method introduces load into the specimen via shear along the specimen ends.

TABLE 3.20
Nominal Mechanical and Physical Properties of Various Vacuum-Infused Quasi-Isotropic Composites[25]

Nominal Matrix/Fiber and Fabric Designations	Fiber Volume Fraction (%)	Tensile Strength		Compression Strength		Shear Strength	Young's Modulus		Shear Modulus	Poisson's Ratio	Density
		F_{1t} (Mpa) 0°	F_{2t} (Mpa) 90°	F_{1c} (Mpa) 0°	F_{2c} (Mpa) 90°	F_{12} (Mpa) 0°/90°	E_1 (Gpa) 0°	E_2 (Gpa) 90°	G_{12} (Gpa) 0°/90°	ν_{12} 0°/90°	ρ (g/cm³)
UPR/E-Glass NCF	50	393	393	386	386	148	20.7	20.0	7.9	0.30	1.88
UPR/S-Glass NCF	53	476	476	476	476	172	24.1	24.1	9.2	0.31	1.85
UPR/P-Aramid NCF	53	407	407	62	62	41	26.9	26.8	10.3	0.27	1.33
VE/SM Carbon NCF	54	510	510	324	324	260	42.1	42.1	17.0	0.27	1.52

This greatly reduces end brooming/splitting failure modes, but the test fixture can be bulky and expensive to produce. The current industry standard is ASTM D6641 "Standard Test Method for Compressive Properties of Polymer Matrix Composite Materials Using a Combined Loading Compression (CLC) Test Fixture", which, as the name implies, introduces load both by end and shear. For more multidirectional and larger unit-cell size laminates, ASTM D8066 "Standard Practice Unnotched Compression Testing of Polymer Matrix Composite Laminates" may be preferred as the specimen sizes are larger.

The compressive properties of composite laminates can be difficult to ascertain given the multitude of issues that can crop up during testing. Apart from the premature end brooming/crushing failure, buckling is also a concern. Placing strain gauges on either side of the specimen can indicate if any buckling occurs, and the D6641 method states a minim amount of buckling that is acceptable before failure. Proper specimen preparation is key to generating good data, which includes level and flat ends and surfaces.

3.4.4.3 In-Plane Shear Strength and Modulus

As with the previous property test methods, several options are available for measuring in-plane shear strength and modulus of composite laminates. The simplest is ASTM D 3518 "Standard Test Method for In-Plane Shear Response of Polymer Matrix Composite Materials by Tensile Test of a ±45° Laminate" which is similar in execution as the D3039 and D5083 tensile tests, but with the specimen required to be biaxial and the fibers running at ±45°. For quantitative shear data, however, methods ASTM D5379 (informally known as the "Iosipescu" test) and ASTM D7078 are much more appropriate. Other methods exist such as the D4255 rail shear and D5448 hoop-wound cylinder but are difficult to run and can have issues producing reliable data.

As with compression testing, in-plane shear test methods require high precision specimen machining and proper strain gauge application. In particular, the gauge section must be properly machined to not produce poor load introduction and response.

3.4.4.4 Flexural Strength and Modulus in Sandwich

In a sandwich construction, it is generally accepted that the face sheet materials take up most of the in-plane bending loads (tension and compression), while the core material response via shear deflection. The face sheet properties can usually be tested using the methods described in Sections 3.4.4.1 and 3.4.4.2 and combined with nominal core material data, the full sandwich structural response to load can be estimated. However, there are certain variables that are not considered with this process, most notably how well the face sheets adhere to the core material. As such, there have been several test methods developed to test the face sheets, core, and total sandwich properties from sandwich panel specimens.

To test the unidirectional compression response of a composite face sheet more accurately in a sandwich construction, the ASTM D5467 "Standard Test Method for Compressive Properties of Unidirectional Polymer Matrix Composite Materials Using a Sandwich Beam" was developed using a 4-point loading scenario. A more encompassing method, ASTM D7249 "Standard Test Method for Face sheet Properties of Sandwich Constructions by Long Beam Flexure", can be used to determine face sheet tension and/or compression strength or overall sandwich stiffness using the offshoot method, ASTM D7250. For determining core material response, Method ASTM C393 "Standard Test Method for Core Shear Properties of Sandwich Constructions by Beam Flexure" has been used with full sandwich construction specimens using either 3-point or 4-point testing configurations. To determine 2D panel response instead of a 1D beam configuration, test method ASTM D6416 "Standard Test Method for Two-Dimensional Flexural Properties of Simply Supported Sandwich Composite Plates Subjected to a Distributed Load" was created. This method does require special bladder and fixture equipment but can be useful in determining flexural response especially for marine industry composite hulls.

3.4.4.5 Interlaminar Shear Strength

Knowing the value property of a laminated composite is often desired as it indicates how well the matrix adheres to the reinforcement and how it would react in more dynamically loaded situations. The most popular test method, ASTM D2344 "Standard Test Method for Short-Beam Strength of Polymer Matrix Composite Materials and Their Laminates", does not actually measure shear strength, hence the test name calling it a "short-beam" strength. This is due to a good amount of bending present in the test specimen during loading, making it not a "pure" shear test. It is, however, still used extensively due to its ease of use and comparative value. Another test method which utilizes notched specimens and a compressive load introduction, ASTM D3846 "Standard Test Method for In-Plane Shear Strength of Reinforced Plastics", despite its misleading name, is a true out-of-plane shear test but can be difficult to perform as shear area calculation depends on the post failure analysis.

As with in-plane shear strength and modulus, ASTM D5379 and D7078 can be used for determining quantitative data, via using very thick laminates and specimens machined in the proper plane for interlaminar stressing.

3.4.4.6 Damage Tolerance and Impact Properties

Damage tolerance of laminate composites are directly related to their fracture toughness properties, which are separated into three modes (Mode I – tensile or "opening" mode, Mode II – shear or "sliding" mode, and Mode III – out-of-plane shear or "tearing" mode). Test methods have been developed to determine fracture toughness values for these modes (with a Mode III method still in development), as well as mixed mode tests, and all generally require some initial crack length and growth measurement during load. Specific methods include ASTM D5528 (Mode I), ASTM D7905 (Mode II), and ASTM D6671 (Mixed Mode I/II).

Impact testing can encompass several different modes of dynamic loading, and here will include pretesting impact conditioning for testing residual strength. One of the first impact tests modified for composites is the "pendulum" test or ASTM D256 "Standard Test Methods for Determining the Izod Pendulum Impact Resistance of Plastics". Specimens for this test are generally notched for accurate gauge failure but can be run without notching, which conforms to the ISO variant of this test, ISO 180. The ASTM D7136 "Standard Test Method for Measuring the Damage Resistance of a Fiber-Reinforced Polymer Matrix Composite to a Drop-Weight Impact Event" method, often call the "Dynatup" test, utilizes a drop tower where a weighted impactor is dropped from a specific height onto the composite specimen. Residual compression properties of laminates subjected to the D7136 method are tested via the D7137 method, which is often referred to as the "CAI" or "compression after impact" test. Flexural residual strength is determined using the D7956 method using a 4-point loading configuration. A newer method, ASTM D8101 "Standard Test Method for Measuring the Penetration Resistance of Composite Materials to Impact by a Blunt Projectile", provides dynamic response using higher velocity and smaller impactors and is more relevant for testing for ballistic properties.

REFERENCES

1. MMPDS-04: Metallic Materials Properties Development and Standardization (MMPDS). Washington, D.C.: Federal Aviation Administration, 2008. Internet resource.
2. Engineered Thermoplastic Polymers (ETP). https://biomerics.com/products/medical-polymers/custom-compounds/engineered-thermoplastic-polymers-etp/
3. Structural Plastics Research Council. *Structural Plastics Selection Manual*. New York: American Society of Civil Engineers, 1985.
4. Solvay Technyl® C 206 Nylon 6, EH0. http://www.matweb.com.
5. Solvay AvaSpire® AV-722. http://www.matweb.com.
6. SABIC ULTEM™ 1000 PEI (Americas). http://www.matweb.com.
7. Solvay Specialty Polymers Torlon® 4200 Polyamide-imide (PAI). http://www.matweb.com.
8. Mitsubishi Chemical Advanced Materials Duratron® D7000 PI Unfilled Polyimide. http://www.matweb.com.
9. "Wood Handbook Wood as an Engineering Material," Madison, WI: U.S. Dept. of Agriculture, Forest Service, Forest Products Laboratory, 2010.
10. APA – The Engineered Wood Association, *Panel Design Specification*. Tacoma, WA, May 2012. http://www.apawood.org.
11. F. T. Wallenberger, J. C. Watson, and H. Li, "Glass fibers." In *ASM Handbook*, D.B. Miracle, and S. L. Donaldson, eds. Geauga County: ASM International, vol. 21, 2001.
12. R.D.S.G. Campilho, ed., *Natural Fiber Composites* (1st ed.). Boca Raton, FL: CRC Press, Taylor & Francis, 2015.
13. "3M Nextel™ Ceramic Fibers and Textiles Technical Reference Guide," https://multimedia.3m.com/mws/media/1327055O/3m-nextel-technical-reference-guide.pdf.
14. American Composites Manufacturers Association. *CCT: Basic Composites Study Guide*. Arlington, VA, ACMA, 2006.
15. G. Lubin, *Handbook of Composites*. New York: Van Nostrand Reinhold, 1982.
16. S.H. Qureshi, *Phenolic Resins: An Overview in Composites Applications*. Albany: Georgia-Pacific Resins, Inc.
17. F. I. Von der Fakultät, "Thermoplastic composite sandwiches for structural helicopter applications," doctor degree dissertation, Lehrstuhl für Polymere Werkstoffe Universität, Bayreuth, Germany, 2018.
18. Stress-Strain Relations for Principal Directions," https://www.efunda.com/formulae/solid_mechanics/composites/comp_lamina_principal.cfm
19. J. R. Vinson and R. L. Sierakowski, *The Behavior of Structures Composed of Composite Materials* (2nd ed.). Dordrecht: Springer, 2008.
20. NCAMP, *CAM-RP-2009-001 Rev C: Advanced Composites Group, ACG MTM45-1/Style 6781 S2 Glass, Qualification Material Property Data Report*. May 1, 2013. https://www.wichita.edu/research/NIAR/Research/cytec-mtm45-1/Style-6781-S2-Glass-2.pdf.

21. NCAMP, *CAM-RP-2012–017 Rev NC: Cytec Cycom 5320–1 T650 3k-PW Fabric Qualification Material Property Data Report.* Oct. 13, 2015. https://www.wichita.edu/research/NIAR/Research/cytec-5320-1/T650-3k-PW-2.pdf.
22. NCAMP, *CAM-RP-2011–005 Rev A MPDR TenCate TC250 HTS 12K SFP OSI* Sept 26, 2016. https://www.wichita.edu/research/NIAR/Research/tencate-tc250/12k-HTS40-2.pdf.
23. Solvay, *Technical Data Sheet CYCOM® 919.* Dec. 18, 2017 https://www.solvay.com/en/product/cycom-919.
24. Cytec Engineered Materials, *Cycom® 950–1 Epoxy Prepreg.* Dec 2002. https://ecm-academics.plymouth.ac.uk/jsummerscales/mats347/safety/Cycom_950-1_epoxy_prepreg.pdf.
25. Vectorply Corporation, *Technical Data Sheets.* https://vectorply.com/database-search/.
26. ASTM Standard D4762-18, 2018. *Standard Guide for Testing Polymer Matrix Composite Materials* ASTM International, West Conshohocken, PA, 2018. http://www.astm.org.

4 Laminating Processes of Thermoset Sandwich Composites

Wenguang Ma

Russell Elkin

CONTENTS

4.1 Dry Laminating Process 126
    4.1.1 Facing Materials 126
    4.1.2 Adhesives 128
    4.1.3 Laminating Processes 130
4.2 Wet Laminating Process 132
    4.2.1 Hand Layup Process 133
    4.2.2 Spray Layup Lamination 134
    4.2.3 Prepreg Lamination, Press or Autoclave Curing 135
    4.2.4 Closed Mold Lamination 138
        4.2.4.1 Vacuum Infusion 139
        4.2.4.2 Resin Transfer Molding (RTM) 147
    4.2.5 Pultrusion 149
    4.2.6 3D Printing and Additive Manufacturing 151
References 156

A structural sandwich is a layered composite formed by bonding two relatively thin facings to a relatively thick core. The core increases the spacing between the two facings and transmits forces imposed on one facing to the other in shear, so that they are effective about a common neutral axis. The facings resist nearly all of the applied edgewise (in-plane) loads and flatwise bending moments. For low-density core, separated facings provide nearly all of the bending rigidity to the construction. The core also provides most of the shear rigidity of the sandwich construction. By proper choice of materials for facings and core, constructions with high ratios of stiffness to weight can be achieved.

The facing materials can be almost any structural material which is available in the form of thin sheet and can be used to form the facings of a sandwich panel. The properties of primary interest for the facings are as follows: high axial stiffness

DOI: 10.1201/9781003035374-4

giving high flexural rigidity, high tensile and compressive strength, impact and temperature resistance, environmental resistance, wear resistance, and optionally an attractive surface finish.

The commonly used facing materials can be divided into two main groups: metallic and nonmetallic materials. The metallic group is comprised of primarily steel, stainless steel, and aluminum alloys. Within each type of metal, there are a vast variety of alloys with different strength properties, whereas the stiffness variation is very limited. The second group is the larger in the two groups, including plywood, cement, veneer, reinforced plastic, and fiber reinforced composites.[1]

The facing can also be divided into another two groups: rigid sheets before laminating on to the cores, which will be constructed with the cores by using adhesives and by a dry consolidation process; and facings formed during the laminating process on to the cores by a wet process. Most large sandwich parts are made by using composite fabrics and thermosetting liquid resin in wet laminating processes.

Dry fabrication refers to lamination with non-FRP or precured facing materials. The process is completed by using face sheets, core, and adhesive. The faces are metallic or nonmetallic rigid sheets, cured thermoset or thermoplastic FRP sheet, or prepreg. The adhesives could be thickened resin, liquid adhesives, or thermosetting or thermoplastic adhesive films. The fabrication processes are hot pressing, vacuum bagging, pinch roller pressing, and continuous pressing, with many variations for all of these.

It is important to consider the core format for dry lamination since bonding will occur between two semi-rigid sheets. The thickness tolerance of the sheet may need to be improved for flat panel fabrication and autoclave cure.

Wet lamination combines the processes of wetting out the fiber reinforcement and bonding all the various layers together. The typical facing consists of dry fabrics; however, precured and noncomposite materials may be incorporated into the laminate. The matrix materials are the thermosetting resins primarily and to a lesser extent, thermoplastic resins that are gaining attention because of their recyclable properties. Wet lamination is any process where liquid resin is transferred to the part. This process may be conducted in the open by hand using a bucket of resin and a brush, machine, chop and spray lamination, or with a or closed to the atmosphere whether vacuum (VARTM, infusion) or with positive pressure, Lite-RTM, RTM, etc. and pultrusion. Prepreg may be considered wet or dry and may be combined with other processes such as infusion if the laminate will be cured at elevated temperatures.

4.1 DRY LAMINATING PROCESS

4.1.1 FACING MATERIALS

Metal sheets are the primary facing materials for producing sandwich panels used in the rail transportation market. The young's (elastic) modulus, thickness, surface treatment for bonding, cosmetic requirements, and environmental protection are the most important factors. Stainless steel and aluminum sheets, with a thickness of 0.4–0.7 mm, primarily are laminated using a continuous process.

Thermosetting fiber reinforced plastic sheets, referred to as "FRP" in the composites industry, mainly are used as facings for making sandwich wall panels for

FIGURE 4.1 FRP and RV with the wall made with FRP.

recreational vehicles, delivery trucks, and semitrailer trucks. FRP sheets as shown in Figure 4.1, about 1–3 mm thick, generally are made with unsaturated polyester or vinyl ester as a matrix, chopped glass fiber or fabric as reinforcing materials, and made in a continuous process. The surface of the FRP is coated with gel coat that can be of a variety of colors and glossy and has the function of weather and impact resistance.

Continuous fiber reinforced thermoplastic (CFRTP) sheet is made by impregnating nonwoven continuous fibers with thermoplastics.[2] The fibers can be glass, carbon, and aramid materials, and the plastics can be polypropylene, polyethylene, PET, polyamide 6, and polyvinylidene fluoride. For oriented unidirectional or biaxial CFRTP as shown in Figure 4.2, the unidirectional 0° direction tape is produced first and then optionally slit into ribbons, woven, and heat-laminated together to create a 0°/90° bi-directional structure that has balanced (or intentionally unbalanced) physical and mechanical properties. The CFRTP product resists rot, corrosion, and mildew. Some CFRTP sheets with a thin scrim liner material on one side promote adhesion for sandwich composite laminating. The scrap, waste, and recovered product after serving its useful life can be recycled into new products by traditional hot melting processes because the matrix is thermoplastics. Also, the sheets can be laminated by applying heat without adhesives and emit no volatile organic compounds (VOCs). The products are primarily used for producing sandwich panels in such applications as truck bodies, shipping containers, and building construction wall panels.[3]

FIGURE 4.2 CFRTP sheet and sandwich lamination.

4.1.2 Adhesives

There exist a variety of adhesives for dry laminating sandwich composites. The adhesive must meet the mechanical requirements of the structure: providing a good bond between the facing and core in the environment that the structure is to work and performing satisfactory against such considerations as static strength and fatigue, heat resistance, aging, and creep. The following are the most frequently used adhesives:

Hot-melt adhesives provide optimal adhesion and superior manufacturing flexibility compared with waterborne or solvent-based adhesives and are widely used in a variety of applications, including product assembly, woodworking, and pressure-sensitive tapes and labels. The products are solvent-free, thermoplastic-based materials that are specially formulated and are applied in the molten state at temperatures varying from 120°C to 180°C. Most hot-melt adhesives are composed of three main components: high molecular weight polymer (e.g., EVA or synthetic rubber and polyurethane), which acts as a backbone and provides the primary mechanical properties of the adhesive; tackifying resin, which provides wetting and adhesion properties; and plasticizer, such as an oil or wax, which controls the viscosity of the blend and enables the adhesive to be handled by simple machinery.

There are rod and pellet types of hot-melt adhesives that can be dispensed by directly laminating and melting using a hot press, and by glue gun, or by roller coating and spraying after melting. The hot-melt adhesives can be solidified at short time and with very little VOCs released. Generally, sandwich panels are fabricated using hot-melt or pressure-sensitive adhesives for applications with low structural requirements.

Epoxy resins and adhesives can be formulated to cure at a wide range of temperatures, but for most consumers, sandwich applications are low-temperature curing resins, normally between 20°C and 90°C. For aerospace and other demanding applications, some epoxy formulations are formulated for high temperature curing (130°C–220°C). In general, epoxy resins and adhesives have the advantage of being used without solvents, cure without creating volatile by-products, and have a relatively low volumetric shrinkage rate. The absence of solvents makes epoxies usable with almost every type of core material. Epoxy is available as pastes, films, powders, or solid adhesives. The typical shear strength of epoxy is about 20–25 MPa.

Polyurethane adhesives can be one-part adhesive or two-part adhesive. One-part adhesives are most widely used for bonding sandwich elements that cure by reacting with moisture in air. This is because they provide excellent adhesion to most materials. They can be used as a paste or liquid in a wide range of viscosities, may have long or short cure times, and can be made fire-retardant and water-resistant. Most PUR adhesives contain virtually no solvents and are thus environmentally friendly and once cured are the least toxic of all the adhesives.

Urethane acrylate adhesives are resins which are compatible with polyesters and vinyl esters. In fact, urethane acrylates are so compatible that they can be incorporated, in, e.g., a wet polyester laminate. Urethane acrylates are very tough and exhibit almost no curing volumetric shrinkage.

While it is possible to use commonly available polyester and vinyl ester resins, these are not generally optimized for dry bonding. Problems are low viscosity, lack of gap filling, and their curing volumetric shrinkage often creates very high interfacial shear stresses. One way of decreasing the effect of shrinkage is to add a low-profile

Laminating Processes

additive (LPA) or some fillers to the resin for reducing the shrinkage and increasing viscosity. Specific core-bonding adhesives are available, expressly designed for this purpose. The supplier of the adhesive will work with the sandwich core suppliers to develop these products and certify them for use.

Adhesive films are solid resin sheets that are thermally activated. They do not depend on the experience of the person mixing as in the case of glues with two components and eliminate waste, and the properties are guaranteed by the manufacturers. The adhesive films are made majorly by two types of polymer materials. One type is thermal plastics: polyolefin, such as polyethylene vinyl acetate (EVA), polyethylene acrylic acid (EAA), LDPE, HDPE, and PP. Bonding can only occur by heat and pressure and then gives the bonding strength after cooling down to room temperature. Typical physical and mechanical properties and processing conditions of polyolefin films are listed in Table 4.1. These films are suitable to assemble aluminum face to honeycomb core into a sandwich structure.

The second type of adhesive film is stage B thermoset resin, such as epoxy and bismaleimide (BMI), which activates by heat and finishes crosslinking reaction to give bonding strength between skins and core. The epoxy film is a high-strength adhesive formulation supplied in the form of a lightweight flexible film. It is intended for metal-to-metal or sandwich core-to-skin bonds and has a strong self-filleting action in honeycomb-to-skin bonds. The features of epoxy film are flexible low-to-medium cure schedule 70°C–130°C; accurate control of adhesive distribution ideal for honeycomb sandwich construction; bonding in both metallic and composite structures; suitable for press molding, autoclave, and vacuum bag cure; no solvents; low volatile content. The properties and processing conditions of one type of epoxy adhesive film are listed in Table 4.2. Also, there is a foaming type of adhesive films that expand when cured at elevated temperature, so that makes them ideal for gap filling, honeycomb core edge bonding, and core splicing.

The choice of adhesive is determined by required mechanical performance, compatibility with the materials and process, ease of use, availability, and cost.

TABLE 4.1
Typical Physical, Processing, and Mechanical Properties of Polyolefin Adhesive Film[4]

Properties	Unit	Material Basis					
		EVA	EAA	LLDPE	LDPE	HDPE	PP
Area weight	g/m^2	100	95	100	100	100	100
Heat deflection temperature (DMTA)	°C	75	80–90	105	110	120	135
Minimal bond line temperature	°C	100	115	130	140	140	165
Pressing temperature	°C	105	120	155	155	155	170
Pressing pressure	Bar	<1	<1	<1	<1	<1	<1
Pressing time	Min	5	3	3	5	5	5
Lap shear strength on aluminium at 23°C	Mpa	10	10.2	11	12	12.8	11
Lap shear strength on aluminium at 47°C	Mpa	5	6	6.5	10	10.6	9.6

TABLE 4.2
Typical Properties and Processing Conditions of One Type of Epoxy Adhesive Film[5]

Properties or Conditions	Unit	Typical Value
Available weight	g/m^2	100,200,300
Recommended cure time at 70°C	hour	8
Recommended cure time at 80°C	hour	5.5
Recommended cure time at 100°C	hour	2
Recommended cure time at 120°C	hour	0.5
Shelf life at 20°C	Day	30
Shelf life at −18°C	Month	12
Lap shear strength	Mpa	30–35
Maximum service temperature after curing	°C	121

4.1.3 LAMINATING PROCESSES

Dry laminating is an assembling process. The process is more often used for producing small and simple parts but can also be used for making large and complicated products. Some lamination processes may just employ a hot press or a vacuum bag, while others require large and complicated equipment. In the process, adhesive layers are interleaved between the faces and the core, and the whole stack is subjected to elevated temperature and pressure as required by the adhesive or resin until it cures, and then the sandwich product is cooled. For high-performance applications, the bonding process likely takes place using a vacuum bag and an autoclave, whereas for less-demanding application, it may be sufficient to use a vacuum bag and/or weights or a hydraulic press. Since, ideally, there should be little or no resin bleeding if the bonding is correctly performed, the vacuum-bagging arrangement is the simplest when compared to wet laminate manufacture.

It is normally necessary to prepare the surfaces to be bonded in order to achieve a high-quality adhesion. Unless already done, cores are typically vacuum cleaned (compressed air contains water & lubricating oil) to remove all loose particles, and they may also be primed (coated with resin) if necessary. Face materials must also be cleaned. Metals may need to be etched and/or primed for bonding. Thermoplastic resins may need surface activation such as corona or plasma treatment.

The typical processing sequence for bonding of composite laminates to a core using an epoxy or polyurethane adhesive is shown in Figure 4.3. Figure 4.4 shows sandwich structural insulated panels (SIPs) manufactured by using adhesive and compression laminating. In this case, the open time of the adhesive is long enough to produce many panels in a single, large opening press. This greatly reduces production time but makes it more challenging if heat is required for proper cure.

The dry lamination process has been used in automatic production for making large-scale sandwich panels for the last 30 years in various industries. Several technologies have been developed for continuous or semi-continuous production. A pinch roller laminating system and double belt press system are rapid fabrication

Laminating Processes

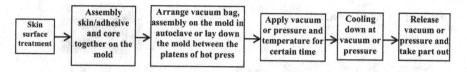

FIGURE 4.3 Dry lamination procedures by using heat and pressure.

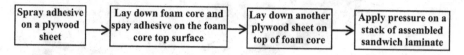

FIGURE 4.4 Sandwich structural insulated panel (SIPs) manufactured by compression lamination.

processes using hot-melt polyurethane adhesive. The adhesive is initially melted, and then coated to the foam core panel using a specially designed rubber roller. The face sheets are laid below and above the foam panel, and then the assembly is sent into the pinch roller system or double-belt press for consolidation, as shown in Figures 4.5 and 4.6. The adhesive can cure and then cool down to give enough strength (green strength) for holding the three-layer assembly together after pinch rolling and will have a maximum strength after just 24 hours.[6]

Moisture curing polyurethane adhesive is used in many large-scale lamination processes for making RV and truck sandwich wall, floor, and roof panels. A multi-opening platen/vacuum powered press for producing a sandwich panel with FRP facings up to 15 m long and 3 m wide. Each platen opening can operate by drawing vacuum independently, while other platens open for moving their laminate in or out. The laminating processes are shown in a flow chart shown in Figure 4.7.

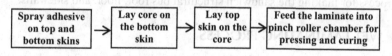

FIGURE 4.5 Pinch roller lamination process.

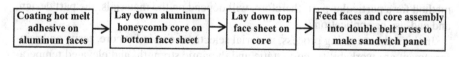

FIGURE 4.6 Aluminum face and honeycomb core sandwich panel by using hot-melt adhesive and double belt press.

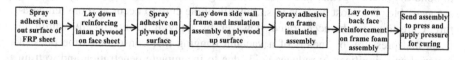

FIGURE 4.7 Four-level vacuum powered press for making RV sandwich sidewall.

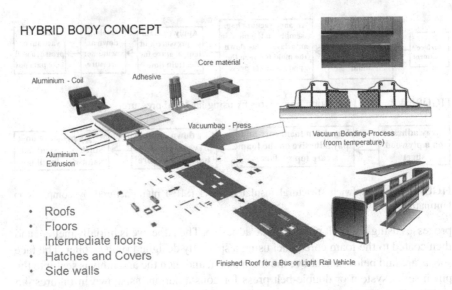

FIGURE 4.8 Vacuum bagging laminating process for making the sandwich structural bus roof, floor, and sidewalls.

Adhesive films and a double belt press are commonly used in continuous laminating processes for making sheet metal or composite-faced sandwich panels. Hot-melt adhesive films are comprised mainly of polyethylene copolymer or thermoplastic polyurethane (TPU). The double belt press may be set up for heating and cooling in different zones so that the laminate is heated first under pressure and then cooled under pressure until leaving the press. Figure 4.8 shows a vacuum bagging laminating process for making the sandwich structural bus roof, floor, and sidewalls.

4.2 WET LAMINATING PROCESS

There are numerous methods for fabricating sandwich composite products. Some methods have been borrowed (injection molding, for example), but many were developed to meet specific sandwich design or manufacturing challenges. Selection of a method for a particular part, therefore, will depend on the materials, the part design, and end-use or application.[7]

In the wet laminating process, material selection criteria for core, resin, and reinforcement are workability, cost, static and dynamic strength, and elevated temperature performance. The wet laminating processes involve some form of molding, to shape the core, resin, and reinforcement into a specially designed product prior to and during cure. After wet laminating, the resin in the facings needs to cure in order to bond the fibers together and bond the facing to the core. Several curing methods are available. The most basic is to simply allow the cure to occur at room temperature. Cure time can be accelerated, however, by applying heat, typically within an oven, and pressure, by means of a vacuum bagging that consolidates the plies of material and significantly reduces voids due to incomplete penetration and wetting of the matrix through the reinforcements and core surface.

4.2.1 Hand Layup Process

The most basic fabrication method for making sandwich composites is hand layup, also called wet layup, which typically consists of wetting dry fabric layers with liquid resin, and prewetting the core surfaces with liquid resin, while assembling these onto a mold to form a laminate stack by hand. The trapped air in the fabric is rolled out by a hand roller as shown in Figure 4.9. The laminate also can be consolidated by vacuum bagging or by the force of a platen press. Following the de-airing, the resin is allowed to cure either in an open condition or under vacuum bag or press, at room temperature or in a hot oven. Curing time may be accelerated by curing in a heated oven or using a hot platen press. Figure 4.10 shows a process to make a sandwich composite panel with foam core and carbon fiber face by hand layup and vacuum consolidation.

Resin used for hand layup can be epoxy, polyurethane, vinyl ester, and unsaturated polyester. The resin can be formulated to cure at room temperature. For achieving maximum strength, the part can post cured at a higher temperature. Hand layup has been widely used for many years, is easy to learn and practice using a low-cost mold if curing is accomplished at room temperature, and can accommodate a wide choice of cores, resins, and fiber materials. This method is often used for producing small-size components, such as a prototype, a sample for testing, and parts for large products. The disadvantages of hand layup are that the resulting quality strongly depends on the skill of the operator, limits the various types and weight of fabrics used, and results in a lower fiber/resin ratio, and volatile components of the resin system more readily evaporate into the air. Laminates must also be fabricated in a

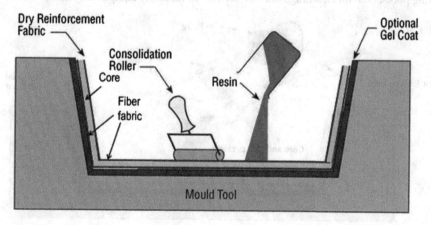

FIGURE 4.9 Hand layup laminating.

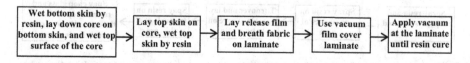

FIGURE 4.10 Hand layup and vacuum bag curing for making sandwich panel.

number of separate steps as only a certain number of plies can be molded at once. Wait time is needed between these "turns", but if too much time passes, secondary bond preparation must occur before a new laminate is added.

4.2.2 Spray Layup Lamination

Spray layup processing is a method to apply the liquid resin by use of a spray gun instead of hand. The fiber may be simultaneously applied with sprayed resin and broadcast on to the mold by use of a chopper gun and mixing process or separately applied to the mold as a fiber fabric before spraying resin as shown in Figure 4.11. For laminating a sandwich structure, the first facing must be prepared by spray layup processing and rolled out to remove trapped air, and then, the prewetted core material is rolled over that. After the mold side facing cures, or before the facing cures, a second facing is spray layup processed over the back side of the core and rolled out to remove trapped air, which is shown in Figure 4.12. Vacuum bagging may be used for consolidating the laminate and drawing out the air after finishing the spray layup processing. The product usually cures at room temperature or alternatively cures in a heated room.

The resins used in the spray layup process primarily are unsaturated polyester and vinyl ester. All types of cores can be used in this process, provided they are compatible with the resin. Chopped strand mat and all woven and knitted fabrics are suitable for this process. Spray layup processing can be performed in a continuous process for making flat sandwich or solid FRP panels, but mostly it is used in the intermittent process for making custom parts in low to medium volume quantities, such as bathtubs, swimming pools, boat hulls, storage tanks, duct and air handling equipment, and the like.

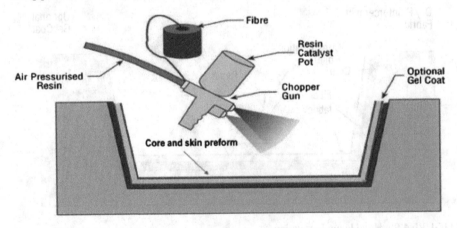

FIGURE 4.11 Spray layup process.

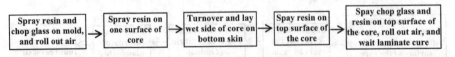

FIGURE 4.12 Process to make honeycomb core sandwich composite panel by the spray layup process.

Advantages of the spray layup process include a very economical process for making small to large parts, using a low-cost mold as well as low-cost materials. The limitations of the process are that it is not suitable making components with a high glass/resin ratio, it is difficult to control the fiber volume fraction as well as the part thickness, and the quality of the component highly depends on operator skill. Only a certain weight of fabric can be wet out by hand (generally a maximum of 36 oz/yd2, 1,400 g/m^2) and a limit on the amount of laminate which can be applied in one step. Often there is a high volatile chemical emission, and the process cannot produce a product with a good finish on both surfaces.

4.2.3 Prepreg Lamination, Press or Autoclave Curing

"Prepreg" is the common term for a reinforcing fabric which has been preimpregnated with a resin system. This resin system (typically epoxy) already includes the proper curing agent. As a result, the prepreg is ready to lay into the mold without the addition of any more resin. In order for the laminate to cure, it is necessary to use a combination of pressure and heat.

The products made by the prepreg process have high mechanical properties compared to the products made by the hand layup and spray layup processes because of using less resin in the laminates. In a hand layup, the products have 50%–70% resin content in most situations. This means that the finished laminate weight is 50% fabric and 50% resin. Excess resin increases brittleness and reduces overall properties. On the other hand, most prepregs contain around 35% resin. This is ideal for maximum cured properties and generally impossible to achieve in normal hand lamination.

Alternatively, semi-preg is a fiber reinforcement that is partially impregnated with a resin in film form, which can be chosen by a fabricator to allow air and voids to escape the fabric as vacuum is drawn, before full consolidation is complete.

Use of prepreg laminating can make composite parts uniformly and repeatedly, without neither resin-rich areas nor dry spots and with good cosmetics. Thickness will be uniform, and every part that comes out of the mold has a theoretical likelihood of being identical. Properties can be assured since the product can be supplied with certification, verifying the fiber/resin ration and mechanical performance. The prepreg process generates less mess and less waste, which are handled at room temperature so not fighting a clock trying to avoid resin setting up before everything ready. The prepreg laminating also needs less curing time because the heat curing cycle is completed, and the part is ready for service.

However, the prepreg fabrics have higher price than hand layup and vacuum infusion process. Even adding up the cost of the resin, cure, and fabric for other processes, prepregs still cost more. The prepreg fabrics also have a limited shelf time. Most prepreg fabrics have 6 months shelf time at room temperature (25°C). Nonetheless, heat cures prepreg, and storage at warmer temperatures will reduce the shelf life. Keeping the material cooler or freezing will extend the life significantly. The prepreg laminating needs heating and pressure equipment, like autoclave oven or hot press, which will increase the processing cost too. Prepreg laminate can be cured by the following processes: autoclave, out of autoclave (OOA), and heat-press compression molding.

Maximizing the performance of thermoset composite materials requires, among other things, an increase in the fiber to resin ratio and removal of all air voids. This can be achieved by subjecting the material to elevated pressures and temperatures. As described in the vacuum bagging section, some pressure can be exerted by applying a vacuum to a sealed bag containing the resin/fiber layup. However, to achieve three dimensional uniform pressures of greater than 1 bar, additional external pressure is required. The most controllable method of achieving this for an infinite variety of different shapes and sizes is by applying a compressed gas into a pressure vessel containing the composite layup. In practice, this is achieved in an autoclave.

The autoclave is a device that can generate a controlled pressure and temperature environment.[8] In this process, the prepregs are laid-up to produce a laminate of the required shape and dimensions, either manually or by using layup machines. The laminate is then vacuum-bagged and placed in the autoclave. Since, the pressures in the bag and autoclave are different, a compacting force is exerted on the laminate. This compacting force combined with the flexibility of the bag exerts pressure normal to the laminate surface resulting in uniform pressure distribution over the laminate surface. This uniform pressure distribution reduces the risk of void formation in the laminate. The autoclave is usually pressurized with nitrogen in order to avoid any fire hazards, which can be imposed by exothermic reaction of resin materials. Figure 4.13 shows two large autoclave equipment in working state.

Composite fabricators in aerospace and other large industries, such as automotive and wind energy, are in search of out-of-autoclave (OOA) manufacturing processes that can cross the mandated 1% void content threshold with less expensive, more efficient equipment yet achieve autoclave-quality composite parts for critical structural applications. OOA usually means vacuum-bag consolidation of the prepregs, followed by oven cure. The term OOA implies that the process is simply a nonautoclave version of the autoclave-based original. Indeed, OOA processing is attractive because it offers a more sustainable manufacturing pathway, based on the cost savings that accrue when foregoing the use of the autoclave, while producing autoclave-quality composites.

But OOA processing of aero structures forces the fabricator to cope with variables that were, for the most part, obviated by the use of the autoclave. Because entrapped air extraction is a time-dependent process, OOA cure cycles are typically longer.

FIGURE 4.13 Autoclave equipment.

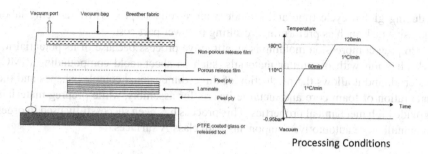

FIGURE 4.14 Typical OOA vacuum laminating setup and one processing condition.[9]

Microwaves reportedly heat only the composite structure, while the tooling and oven chamber remain cool, which drastically cuts heating and cooling time, as well as energy consumption. Figure 4.14 shows a typical OOA vacuum setup and curing schedule.

Compression molding of composite materials has been used for many years in composites applications. The main advantages of this process are very little material waste, its potential for automation, and its ability to mold large and fairly intricate parts. For structural parts, prepreg materials are preferred due to their high mechanical performance, achieved through a higher fiber volume content and the continuous nature of their fibers. The autoclave process makes it unsuitable for the automotive industry due to long cycle time and low level of automation. However, the latest development in resin technology has made it possible to compression-mold prepreg materials, combining the advantage of both prepreg materials and the compression molding process. Prepreg compression molding allows cycle time as low as 2 minutes while meeting the same structural properties obtained in an autoclave, making it an attractive process for the automotive industry.

Figure 4.15 schematically illustrates the different steps of the prepreg compression process. The prepreg plies are cut in the desired shape with the desired fiber orientation and assembled with core material together either manually or automatically. The laminate is then preformed over a dedicated preforming tool which gives the part its final shape. Preforming is then followed by consolidation and curing of the part under pressure and temperature where the cross-linking reaction occurs. Every step of the prepreg compression molding process can be fully automated, which helps

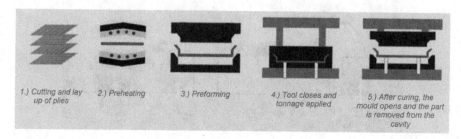

FIGURE 4.15 Prepreg compression laminating process.[10]

reducing global cycle time and provide a more reliable process by reducing labor intensive task such as preforming, resulting in lower part cost.

Prepreg compression molding is an attractive process because of its potential for hybridization with dissimilar materials, such as sheet mold compounding (SMC) material, and it allows the production of three-dimensional complex shapes and the integration of foam core and metallic inserts for assembly, while prepreg materials provide high mechanical properties. This process has been successfully used in order to manufacture automotive components with class A surfaces.

4.2.4 Closed Mold Lamination

There are a variety of ways to impregnate fibers with resin and consolidate sandwich laminates. All involve pressure. Pressure from a human hand or tool, from the atmosphere (vacuum bag, infusion), a machine (resin impregnator, prepreg), a hydraulic press or autoclave, or a pump (VARTM, RTM).

Resin transfer molding, vacuum-assisted RTM, and vacuum infusion are processes where the resin is introduced after dry fiber placement, transferring resin from a reservoir throughout the part.

RTM and the assortment of variations (ZIP, MIT, CARTM, HPRTM, etc.) use pressure above 1 atm to deliver the resin into the composite tool. VARTM and its emulations (Lite-RTM, CCBM) use pressure between 0 and 1 full atmosphere. Vacuum infusion uses only vacuum pressure as shown in Figure 4.16.

Closed molding was originally invented during the early years of composite lamination since resins of that time suffered from cure inhibition when exposed to air. Later, resin suppliers added paraffin wax to their products, which eliminated the need for complicated tooling and the associated setup. This allowed hand/spray layup to become the dominant molding process for a number of industries. While prepreg became the regimen for aerospace, there remained a void in other markets to find a way to both improve performance, quality, and repeatability. A suitable method was developed in the 1980s with the invention of SCRIMP, the Seeman Composite Resin Infusion Molding Process.[1] First patented in 1990, the method used the resins, fabrics, and cores available; resin distribution was achieved using surface

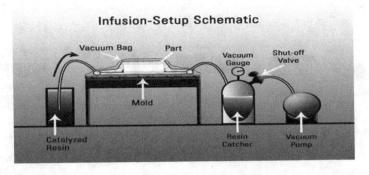

FIGURE 4.16 Vacuum infusion setup.[11]

media, and feed and vacuum lines were wire coils equally spaced in an alternating pattern. As the process became more popular, resins, fabrics, core finish (perforations, grooving), and other materials were developed specifically for resin infusion, and improvements to the infusion process were cultivated and adopted. Today, infusion cannot be defined as a specific method as there are dozens of varieties, materials, and design alternatives.

For performing closed molding, three main parts of the sandwich structure each have individual requirements. The fiber reinforcements (fabrics and mats) should be products that can be dry-laid on the mold and do not restrict resin flow, either by pressure or vacuum. Resins should be low in viscosity and free from too much filler or thickening agents so that they can easily flow and saturate all dry fiber bundles and completely wet all surfaces of the core. The catalyst or hardener should create a cure profile that achieves full cure at a lower peak exotherm. This is important because all the resin is introduced into the part at once. The core should be finished into a particular shape to meet the design requirements and fit of the mold, perforated, and optionally cut with flow channels that allow the infused liquid resin to readily flow through and across the sheet.

Like vacuum bagging, certain closed mold processes also require auxiliary materials such as flow media, release film, peel ply, resin feed lines, vacuum tape and film, vacuum system, or resin pressure injection equipment, and the like.

Closed molding will also involve specialized tooling or modification of existing molds to possess the proper flanges and be fully sealed against leaks.

4.2.4.1 Vacuum Infusion

Vacuum infusion (aka resin infusion) is a closed mold process where the second or B-side mold is a flexible membrane, typically a vacuum bag. The benefits of this specific process are low capital costs and unlimited part size, as well as molding parts with negative draft.

Vacuum infusion follows a specific equation which defines how far or how fast the resin will flow through the part. Darcy's Law, named after Henry Darcy, defines how a fluid flows through a porous medium.

$$\text{Darcy's Law } q = -\frac{k}{\mu} * \nabla p$$

where
 P = Pressure
 μ = Fluid viscosity
 q = Flow rate
 k = Porosity/permeability

Because the pressure differential is always less than 1 atm, there is a maximum distance the resin can flow. There are generally two directions to consider, resin flow through the thickness of the medium and flow across the thickness or surface (porosity vs. permeability).

FIGURE 4.17 Special core formats for vacuum infusion process.

Fabrics and core materials for vacuum infusion do not require good flow across the surface. They must however be porous through the thickness. Veils, cloth, and carbon fiber will all limit resin flow, but all may be used. They merely require another method for resin distribution, either surface media, the addition of a flow-friendly fabric such as continuous filament mat (unifilo) or core channels (surface grooves). This creates good flow in one direction.

Making a sandwich composite product by the vacuum infusion process needs special format of core materials. Porosity through the thickness is achieved by perforations. Although balsa core is porous and will transfer air easily through the core sheet, the rate is not good enough for vacuum infusion. The purpose of the perforations is to ensure a good transfer of resin to the laminate on both sides of the core. Figure 4.17 shows a core product grid scored, grooved, perforated, and laminated with scrim on one side for fitting curved geometries. The design of the grooves (width, depth, and spacing) can create fast resin flow, but this must be balanced with resin absorption and final part weight as well as keeping the fibers straight. If too wide, the face sheet can dip into the groove resulting in weak spots, and excessive resin volume can result in a poor "checker-board" surface due to shrinkage during cure.

A final consideration for fabrics and core is their combination. It is important to understand how individual materials wet out with resin when infused in order to recognize how they will fill together. If a fabric which is very difficult to wet out is combined with another material which creates high resin flow, the "slower" fabric may not get fully impregnated. It will look wet out, but the fiber bundles (tows) can be dry in the center. Laminate thickness is also a consideration. Very thick laminates become a two-dimensional problem. The time for the resin to flow through the thickness of the laminate must also be measured. Care should be taken to balance resin flow rates. This may be as simple as restricting the flow at the beginning of an infusion, until the feed rate drops for complete, 100% impregnation. The feed may then be fully opened afterward.

Resins for the vacuum infusion process should have a viscosity less than 400cps so they can flow through the reinforcing fabric quickly and uniformly. Principal parameters for selecting a suitable resin system are viscosity, gel time, cure schedule, and glass transition temperature (T_g). Useful technical data, unique characteristics, and applications of polyester and vinyl ester resins are listed in Table 4.3.

Resins can be formulated to the laminator's requirements ready simply for the addition of the catalyst prior to molding. While thixotropic agents should be avoided,

Laminating Processes

some additives and fillers can be used with vacuum infusion. The level of filler should be set so fill is balanced; if too high, the fabrics in the laminate can filter them out after a specific distance away from the resin feed line.

TABLE 4.3
Vacuum Infusion Grade Unsaturated Polyester and Vinyl Resin General Properties[12]

Resin Type	Viscosity at 25°C (cps)	Gel Time Adjusted by Initiator Level(min)	Gel to Peak Time (min)	Tensile Elongation ASTM D638, 3.2 mm Casting (%)	HDT ASTM D648, 3.2 mm Casting (°C)	Unique Characteristics	Applications
Isophthalic	220–275	20–25	8–15	2.4	108	Excellent dimensional stability, FDA-approvable	Components requiring good corrosion resistance and high-temperature applications
Orthophthalic	140–180	30–36	10–20	1.9	82	Economical, high reactivity with low exotherm	Wide range of applications, including marine
Dicyclopentadiene (DCPD)	80–100	8–12	5–10	1.4	108	Most economical, high reactivity in polymerization	Rigid composite and high-temperature parts needing low shrinkage
Vinyl ester resin	80–120	70–90	10–25	4.5	99	High-performance ingredients conform to contact food, moderate requiring high laminate exotherm allows for flexible timing of manufacturing	Marine and component requiring high corrosion resistance and high-temperature applications

Epoxies are fine-tuned for the desired application and manufacturing process. For example, protrusion and compression molding epoxy resins are heat-activated, while an infusion resin has a lower viscosity and can easily and quickly cure at any temperature from 5°C to 150°C, depending on the choice of curing agent. The low shrinkage during cure is the most advantageous property for closed molding, one that will minimize fabric and core print-through.

Table 4.4 shows a low-viscosity epoxy system designed for the fabrication of parts and structures by the resin infusion method. The epoxy system is available with two or more hardener options, slow, medium, or fast pot life. This is advantageous for infusion since the slow hardener may be used at the start, when the mass of resin in the supply reservoir is high and then changed as the part is filled. Combinations may also be used.

Any type of fabric may be used for vacuum infusion. This even includes 3-d spacer fabrics; however, the vacuum must be released at the proper time to permit the fabric to "snap" open. Most suppliers will offer a group of infusion or flow-friendly products that can be used without surface media. A major benefit of vacuum infusion is that the process

TABLE 4.4
Properties of One Infusion Grade Epoxy Resin[13]

Product Specifications	Epoxy Resin	Hardener One	Hardener Two
Color	Amber clear	Amber	Amber
Viscosity at 25°C (cps)	1,000	60	25
Specific gravity (g/cm^3)	1.0	0.99	0.97
Mix ratio by weight		100:24	100:22
Mix ratio by volume		4:1	4:1
Pot lift, 0.03 liter at 25°C (min)		40	70
Color		Mixed Resin and Hardener One or Two	ASTM Test Method
		Light Amber	Visual
Mixed viscosity		320	D2393
Cured hardness		86	D2240
Specific gravity (g/cm^3)		1.08	D1475
Tensile strength of cast bar (Mpa)		76	D638
Elongation at break of cast bar (%)		6.3	D638
Tensile modulus of cast bar (Mpa)		3,356	D638
Flexural strength of cast bar (Mpa)		131	D790
Flexural modulus of cast bar (Mpa)		3,645	D790
Compressive strength (Mpa)		110	D695
Compressive modulus (Mpa)		3,390	D695
Izod impact strength method A (ft-lbs/in)		1,093	D256
Glass transition temperature DMA Tg Onset (E) (°C)		72–96	D4060
Coefficient of thermal expansion, range of 40°C–60°C		5.42×10^{-5} in/in/°C	D696

Laminating Processes

permits the use of very heavy fabrics. These could not be wet out properly by hand or available as prepreg. Furthermore, multiple types of fabric can be stitched together and combined with surface veils and bulker mats. This greatly reduces the number of plies placed into a part as well as the time and labor associated with fabric placement.

Auxiliary materials for vacuum infusion are flow media, peel ply, resin feed and vacuum lines, sealant tape bagging film, fittings and tubing, vacuum system, resin injection equipment, and the like as shown in Figure 4.18.[14] There are a wide variety of peel ply fabrics available, and not all of them are suitable for infusion. First make sure the peel ply is suitable for use with the specific resin system. This is not a question of compatibility but time. After a certain period of time, it becomes more difficult to remove the peel ply from the laminate. For epoxy resin systems, with the wrong peel ply, this time is reduced to mere hours. This is a particular problem with resin infusion underneath resin feed lines and inside corners. Adding perforated release film can or using a coated peel ply eliminates this problem. Some peel plies have a tight weave which will restrict resin flow and should not be used under resin feed lines.

These products can be perfect for vacuum lines, however. Others are designed with a coarse weave for secondary bonding. This can increase the part's overall resin absorption and weight. Choosing the proper bag film depends on type of resin system and process temperature. This is also important for the fittings and tubing; they should both be able to handle the resin chemistry and temperature. All fittings should have multiple barbs, as single barbs have a much higher potential to leak. For extremely critical or large parts, it is prudent to use hose clamps on all resin feed lines to prevent leaks during resin cure where the elevated temperature will soften the

FIGURE 4.18 Some auxiliary materials for vacuum infusion.

polymer tubing and fittings. The tubing should also not collapse easily under vacuum. Figure 4.19 shows how to use the auxiliary materials with a vacuum infusion setup.

Proper vacuum pumps are the ones which can provide high vacuum >900 mBar and can run for long periods of time, even days without interruption. Oil-cooled, rotary vane pumps are the best option.

Finally, the sealant tape must be designed for composite lamination. Double-sided tape will not work. Like all other ancillary materials, the tape must also be suitable with the process or exotherm temperature. Note that sealant tape does have a shelf life. If too old, it will lose seal at high vacuum levels.

Vacuum infusion provides the opportunity to precisely position the core and reinforcements. If demanded by the part design, the mold may be first coated with a gel coat. For larger parts or if it must be walked on during dry fiber placement, a skin coat is needed to help fully cure the gel coat and prevent prerelease. The skin coat is applied by open molding and properly prepared for secondary bonding.

Fabrics and core are kept in the proper position using a "tackifier" either preapplied or as the materials are installed. Care must be taken to not overuse the tackspray as most of these products are not soluble in resin and too much will affect laminate strength.

It is vital to consider ply drop-offs and core transitions. In open molding, these thickness changes may be smoothed out either by grinding between steps or with the proper filler-putty. This is not possible for infusion. All gaps will fill with resin. Care should be taken to make sure multiple layers of fabric do not terminate or are cut to eliminate wrinkles at the same location. Most of these issues can be solved by using a precut kit. Kitting cannot solve the potential problem of fabric bridging, gaps between plies, particularly at inside corners. Bridging will lead to excessive resin absorption in the corner, a weakened laminate, and race-tracking. Race-tracking is a term for uncontrolled resin flow.

Proper core fit is also important. In open molding, gaps between core sheets can be repaired before the second skin is laminated. Poor fitting core sheets will result in higher part weight with excessive resin absorption, but more critical is race-tracking. Resin which races through gaps between fabric plies or around a sheet of core can cut off a section of the laminate from the vacuum and result in dry spots. Proper core fit is best achieved by using precut kits. If gaps remain, they should be filled with scraps of dry fabric; larger gaps should be filled with strips of core.

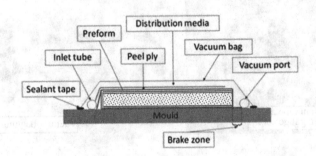

FIGURE 4.19 Vacuum infusion setup with auxiliary materials.[15]

Once all of the fabric and core is installed into the part, peel ply fabric covers all surfaces, and if needed, flow media is added. Peel ply will define the inside surface of the part, so careful placement is recommended.

The spacing of resin feed lines is determined by testing. Software is available to simulate part fill, but the material parameters for the software can only be found by direct testing. Flow testing should be conducted on a large enough panel or section of the part, so the resin begins to cure before it reaches the vacuum line or the longest distance required for the part is covered. Intervals of time are marked on the laminate to track progression. A flow test is shown in Figure 4.20.

The vacuum bag is then positioned and sealed around the mold perimeter, and all the vacuum lines are attached. It is important to use enough vacuum ports throughout the part. This ensures a constant vacuum level and helps prevent the vacuum from getting cut off during part fill. Vacuum may be divided into sectors if some areas of the part will fill before others.

The vacuum pump begins to draw air out from under the bag. For large parts, this is best achieved by a centrifugal pump, even a vacuum cleaner. Vacuum pumps are designed to build pressure, and their service life can be reduced by sucking a large volume of air. They are also slow. A very large part such as a wind turbine blade or boat hull will take many hours to evacuate with only a vacuum pump.

As the air is evacuated, it is important to make sure the bag is properly secured against the part with enough slack, so no gaps are left. If the bag bridges over a corner, it can stretch until it breaks. It is best to locate tabs or wrinkles at any section with the potential for bridging. Full vacuum is not needed at this point. Once the bag is fully in place and inspected for potential bridging, the vacuum level is increased until the lowest level of vacuum is achieved, and any leak is detected and then sealed. While one's ear and a simple stethoscope tube can be used, checking for leaks requires an ultrasonic leak detector and a drop-down test. For critical parts, an infrared camera can also be used in combination with cold spray.

To conduct a drop-down test, the part is cut off from the vacuum and the pressure drop is recorded over time. A maximum level of drop is permitted. If an acceptable drop is not achieved, additional checks for leaks are conducted. In the rare

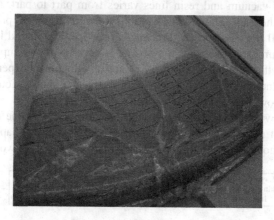

FIGURE 4.20 Vacuum infusion flow test setup.

occurrence, it may take many hours to find and fix all of the leaks. Lost time is far better than a lost part.

For larger parts, the drop test is conducted before resin feed tubes are connected between the vacuum bag and the resin container. Since leaks at the feed lines are more important, a second drop test is conducted to make sure the part is ready to fill. Also for large parts, it is good to use a manifold both for vacuum and resin. This allows for better control and faster response to potential problems.

Finally, the valve between the bag and the resin container is opened, and the resin enters the bag and wets out all dry reinforcements and core surfaces. Full vacuum is maintained until the resin has wetted out the entire laminate, and then the resin valve is closed to stop drawing resin in. After that, the laminate is kept under vacuum until the resin cures.

A setup procedure for infusing a boat hull is displayed in Figure 4.21. For making a large part, multiple hoses are installed to draw air out and resin into the bag.[16,17]

By using vacuum infusion, one can make products with very high fiber-to-resin ratio, a minimum amount of resin, and with superior mechanical properties. This is so because when resin is pulled into the mold, the laminate is already compacted; there is no room for extra resin. Parts can be lighter than other processes, but infusion will not produce the lightest parts.

Vacuum infusion offers the potential for a substantial reduction in emissions compared to either open molding or wet layup vacuum bagging processes. The limitations are a greater level of skill for producing quality parts, the risk of air leaks, the extra expense on auxiliary disposable materials, and the capability for producing only one cosmetic side of the product. The "extra" expense will almost always be offset by the saving in PPE and other cardboard or plastic used in open molding.

Like any laminating process, VIP is not without its drawbacks. It is highly recommended to not attempt to infuse a part without a lot of practice. There is no need learn by trial-and-error where the first few parts may be scrapped. Training, attention to detail, and good-quality control documentation is the path to success and learn from one's mistakes. The time before the resin is fed into the part is unlimited. Mistakes can be corrected.

Placement of vacuum and resin lines varies from part to part, and there is no one way to set them up. These considerations must be evaluated prior to layup, or else the part will not fill properly. Practice is used to observe and learn how resin flows through laminates. Remember viscosity is dependent on temperature. A cold part will reduce the flow rate significantly. Carefully document experiments and the infusion of production parts in order to learn from each one. Various types of feed geometry are shown in Figure 4.22.

Eccentric flow is best for smaller parts where the vacuum line must be placed close to the part perimeter. Concentric flow is best for closed shapes or when the resin feed line should not be in direct contact with the final part (it would be trimmed

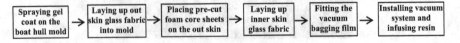

FIGURE 4.21 Setting-up process for vacuum infusing a boat hull.

Laminating Processes

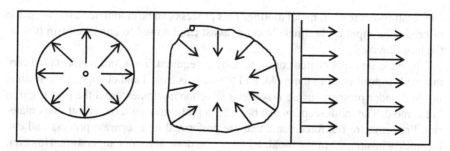

FIGURE 4.22 Eccentric, concentric, and linear (sequential) flow.

away). Both are also used for parts with an irregular shape. Branch feed systems use a combination of linear and concentric (inside L shape) or linear and concentric (outside L shape). Linear and branch flow are best for larger parts; sequential flow is an array of linear flow.

A further advantage is that infusion feeder lines can be positioned where it is most strategically appropriate, independent of the core sheet position. As mentioned earlier, another important benefit with perforations is that it yields less print through from resin shrinkage compared to having grooves facing the outer surface.

When resin is first introduced into the laminate, an associated drop in vacuum level will occur. This will cause the laminate close to the feed line to decompress slightly. The result is a higher resin content at the location adjacent to the feed vs. close to the vacuum line. This can be countered by changing the feed line to a vacuum line once the part is full or with a "double bag". Double (flexible) bag infusion is a method where one vacuum level is applied to the reinforcement positioned between the mold and an inner flexible bag and a second vacuum level (usually a higher vacuum level) is applied to a porous medium held between the inner vacuum bag and a second flexible vacuum bag. A second process was developed by NASA named the "double vacuum bag (DVB)" process, where a rigid perforated shell is interposed between two flexible bags. Separately controllable levels of vacuum may be applied to the inner bag or to the void between the two bags. The DVB process is particularly suited to the manufacture of composite parts involving resins which release water or solvent vapors during curing.

The specific feed geometry and infusion process variant permits close control of how the resin flows through the part and the fiber/resin ratio.

If race tracking or another problem leads to dry spots in the laminate, it is best to leave them, let the resin cure and re-infuse to fix. The critical time is when the part is completely wet out (full), but the resin is not yet cured. It is the only time a leak can ruin a part. If the instructions are followed and all leaks are found beforehand, the potential for a leak after part fill is extremely remote. In a manufacturing environment, anywhere from a few months to a number of years may be needed for training, testing, and preparation.

4.2.4.2 Resin Transfer Molding (RTM)

RTM is an intermediate volume molding process for producing composites. The RTM process involves injecting resin under high pressure into a mold cavity. RTM can use a wide variety of tooling, ranging from low-cost, composite molds to temperature-controlled, metal tooling. This process can be automated and is capable of producing rapid cycle times. Vacuum assist can be used to enhance resin flow in the mold cavity.

The mold set is gel-coated conventionally, if required. The reinforcement (and core material) is positioned in the mold, and the mold is closed and clamped. The resin is injected under pressure, using mix/meter injection equipment, and the part is cured in the mold. The reinforcement can be either a preform or pattern cut roll stock material. Preforms are reinforcements that are preformed in a separate process and can be quickly positioned in the mold. RTM can be done at room temperature. However, heated molds are required to achieve faster cycle times and greater product consistency. Clamping can be accomplished with perimeter clamping or clamping in a press.

The process can give a high rate of production, is easy to control, ideal for medium to large series of small to medium parts, has a high level of consistency, and provides good dimensional stability. However, the cost of the equipment is high, requires high precision molds, and requires time to learn. Applications are abundant in the production of truck, bus, and car, and structural and industrial products.

Since RTM incorporates resin under high pressure, the use of sandwich core materials is more limited. The core must withstand hydrostatic forces, and these will be in all directions. Often, there are extreme conditions right at the injection ports. Sometimes, the process must be adjusted to incorporate a sandwich core or the core reinforced in the proper area to prevent cracking or crush.

Vacuum-assisted resin transfer molding (VARTM) refers to a variety of related processes, which represent the fastest growing new molding technology. The salient difference between VARTM-type processes and standard RTM is that in VARTM, resin is drawn into a preform through use of a vacuum, rather than pumped in under pressure. VARTM does not require high heat or pressure. For that reason, VARTM operates with lower-cost tooling, making it possible to inexpensively produce large, complex parts in one shot.

In the VARTM process, fiber reinforcements are placed in a one-sided mold, and a cover (rigid or flexible) is placed over the top to form a vacuum-tight seal. The resin typically enters the structure through strategically placed ports. It is drawn by vacuum through the reinforcements by means of a series of designed-in channels that facilitate wet-out of the fibers. Fiber content in the finished part can run as high as 70%. Current applications include marine, ground transportation, and infrastructure parts. RTM is frequently used for making parts of airplanes. Within thermosetting resins, several classes are suitable for aerospace applications of RTM, such as epoxy, phenolic, cyanate, and bismaleimide.

Resins used for RTM and VARTM are similar to the resin used for vacuum infusion mentioned in the last section. However, polyurethane resin is also suitable for the processes. Polyurethane-based matrix resin for the RTM process contains three-component resin: resin, hardener, and release agent, which has lower viscosity for shorter injection times than epoxy products. Polyurethane resins that cure significantly faster than the epoxy products usually are employed for the cost-effective RTM production of composite components under mass production conditions.

The selection of a reactive thermosetting resin for RTM application forces to deal with a large number of choices of chemical engineering, mainly due to a strong relationship between the chemistry and process engineering. In fact, the process parameters, such as temperature and pressure, cannot be selected without considering the chemistry of the resin to be used. Factors to take in consideration for an RTM system can be divided into two broad categories: processing and performance. Initial viscosity and molding life are functions of the temperature, and they determine the operational temperature range of a process. The molding time is a function of the rate at which the reaction occurs between the resin and the curing agent, and the rate is directly proportional to the temperature. The viscosity depends on the chemical-physical characteristics of the matrix.[18]

RTM produces parts that do not need to be autoclaved. However, once cured and demolded, a part destined for a high-temperature application usually undergoes post-cure. Most RTM applications use a two-part epoxy formulation. The two parts are mixed just before they are injected. Bismaleimide and polyimide resins are also available in RTM formulations. "Light RTM" is a variant of RTM that is growing in popularity. Low injection pressure, coupled with vacuum, allow the use of less-expensive, lightweight two-part molds.

Light-RTM eliminates the challenges for sandwich core materials. Mostly any product may be used, even honeycombs which have been sealed for closed molding. Generally, the sheet should be perforated to equalize the pressure between the two face sheets.

4.2.5 Pultrusion

Pultrusion is a continuous process for the manufacture of products having a constant cross section, such as rod stock, structural shapes, beams, and channels. After modifying the die and installing core aligning and feeding system, sandwich composite panels can be processed by the pultrusion process as shown in Figure 4.23. Pultrusion produces a sandwich panel with high fiber content faces and high structural properties.

For producing sandwich panel by pultrusion, continuous glass strand mat, cloth, or surfacing veil is impregnated in a resin bath, and then, the core is fed between two sets of the impregnated reinforcements and pulled through a steel die by a powerful tractor mechanism or continuous puller. The steel die consolidates the saturated reinforcement, sets the shape of the stock, and controls the fiber/resin ratio. The resin for pultrusion is specially formulated to cure at a high temperature and thus at a fast speed. The die's platen is designed with multi-heating zones that are controlled by computer. The reinforcements are positioned into the die by tensioning devices and guides. The hardened steel die system is machined and chromium-coated and includes a preform area to do the initial shaping of the resin-saturated reinforcements. The latest pultrusion technology uses direct injection dies, in which the resin is injected into the die, rather than applied through an external resin bath.

The process is a continuous operation that can be readily automated with high efficiency and produces products with consistency properties. The composite skins of a sandwich panel are cured under tensile stress and at a high temperature, so the panel has greater flexural properties and a high heat resistance. The pultrusion process has been used for producing sandwich panels from 5 to 100 mm thick, and up to 3 m width, with different foam, honeycombs and balsa cores, and different facing materials.

Reinforcement materials used for sandwich panel pultrusion usually are glass or carbon fiber fabrics, especially multiaxial stitched mats, such as biaxial and quadaxial with chop strand mat on one side. The multi-directional stitched fabrics give flat and straight reinforcement and retain less resin in composite skins of the sandwich panel. And also stitched fabric can be pulled uniformly and symmetrically through the die in the continuously manufacturing process.

Surface veils are added to the laminate construction to displace the reinforcement from the surface adding a resin-rich finish to the product; since pultrusion is a low-pressure process, fiberglass reinforcements normally appear close to the surface of the product if no surface veil is being used. These can affect the appearance, corrosion resistance, or handling of the products. The two most commonly used veils are made by fiberglass and polyester fiber.

Core materials used for the pultrusion process need high-temperature resistance because they will go through the hot die platens. Most useful cores are polyethylene terephthalate (PET) foam, polyisocyanurate (polyiso) foam, polypropylene honeycomb, balsa wood core, etc.

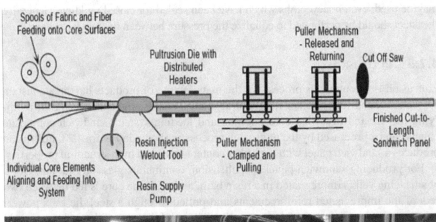

FIGURE 4.23 Sandwich panel pultrusion process.[19]

Laminating Processes

There are some polymer matrices that have been used in the pultrusion process such as polyester, vinyl ester, phenolic, and epoxy. Each polymer matrix has its own specific characteristic advantages for processing, performance, and application. Polyester resin has been widely used to produce pultruded profiles. Excellent performance and cost effectiveness provide the matrix the highest demand in the current pultrusion manufacturing field. However, polyester is poor to resist the high temperature and is quite brittle. Phenolic has the ability to withstand very high temperature compared to other polymers, and it is widely applied in the oil and gas industry. Vinyl ester and epoxy demonstrated better wetting compared to other polymers because of their comparatively low viscosity. In addition, these materials can accommodate higher percentage of fiber, and thus, they offer better mechanical and thermal performance. Epoxy shows much better mechanical performance compared to vinyl ester, but the price of epoxy is higher than vinyl ester. Furthermore, epoxy is advantageous when special mechanical properties, thermal degradation, and high corrosion resistance are required.

The resin formula for pultrusion contains a basic resin and a few additives for curing the resin in short time, binding the fibers and core together, and providing the mechanical and other properties required. The additives include fillers, catalysts, fiber wetting agents, air release agent, mold release agent, UV inhibitors, and color pigments. The polyester or vinyl ester for the pultrusion process is usually not promoted, so the curing reaction must be started by heat and catalysts (initiators in technical term). The catalyst system for polyester and vinyl ester resin is complicated in the pultrusion process. The resin needs to be heated up from room temperature to about 150°C and cured in less 1 minute time under die pressure. Multiple types of peroxide chemicals that can initiate a cross linking reaction of the resin at low, medium, and high temperature, respectively, are needed. Then exothermic reaction can heat the resin itself to peak temperature and finish the cross-linking reaction. The sandwich composite panel also needs to be cooled down under die pressure to below 80°CC before being pulled out of die. Exempt for selecting raw materials and making resin formula, setting up temperature in different die sections and die openings for making certain thickness panel, and using right pulling speed also are important for making a high-quality product with designed properties.[19,21,22]

4.2.6 3D Printing and Additive Manufacturing

The terms "additive manufacturing" and "3D printing" both refer to creating an object by sequentially adding build material in successive cross-sections, one stacked upon another. Although the mainstream media and many in the industry use the two terms interchangeably, additive manufacturing (AM) is the broader and more all-inclusive term. It is commonly associated with industrial applications and involves end-use applications like the mass production of components.

3D printing is a process of building an object one thin layer at a time. It is fundamentally additive rather than subtractive in nature. To many, 3D printing is the singular production of often-ornate objects on a desktop printer. The term was understandably adopted by the mainstream media when fused deposition modeling (FDM) first appeared in the 1980s because the FDM process worked very much like a 2D inkjet printer. Instead of a print head laying down a single layer of ink, the 3D print head deposited multiple layers of build material typically delivered as a thermoplastic filament.

The need for a low-cost, design flexibility, and automated fabrication process has spurred the development of AM for sandwich composites. AM refers to a group of fabrication techniques, including 3D printing, where parts are fabricated layer-by-layer directly from a computer-aided design file. Eradicating the need to have mold makes AM attractive to design and produce complex parts with lower fabrication costs. Development of short fiber reinforced polymers that exhibit superior strength for AM processes has been on-going for at least a decade. Recent research trends are in the direction of developing new fiber reinforcement composites including continuous fibers and cores.

Fused deposition modeling (FDM) technology also known as fused filament fabrication (FFF) is currently the most widespread way to rapid production of items utilizing additive manufacturing. 3D printing by FDM consists in depositing a filament of thermoplastic material. A portion of filament from a reel passes through a hot head, with a temperature higher than the melting point of the filament, then is extruded in the XY plane creating a layer of solid material on the build plate. Creating a model can be done by depositing a layer contour and then filling the inside with plasticized material by zigzag movement of the head. After printing one layer, the head moves along the Z axis initiating the build-up of the next layer. Using this technique, we can create complex shapes with a minimum of preparatory processes. The production process begins with creating a model in the CAD program, and then the model is incorporated into the program enabling control of process parameters such as head movement, feed rate, layer thickness, infill, head and table temperatures, slicing, support application, etc. as shown in Figure 4.24. Such a program generates a G-code, which uploaded in a 3D printer enables making a real model. The model removed from the printer needs finish machining, for instance to remove the supports and imperfections.[20]

For creating a sandwich composite structure, two materials are needed for core and skin in most situations, which can be completed by alternate dual material printing. A diagram illustrating the making of a composite structure by means of alternate printing using two materials is given in Figure 4.25. The process of selective printing with two materials is an advanced 3D printing method, getting increasingly more common. It can be done using a head consisting of two nozzles or newer solutions of one nozzle with a filament change system. The method consists in alternate deposition of different materials in one printing process. The advantage of this method is the possibility to control the properties of a finished item at the design stage because

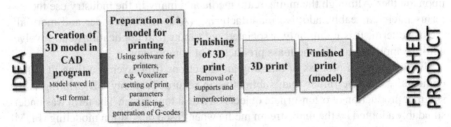

FIGURE 4.24 Diagram of 3D printing procedure by FDM method.[23]

Laminating Processes

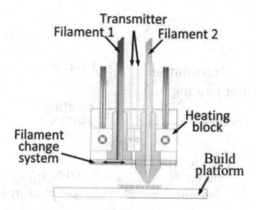

FIGURE 4.25 Diagram of making a composite structure by means of alternate printing using two materials.[23]

we can design and then make any composite structure. Figure 4.26 shows examples of sandwich structure composites made by dual material printing.

The greatest limitations of that manufacturing method stem from dual material printing imperfections, e.g., contamination due to the impossibility of transition from one filament to the other, fragment after fragment, as well as problems in proper configuring of the G-code, and the requirement of using materials with similar thermal characteristics. These problems are being constantly solved by introducing new systems of filament change and software updates by printer manufacturers. Creating a layered composite structure as face sheet using a special nozzle allows introduction of the continuous fiber. Figure 4.27 schematically shows the making of continuous composites by in-nozzle impregnation. Filament being the matrix material, continuous reinforcing fibers are supplied separately to the printer nozzle. In order to increase fiber impregnation with thermoplastic material, before its insertion in the nozzle, the reinforcement phase is heated. Fibers are automatically supplied to the head through filament motion. Filament is plasticized inside the printer head and connected with the reinforcement phase. The subsequent stages of the printing process are similar to traditional 3D printing.

For increasing stiffness of the composite made by 3D printing, some thermosetting polymers, such as epoxy, have been used with or without short fiber with it. These reinforced materials exhibit Young's modulus values that are an order of magnitude

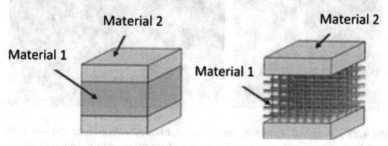

FIGURE 4.26 Example print structures of dual-material composite.[23]

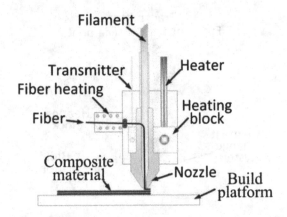

FIGURE 4.27 Creating a layered composite structure using a special nozzle allowing introduction of the continuous fiber.[23]

higher than those obtained by thermoplastics and photo curable resins developed for commercial 3D printing methods, while retaining comparable strength.

An epoxy-based inks that embody the essential rheological properties required for our 3D printing method were used to fabricate lightweight core sandwich composites. Unlike the ink designs that undergo solidification via gelation, drying, or on the fly photo-polymerization, the epoxy resins are reactive materials that initially exhibit a low viscosity, which rises over time as the reaction proceeds under ambient conditions. Next, silicon carbide whiskers (0.65 μm in diameter; 12 μm mean length) and carbon fibers (10 μm in diameter; 220 μm mean length) are added to the base formulation to create fiber-filled, epoxy inks.[24]

These high aspect ratio fillers align under the shear and extensional flow field that develops within the micronozzle during printing as shown in Figure 4.28, resulting in enhanced stiffening in the cured composite along the printing direction. Finally,

FIGURE 4.28 Image of 3D printing of a triangular honeycomb core (a) and schematic illustration of the progressive alignment of high aspect ratio fillers within the nozzle during composite ink deposition (b).[24]

Laminating Processes

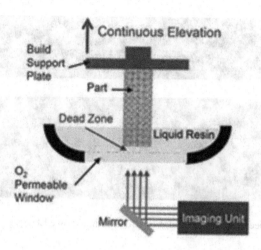

FIGURE 4.29 Diagram of continuous liquid interface production (CLIP).[27]

an imidazole-based ionic liquid was added into the epoxy as a latent curing agent to greatly extend the printing window (i.e., 30-day pot-life) under ambient conditions. Moreover, these inks ultimately require thermal curing at elevated temperatures (100°C–220°C) for several hours to complete the polymerization process. Some special honeycomb cores and sandwich composite samples are made by 3D printing.[24]

Although 3D printing, such as FDM, is now possible using relatively small and low-cost machines, it is still a fairly slow process. This is because 3D printers require a series of steps to cure, replenish, and reposition themselves for each additive cycle. Newly developed continuous liquid interface production (CLIP) is a process to effectively grow solid structures out of a liquid bath as demonstrated in Figure 4.29. CLIP works by projecting light through an oxygen-permeable window into a reservoir of UV curable resin. The build platform lifts continuously as the object is grown without requiring more time-consuming steps. By controlling the oxygen flux, CLIP creates a thin layer of uncured resin between the window and the object. This thin layer is called the "dead zone". This makes it possible to grow the object without stopping. A continuous sequence of UV images is projected, and the object is drawn from the resin tank without interruption.[25, 26]

The key to the process is the creation of an oxygen-containing "dead zone" between the solid part and the liquid precursor where solidification cannot occur. The precursor liquid is then renewed by the upward movement of the growing solid part. This approach made structures tens of centimeters in size that could contain features with a resolution below 100 µm. The CLIP process is 25–100 times faster than usual 3D printing techniques. An additional advantage of the CLIP technology is the resistance of the 3D printed objects, with no weaknesses between layers.[28] Special designed core and sandwich structure can be made by CLIP as shown in Figure 4.30.

FIGURE 4.30 Special core and sandwich structure made by CLIP.[26]

REFERENCES

1. "Manufacturing of Sandwich Composites." https://www.angelfire.com/ma/ameyavaidya/F_sandwch3.htm.
2. "Overmolding with Polystrand™ Continuous Fiber Reinforced Thermoplastic Composites," https://www.polyone.com/sites/default/files/resources/Composites_Overmolding_Overview.pdf.
3. "Conventional fiberglass Versitex used for Truck/Trailer Vehicles," http://www.uslco.com/versitex.php.
4. "Adhesives containing olefin polymers for safe and efficient functionality at low application temperatures," https://www.bostik.com/global/en/smart-adhesives/lead-technologies/polyolefin-adhesive/.
5. "Film Adhesives," https://www.toraytac.com/products/adhesives-and-core/film-adhesives.
6. "Aluminum skin plastics core sandwich panel laminating system," https://www.shsinopower.com
7. E. Greene, "Marine Composites," ACMA Composites 2004 Conference, Tampa, October 2004.
8. Q. Guo, Ed., P.J. Halley, "*Thermosets Structure, Properties and Applications,*" Woodhead Publishing Limited, Sawston, 2012, pp. 92–117.
9. S. Mortimer, M. J. Smith, and E. Olk, "Product Development for Out of-Autoclave (OOA) Manufacture of Aerospace Structures," https://www.hexcel.com/user_area/content_media/raw/ProductProcessDevelopmentForOutofAutoclave.pdf.
10. C. Pasco and K. Kendall, "Characterization of the Thermoset Prepreg Compression Moulding Process," SPE Conference, Sept. 2016.
11. "Vacuum Infusion set up," www.mid-composties.com.
12. "Vacuum Infusion Grade Unsaturated Polyester and Vinyl Resin General Properties," https://interplastic.com/category/products/.
13. "Properties of One Infusion Grade Epoxy Resin," https://www.fiberglass.com/a/fg7501.html.
14. "Vacuum infusion start-up kit," http://www.airteck.com.
15. S. Kazmi, Q. Govignon and S. Bickerton, "Control of laminate quality for parts manufactured using the resin infusion process," *Journal of Composite Materials*, vol. 53, pp. 327–343, Jan. 2011.
16. "Setting-up process for vacuum infusing a boat hull," http://www.diabgroup.com.
17. "Wind turbine blade by vacuum infusion," http://www.mathfem.it.

18. S. Laurenzi and M. Marchettihttps, "Advanced Composite Materials by Resin Transfer Molding for Aerospace Applications," http://www.intechopen.com/books/composites-and-their-properties.
19. "Sandwich panel pultrusion process" http://www.vixencomposites.com.
20. "Quadaxial and biaxial stitched chop strand mats used as skin reinforcement of sandwich composite pultrusion," https://www.vectorply.com.
21. https://www.strongwell.com.
22. https://www.creativeputrusions.com.
23. K. Bryl, E. Piesowicz, P. Szymański, W. Slączka and M. Pijanowski, "Polymer composite manufacturing by FDM 3D printing technology", *MATEC Web of Conferences*, (Open access), 237, 02006, 2018.
24. B. Compton and J. Lewis, "3D-printing of lightweight cellular composites," *Advanced Materials*, vol. 26, pp. 5930–5935, 2014.
25. V. Dikshit, et al., "Investigation of out of plane compressive strength of 3D printed sandwich composites," *IOP Conference Series: Materials Science and Engineering*, (open access), vol. 139, no. 1, pp. 012017, 2016.
26. J. Austermann, A. Redmann, V. Dahmen, A. Quintanilla, S. Mecham and T. Osswald, "Fiber-reinforced composite sandwich structures by co-curing with additive manufactured epoxy lattices," *Journal of Composites Science*, (open access), vol. 3, no. 2, p. 53, 2019.
27. "Hardware M2 Printer," Carbon, Inc.: Redwood, CA, USA, https://www.carbon3d.com/products/.
28. J. Tumbleston, D. Shirvanyants and N. Ermoshkin, "Continuous liquid interface production of 3D objects," *Science*, vol. 347, no. 6228, pp. 1349–1352, 2015.

5 All-Thermoplastic Sandwich Composites

Wenguang Ma

Russell Elkin

CONTENTS

5.1 Concept and Specialties 160
5.2 Thermoplastic Core Materials 161
5.3 Thermoplastic Face Materials 164
5.4 Sandwich Structure Process Methods 169
    5.4.1 Compression Molding 171
    5.4.2 Continuous Laminating – Double Belt and Pultrusion ... 172
    5.4.3 In-situ Core Foaming 173
    5.4.4 Diaphragm Forming 176
    5.4.5 Manufacturing of 3D Thermoplastic Sandwiches by One-Step Forming 176
5.5 Postprocessing and Recycling 177
References 182

The sandwich composites that are made using thermoplastic foam, honeycomb or composite mat cores with the solid thermoplastics or fiber-reinforced thermoplastics skins, and by thermal bonding or welding process are called all-thermoplastic sandwich composites (TSCs). Features of all-TSCs are high fracture toughness and impact resistance and short laminating time in a clean and healthy work environment, as well as the possibility to highly automate fabrication. Scraps and used finished goods can be re-formed, re-processed, repaired, or recycled all by the thermal melting process. Other features include the lower-cost raw materials and long raw material shelf life to ensure the lower processing cost. Furthermore, it is possible to use a single polymer throughout the core and skins which brings on optimal face–core compatibility due to the absence of a separate adhesive layer. This chapter will introduce the concept and specialties of all-TSCs firstly and then present thermoplastic core, face sheet, and processing methods that include compression molding, continue laminating, in-situ core foaming, diaphragm forming, and 3D one-step forming. Finally, the thermoforming, thermal welding, and other post-processing methods, and recycling by melting/molding process are unveiled.

DOI: 10.1201/9781003035374-5

5.1 CONCEPT AND SPECIALTIES

Thermoplastic materials are characterized by molecule chains that associate reversibly through intermolecular forces (such as van der Waals' forces), which allow disconnection and thereby softening (melting) of the material by the input of heat or mechanical forces. Material softening offers the potential to be thermoformed or joined by melting. Cooling returns the polymer to a solid. With this concept, the problems traditionally associated with thermoset sandwich structures may be alleviated. The fact that the polymer is thermoplastic means that mere physical processes are required to fabricate the structure.

All-TSC has the property of high impact resistance. Processing time can be quite short, in a clean and healthy work environment, because of no crosslinking process, as well as the possibility to highly automate fabrication. Scraps and used finished goods can be re-formed, re-processed, repaired, or recycled all by the thermal melting process. These features along with lower cost raw materials and long raw material shelf life ensure the lower processing cost. Furthermore, the use of a single polymer throughout the sandwich brings on optimal face–core compatibility due to the absence of a separate adhesive layer. Figure 5.1 shows the structure of thermoplastic sandwich panel (left, thermoplastic adhesive film is not used for most situation) and curved thermoplastic sandwich products made by curvature mold or by post thermoforming.

Cores for all-TSCs include thermoplastic honeycombs and foams and fiber-reinforced thermoplastic rigid porous or nonporous boards. Honeycombs can be made by polypropylene (PP), polyethylene terephthalate (PET), polycarbonate (PC), and polyetherimide (PEI); foams can be made by extrusion and bead infusion. Thermoplastic foam polymers are PET, polystyrene (PS), PP, PEI, polymethacrylimide (PMI), polyamide (PA), polyphenylsulphone (PPSU), and polyethersulfone (PES). The most useful bead foams are PS and PP bead foams. The light density fiber-reinforced thermoplastic rigid boards are made by glass, carbon, aramid and polyester fiber, or natural fiber with thermoplastic polymer PP and PET mostly.

Skins for all-TSC sandwiches consist of fiber-reinforced laminates with a thermoplastic matrix in which either short or continuous fibers are used. Thermoplastic composite skins are available as fully consolidated laminates, semi-prepregs, prepregs, or commingled fabrics. The thermoplastic matrix polymers in general range

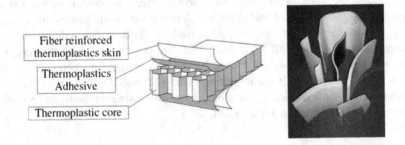

FIGURE 5.1 Thermoplastic sandwich structures and products with special shapes.

TABLE 5.1
Typical Skin Materials for All-Thermoplastic Sandwich Composites

Matric Polymer	Reinforcement	Fiber Content	Product Format
Polypropylene (PP)	Glass fiber unidirectional	60–75 wt%	Unidirectional tap or 0/90 multilayer sheet
Polypropylene (PP)	Glass fiber fabric	60–75 wt%	Commingling of glass fiber and PET fiber woven roving
Polyethylene terephthalate (PET)	Glass fiber unidirectional	65 wt%	Unidirectional tap or 0/90 multilayer sheet
Polyethylene terephthalate (PET)	Glass fiber fabric	65 wt%	Commingling of glass fiber and PET fiber woven roving
Polyamide (PA)	Glass, carbon, or aramid fiber fabrics	40–60 wt%	Dry prepreg fabrics
Polyetherimide (PEI)	Glass, carbon, or aramid fiber fabrics	60–70 wt%	Dry prepreg fabrics
Polyether ether ketone (PEEK)	Carbon fiber unidirectional	60–70 wt%	Impregnated tape or 0/90 sheet

from commodity plastics such as PP and PET or engineering polymers such as PA to high-performance polymers such as PEI or PEEK. Table 5.1 gives an overview of thermoplastic composite sandwich skin materials. Commodity plastics in combination with reinforcing glass fibers are mostly employed as skin materials for TSC sandwich structures. Especially, PP-based composites are applied. This can be explained by the low price of the material and the ease of process ability of PP since many commercial products are available in the market, and most investigation has been performed for many applications where manufacturing times need to be extremely short, less than 2 minutes. Of interest to the aviation industry is PEI, often combined with reinforcing glass fibers. Glass fibers are mostly used for reinforcement, but for some high-performance applications, such as aerospace and military, carbon fiber and aramid fiber also are used.

Due to this unique combination of materials, sandwich structures can easily be obtained by a rapid process comprised of two basic steps, manufacturing of all-TSC sandwich components and forming of intermediates into sandwich parts. Laminating and forming processes of all-TSCs include continuous thermal pressing and pultrusion, vacuum infusion in-situ polymerization, vacuum diaphragm, and compression molding. Post-thermal forming is used to produce 3D products, and thermal welding is used to make large and complex shapes.

5.2 THERMOPLASTIC CORE MATERIALS

Different thermoplastic polymers used for sandwich cores such as PS, PP, PET, PEI, and PES are listed in Table 5.2 and described in Chapter 1. Table 5.2 illustrates that mostly commodity plastics such as PP or PET are applied as foam core structures for ground transportation and industrial applications. In the area of high-performance

TABLE 5.2
Typical Foam and Honeycomb Core for All-Thermoplastic Sandwich Composites

Plastics Material	Density Range (kg/m^3)	Core Format	Commercial Name
Polystyrene (PS)	50, 60	Extruded Foam	COMPAXX
Polypropylene (PP)	60, 96	Extruded Foam	Strandfoam
Polypropylene (PP)	40, 60, 80	Bead Foam	Kaneka EPP
Polyethylene terephthalate (PET)	60, 80, 100, 130, 200	Extruded Strandfoam	Airex T92 and T90
Polyether sulfone (PES)	90	Extruded Foam	Divinycell F
Polyetherimide (PEI)	50, 60, 80, 110	Extruded Foam	ULTEM foam
Polypropylene (PP)	60, 80, 100, 130, 160	Extruded Honeycomb	Plascore
Polycarbonate (PC)	60, 80, 100	Extruded Honeycomb	Plascore

thermoplastic polymers, PES, PMI, and especially PEI cores are favored. With an operation temperature of up to 180°C and excellent FST (fire, smoke, and toxicity) properties, PEI and PES are interesting materials for the aviation industry. In comparison to other core materials, PEI cores can originate during the production (in-situ) of the sandwich structure. Due to the combination of a PEI film with a blowing agent, the material expands under heat treatment. Furthermore, Table 5.2 illustrates cores with a density in the range of 40–240 kg/m^3.

Extruded polystyrene (XPS) foam core material shows an excellent strength-and-stiffness-to-weight ratio. Composite panels using XPS foam core material exhibit high impact strength. The ductile behavior of XPS foam makes this core material particularly resistant to fatigue reducing shear heating. XPS foam core material, with its homogeneous closed cell structure and smooth dust free surface, is an excellent core material for use in combination with thermoplastic skin for making all-TSC product used in automobile and transportation industries. Typical mechanical and physical properties of two types of XPS foam cores are listed in Table 5.3.

PP Strandfoam is moderately strong, low-density closed-cell thermoplastic foam that can be used for making all-TSC products. PP Strandfoam has a unique orientation which offers superior compressive modulus and strength with minimal displacement, which are much higher against strand direction than parallel to the strand direction. The core sheet for sandwich composite application is always made into end strand format so the extrusion direction is perpendicular to the face sheets. The PP strandfoam planks can be welded into a large block by heat, which is sliced to different thicknesses (Table 5.4).[3]

Expanded polypropylene (EPP) bead foam has been widely used throughout the vehicle for the purposes of energy management and low-speed impact protection. EPP has been available in molded densities ranging from as low as 14 kg/m^3 and as high as 300 kg/m^3, which equated to resin expansion ratios of 64 times to 3 times. EPP bead foam is available in grades necessary for a wide range of applications, based on the technical requirements. High density grades are used where energy management is important, such as automotive bumpers and interior passenger safety

TABLE 5.3
Mechanical and Physical Properties of Two XPS Foam Cores[2]

Property	Unit	Typical Value	
		XPS Foam 1	XPS Foam 2
Density	kg/m^3	50	60
Compressive strength at 10% deformation	Mpa	0.7	0.9
Compressive modulus	Mpa	50	85
Tensile strength	Mpa	1.2	1.35
Tensile modulus	Mpa	40	65
Shear strength	Mpa	0.6	0.85
Shear modulus	Mpa	18	22
Shear elongation at break	%	80	60
Thickness	Mm	Up to 100	Up to 100
Coefficient of linear thermal expansion	mm/m-K	0.07	0.07
Processing temperature	°C	95 max	95 max
Service temperature	°C	−50/+75	−50/+75

TABLE 5.4
Mechanical and Physical Properties of Two PP Strandfoams[4]

Property	Unit	Typical Value	
		PP Foam 1	PP Foam 2
Density	kg/m^3	40	64
Compressive strength at against strand direction and 10% deformation	Mpa	0.34	0.76
Compressive modulus at against strand direction	Mpa	9.6	28.4
Thermal stability (linear change at 82°C)	%	<1	<1
Coefficient of linear thermal expansion	mm/m-K	0.04	0.03

components. Low density grades are used for packaging applications, and medium densities find applications in furniture and other consumer products. Recently, developments lead to use medium to high-density EPP bead foam in all-TSC production by in-situ expanding and laminating process (Table 5.5).[5]

Porous fiberglass-reinforced polypropylene rigid boards can be used as thin core sheets for all-TSC products. These low-weight reinforced thermoplastics make up a product group of special lightweight plastic composites with low thermal expansion and excellent mechanical and physical properties. Glass and PP polymer fiber fleeces produced in special textile processes are mixed and supplied as rolls and boards. Another is made with short glass fiber and PP powder by wet process as in paper production.

The rigid boards can be utilized in place of plywood, veneers, fiberboard, and FRP. They can also be used as core for all-TSC production. The porous rigid boards

TABLE 5.5
Mechanical and Physical Properties of EPP Bead Foams[5]

Property	Unit	Typical Value			
Density	kg/m^3	60	82	106	150
Compressive strength at 25% deformation	Mpa	0.39	0.60	0.87	1.50
Tensile strength	Mpa	0.62	0.87	1.05	1.37
Tensile elongation at break	%	14	13	12	11
Flexural strength	Mpa	0.72	1.08	1.40	1.90
Flexural modulus	Mpa	19	28.9	41	73
Coefficient of linear thermal expansion	mm/m-K	0.09	0.075	0.068	0.047

are available in areas per unit weight of 600–2,200 g/m² (with variable shares of PP and glass). Typical thickness ranges from 2.5 to 6 mm. These materials generally come with functional cover materials that provide adhesion, barrier, and ease of handling. The materials consist of a network of randomly oriented fibers coated with, and held together by, thermoplastic resin. A distinguishing attribute of materials with these structures is that they "loft" or expand in the thickness direction when heated. This expansion is due to the softening of the thermoplastic resin which allows the fibers that were previously bent or constrained as a result of the formation process to return to a straightened orientation.[6,7]

One type of porous rigid board called natural fiber composite (NFC) is made by combining natural fibers, kenaf, hemp, and flax are the most popular, with PP or polyester in a dry blending and hot-pressing process. As a thermoplastic base composite, NFC offers realistic opportunities for recovery and reuse of the trim waste and end-of-life components. This will become critical to enact the recycling laws for component manufacturers.

There are three formats of NFC products: flexible nonwoven mat, low-density board, and high-density hardboard. The low-density board is an excellent substitute for many existing wood fibers, wood flour, and fiberglass-reinforced plastics and also is an ideal material as a core for all-TSCs. The low-density board can be made in varying densities and sizes and can be laminated with many fiber-reinforced thermoplastic skins to fit many product possibilities. Typical physical and mechanical properties of the low-density board of the NFC are listed in Table 5.6.

5.3 THERMOPLASTIC FACE MATERIALS

Fiber-reinforced thermoplastic skin materials for all-TSC exist mainly in three formats: towpregs, preconsolidate tapes, and commingled fibers as shown in Figure 5.2. There are fiber fabrics with three prepregs also.

Towpreg is fabricated via a powder coating and consolidation system. A schematic diagram of the powder coating line in reference 9 shows powder coating equipment used to produce glass and carbon fiber-reinforced towpregs.[9] It consists of six main parts: a wind-off system, a fiber spreader unit, a heating section, a coating section, a consolidation unit, and a wind-up section. In order to produce the towpregs,

All-Thermoplastic Sandwich Composites

TABLE 5.6
Physical and Mechanical Properties of NFC Low-Density Boards[8]

Product Composition (%PP/% Natural Fiber)	Weight (g/m²)	Thickness (mm)	Tensile Strength (Mpa)	Flexural Modulus (Mpa)
50/50	2,400	3.0	31.0	3,104
50/50	2,000	2.5	27.6	2,759
50/50	1,800	2.2	26.9	2,552
50/50	1,600	2.0	24.1	2,069
50/50	1,400	1.8	20.7	1,862
50/50	1,200	1.5	20.0	1,793
50/50	1,000	1.2	18.6	1,724
50/50	800	1.0	13.8	1,172

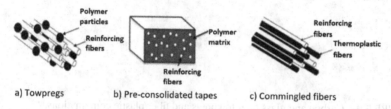

FIGURE 5.2 Formats of thermoplastic fiber prepregs.

the reinforcing fibers are wound-off from their tows and held by a carrier film and then pulled through a pneumatic spreader. After this, they are heated in a convection oven and so made to pass into a polymer powder vibrating bath to be coated. A gravity system allows maintaining the amount of polymer powder constant. The oven of the consolidation unit allows softening the polymer powder, promoting its adhesion to the fiber surface. Finally, the thermoplastic matrix towpreg is cooled down and wound-up on the final spool.

Glass, carbon, and synthetic fiber fabrics can be impregnated by general-purpose plastic like PP and PET and also by high-performance plastics, such as PA, PEI, PEEK, and PPS. Two examples of towpregs and fiber fabric/plastic combinations are shown in Figure 5.3.

Prepreg consolidated tapes are made by melt impregnation techniques. In the case of melt impregnation, the fiber bundle is spread to aid melt penetration prior to entering a crosshead extruder and die which is normally designed to produce a thin flat tape or prepreg. The process, which is shown in Figure 5.4, is capable of offering tape widths up to 300 mm with thicknesses from 0.125 to 0.5 mm. A typical glass or carbon prepreg with a thickness of 0.125 mm will have a fiber volume fraction around 60% and a single ply weight of 210 g/m². The continuous fiber-reinforced tapes could be produced with a variety of fibers and combined with mostly any thermoplastic matrix material.

The prepreg tapes can be weaved into fabrics as shown in Figure 5.5 or thermal-pressed into 0/90 configuration in two or multiple layers as shown in Figure 5.6. For

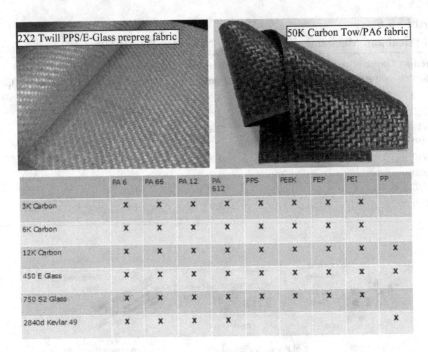

FIGURE 5.3 Carbon and glass fiber towpregs and fiber/plastic combinations.[10]

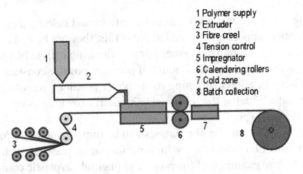

FIGURE 5.4 Schematic of the melt impregnation process.[11]

some special application, a thin layer of polyester fiber liner material is laminated at the top surface for promoting adhesion in the second bonding process (Table 5.7).[13]

Commingled fiber is made by one of the most promising routines which are based on the principle of homogeneous distribution of continuous matrix filaments and reinforcement glass filaments during melt spinning. The homogeneous fiber/matrix distribution of online commingled yarns especially leads to short impregnation paths and low void contents reflected by the high mechanical performance of the thermoplastic composites. Online commingled yarns were spun with three different polymeric matrices PP, polyamide (PA), and PET and glass fibers.

All-Thermoplastic Sandwich Composites

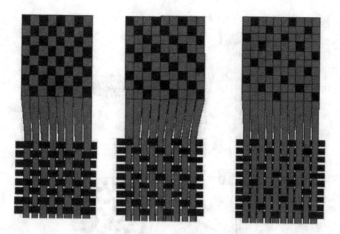

FIGURE 5.5 Weaved fabrics made from prepreg unidirectional tapes (plain-left, twill-center, and satin-right)

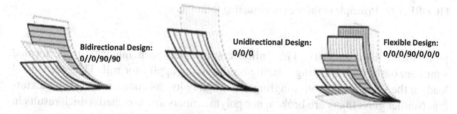

FIGURE 5.6 Fiber-reinforced thermoplastic laminates made from prepreg unidirectional tapes.[12]

TABLE 5.7
Physical and Mechanical Properties of Fiberglass/PP Sheets Made by Thermal Pressing Prepreg Tapes[14]

Property	Units	Typical Values		
Nominal thickness	mm	1.02	1.52	2.03
Normal weight	g/m^2	928	1,661	2,443
Tensile strength	Mpa	231	231	231
Tensile modulus	Gpa	10.3	10.3	10.3
Tensile elongation	%	2.5	2.6	2.6
Specific gravity		1.4	1.4	1.4

The processing of online commingled yarns differs significantly from other commingling techniques. Glass and polymer filaments are simultaneously spun and commingled while passing the sizing applicator as shown in Figure 5.7. In the commingling process, the filament distribution homogeneity is reasonably high since commingling is done at a state where both matrix and glass fiber yarns do not possess

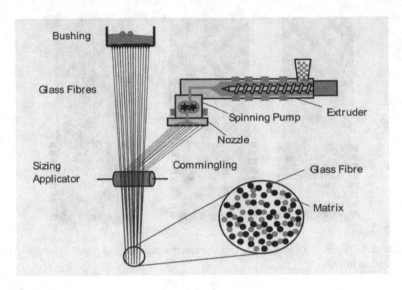

FIGURE 5.7 Principle of online commingling techniques.[15]

pronounced fiber integrity. The online commingled yarn integrity is performed simultaneously with applying a sizing on the sizing applicator roll. The mechanical load on the yarn during commingling is negligibly low as compared to air jet texturing. Neither glass fibers are broken, nor polymer fibers are stretched, which results in high yarn strength and avoids thermal shrinkage during consolidation.

Depending on the matrix polymer, different sizings are applied during the spinning.

The commingled glass fiber yarns can be supplied as roving, and then fabrics can be made by weaving, knitting, or braiding. Consolidation is done by heating above the melting temperature of the polymer matrix and applying pressure, before cooling under pressure. Depending on parts to be produced, vacuum molding, diaphragm, or calendaring process can be used. The states of glass fiber and polymer fiber/metrics of commingled roving before and after heating consolidating are shown in Figure 5.8.

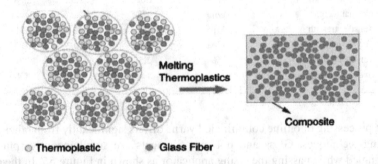

FIGURE 5.8 States of glass fiber and polymer fiber of the commingled roving before and after the melting process.[16]

TABLE 5.8
Physical and Mechanical Properties of Commingled E-Glass/Polypropylene Composite Fabrics and laminates[17]

Fabric Type	Glass Fiber Content (wt%)	Laminate Schedule	Tensile Strength (Mpa) 0° & 90°	Tensile Modulus (Gpa) 0° & 90°	Flexural Strength (Mpa) 0°	Flexural Strength (Mpa) 90°	Flexural Modulus (Gpa) 0°	Flexural Modulus (Gpa) 90°
500 gsm Plain Weave Woven	60	0°/90°	294	13.23	268	242	11.25	11.39
700 gsm Plain Weave Woven	60	0°/90°	310	13.89	171	213	8.79	10.46
600 gsm +45/−45 biaxial stitch-bonded fabrics	60	0°/90°	317	14.41	257	243	11.51	10.84
800 gsm +45/−45 biaxial stitch-bonded fabrics	60	0°/90°	305	14.18	297	291	11.95	11.95

Table 5.8 shows the physical and mechanical properties of commingled e-glass/PP plain woven and biaxial noncrimp fiber stitch-bonded fabrics and their laminates. Tensile and flexural modulus testing demonstrates that the biaxial stitch-bonded laminate has higher tensile and flexural modulus than the woven laminate, which can be attributed to the more highly aligned reinforcement fibers.

5.4 SANDWICH STRUCTURE PROCESS METHODS

In order to produce an all-TSC structure with sufficient quality, a proper bond between skins and core must be achieved. A good bond is defined as one that does not represent the weakest link within the sandwich structure and allows the transition of forces between skins and core reliably. Figure 5.9 gives an overview of different skin-to-core joining techniques. The processes for skin-to-core bonding can be divided into two main groups: adhesive bonding and fusion bonding. Several processes for adhesive and fusion bonding of skin and core have been used in industrial areas. These processes as well as some new developments and products are presented in the following parts.

One approach for laminating thermoplastic skin and core together is thermoplastic hot melt films. The polymer film is placed between skin and core, softened by heat, and finally solidified by cooling resulting in the film functioning similar to a thermoset adhesive. In the case of modified thermoplastic hot melt films, heating techniques such as microwave activation are also possible. The thermoplastic adhesive hot melt concept is considered faster (no curing has to take place) and more eco-friendly (possible recycling by disassembling after re-heating) in comparison to thermoset-based adhesive. Additionally, joining by means of hot melt adhesives offers the advantage that dissimilar polymers/substrates can be combined.

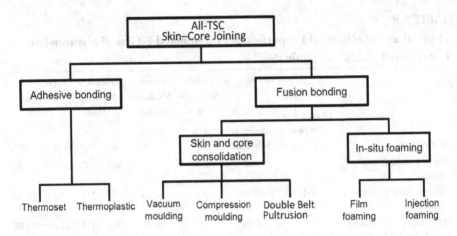

FIGURE 5.9 Processing routes for skin to core joining of all-TSC products.[1]

Usually thermoplastic hot melt films possess a lower softening temperature than core and skins. Skins and core stay in the solid state during the joining process resulting in inter-diffusion of the molecules of skins and core not taking place. The joining process is controlled by the heating temperature, pressure, and consolidation time.

To obtain a fully recyclable thermoplastic sandwich structure, an all-polypropylene (PP) sandwich panel is a good example. A low-sealing-temperature PP copolymer film can be used to adhesively bond PP self-reinforced or glass fiber-reinforced PP laminates to PP foam or honeycomb core. The process temperature is chosen at the melting temperature of the film but less than PP's melting temperature. Along with the appropriate pressure, sandwich skins can be successfully joined to the core without influencing the structure of either core or facing. The sandwich products can be manufactured in a press or with a vacuum setup.

Melting only one component of the sandwich, preferably the skins, is another way to achieve a thermoplastic adhesive bond between skins and core. The bond is based on mechanical interlocking between them. An example can be to make a TSC product by using glass fiber-reinforced PP skins (T_m = approx. 170°C), PET foam (T_m = approx. 250 °C), or PEI foam core (T_g = 215°C [107]). The skin polymer is heated and softened followed by applied pressure, which forces the molten PP polymer to flow into the surface cells of the foam core. After solidification of the skin polymer, mechanical interlocking is obtained, and an adhesive bond is created. Since an increased amount of polymer along the skin-core interface improves the bonding strength, the application of additional polymer at the interface material seems to be beneficial.

Another example is that carbon fiber-reinforced PEEK-plies, separate layers of pure PEEK, and the carbon foam were placed in a mold, which was heated to 380°C for about 10 minutes. Then the mold was transferred to a cold press, which applied a pressure of 1.75 MPa until the mold was cooled down to ambient temperature. The bond between the PEEK-based skins and the carbon foam is created by interlocking of the PEEK polymer and the pores of the cellular core.

One method to join thermoplastic skins and core is fusion bonding. This process is characterized by a joint formed by intermolecular diffusion of the polymers of the components, also referred to as adherents, to be joined. This is opposed to adhesive joining, where the joint is created by mechanical interlocking of the two dissimilar substrates on the surface. Fusion bonding, also called welding, is a well-established joining process for thermoplastics. Due to the inter-diffusion of the molecules, the joint can approach the bulk properties of the adherents. Additionally, fusion bonding can be performed in short cycle times and needs only nominal surface treatment.

Fusion bonding of thermoplastic materials can be explained by the "autohesion" or "self-adhesion" theory. According to this theory, the bonding occurs when the polymer molecules near the surface become mobile and the bond is developed through a combination of surface rearrangement, wetting, diffusion, and randomization of the polymers.[1]

In some applications, a combination of different polymers is required. Fusion bonding of two or more chemically different polymers poses some challenges since it requires miscibility of the components. It is often not even possible to join two dissimilar polymers by means of fusion bonding. Inter-diffusion of the molecules is highly influenced by the temperature, composition, miscibility, molecular weight distribution, chain orientation, and molecular structure of the adherents. With the same polymer in both adherents resulting in inter-diffusion of chemically coinciding molecules, the fusion-bonded interface can achieve the bulk properties of the material. Furthermore, sandwiches containing one single polymer in skin and core offer recycling potential (scrap and components) or even possibilities for postforming. However, if two polymers are miscible, even though they are different, such as PEI and PET, a fusion bond can be created between PEI fiber-reinforced skin and a PET foam core for making a TSC product.

To ensure molecular inter-diffusion between the core and skin more precisely, the core surface will also need to be molten. By applying pressure to achieve intimate contact, the core can be compacted due to its low density of foam or its small area of honeycomb or can collapse due to extensive heating. In order to prevent this, only a small process window exists to establish a good bond between skin and core.

5.4.1 Compression Molding

The term compression molding is used when the sandwich is manufactured in a mold by means of a press. Compression mold molding is an industrially applied method for producing thermoplastic composites and thermoplastic composite sandwich structures. Skins and core are stacked and placed in a mold before pressure is applied by a press. However, the comparatively low mechanical properties of the lightweight cores limit the application of pressure. As with vacuum mold molding, heating can be conducted isothermally (heating the whole stack while applying pressure) and nonisothermally as illustrated in Figure 5.10 (separate heating of the components before skin-core consolidation) depending on the heating method. Similar to vacuum mold molding, the main parameters for isothermal compression mold molding are the heating temperature, mold molding time, and pressure. By separating the heating procedure from the bonding process, the nonisothermal compression mold molding process can additionally be controlled by the skin and core preheating temperature and transfer time.

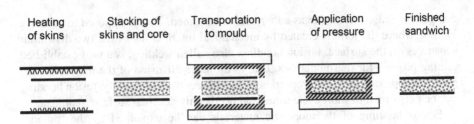

FIGURE 5.10 Nonisothermal compression molding process.[18]

Making all-TSC products by compression molding has two ways in the process: two-step nonisothermal compression mold molding process and one-step nonisothermal compression molding process. In the two-step process, the skins were heated sufficiently high to melt them in the first press, then being stacked with core and consolidated in a second cold press, while preheating stacks of plastic-based preimpregnated fiber fabrics, which were simultaneously consolidated and fusion-bonded to the core, were considered a single-step method. In both cases, mold stops should be applied to control the compaction of the core. Usually, the one-step process led to 20% better tensile strengths than the two-step process, bringing additional advantages such as time and energy saving. Moreover, an extra plastic film layer in the interface on the bond will increase the bonding quality. The pressure and thickness of the extra PP layer are the most significant parameters.

5.4.2 Continuous Laminating – Double Belt and Pultrusion

Continuous production of thermoplastic composite sandwiches based on fusion bonding can be performed by a double-belt laminator and pultrusion process. Since the process enables speeds of several meters per minute, it is considered very efficient and cost-effective.

First skins and core are automatically stacked, before entering a group of contact heating elements. The sandwich stack is heated and fusion-bonded under pressure. Then, a group of cooling elements or section of cooling zone solidifies the sandwich under pressure, see Figures 5.11 and 5.12. During the process, the temperature needs to be monitored to avoid overheating which could lead to core collapse.

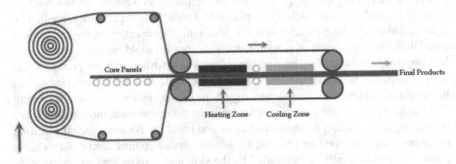

FIGURE 5.11 Double-belt production process for making TSC panel continuously.[19]

All-Thermoplastic Sandwich Composites

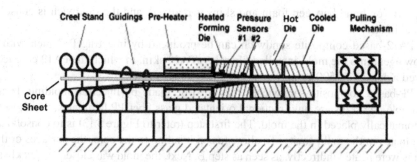

FIGURE 5.12 Making all-TSC panel by pultrusion process continuously.

After the consolidated sandwich leaves the double-belt laminator or pultrusion die, it can be cut into individual pieces based on requirements.

Process variables include material type, thickness, running speed, and pressure, as well as heating and cooling temperatures. The belt for the double-belt process, which ensures the continuous transfer, is made of materials such as polytetrafluoroethylene (PTFE) to avoid sticking of the molten polymer on the belt.

One type of sandwich composite panel is all-PP thermoplastic, with PP honeycomb as core and fiber glass-reinforced PP biaxial tapes as skins. If a layer of polyolefin film with lower melting temperature is used between the skin and the core, and by controlling temperature, belt gap, or die gap for pultrusion and running speed, the sandwich panel can be produced at a temperature lower than PP's melting temperature and also can prevent honeycomb collapsing and increasing bonding strength between skin and core.

Another all-TSC panel can be made by using a fiber-reinforced thermoplastic skin with a lower melting temperature than the core in double belt or pultrusion process, for example, using a fiberglass/PP biaxial laminate as the skin material (Tm is about 170°C) and a PET foam as the core (Tm is about 250°C). The processing temperature should be controlled to melt PP skin only and to bond the PET core together. The bonding mechanism between the skin and core is mechanical. This process is easy to control because there is no need to be concerned about melting and crushing the core.

Porous fiberglass-reinforced polypropylene rigid boards and natural fiber-reinforced polypropylene low-density board can be ideal core materials for double belt and pultrusion processes. Towpreg fiberglass/PP fabric, preconsolidate tape 0/90 laminate, or commingled glass fiber/PP fabrics are options for facing materials in the process to making sandwich panels continuously. The porous rigid board is heated above the melting point of the PP matrix which leads to self-expansion of the core that creates a pressure against skin and generates good bonding between skin and core. The self-expansion is caused by the release of internal stress stored in the fibers during the compression and consolidation.

5.4.3 In-situ Core Foaming

One in-situ foaming method is injection foaming. The core polymer and blowing agent are separately mixed in a batch. The skins are positioned in a mold. Then the foam is injected at a pressure of approximately 20 bar in between the skins. In doing so, a fusion bond between foam and skins is realized, and the sandwich is consolidated in-situ.

PA12-based composite sandwich can be produced by injecting P12 melt with a blow agent into the mold-installed fiberglass/PA12 skin, in which the PA12 core featured a density of $500\,kg/m^3$.

PP-based composite sandwiches may also be made by injection molding. In an example, the skins are unidirectional orientated glass fiber/PP laminates, which were automatically placed in the mold. The first step (refer to Figure 5.13) is to consolidate laminates in the mold cavity. To allow fusion bonding to the core, the surfaces of the skins were heated indirectly, as seen as step B. Next, the mold was closed as quickly as possible in order to minimize heat loss, followed by injection of the gas-loaded PP melt. After filling of the cavity had been completed, a pressure drop was created by expansion of the mold, which led to foaming of the core. The sandwich dimensions were controlled by the expansion of the mold. After cooling, the sandwich was released.

Today, expanded PP particle foams, named EPP, are often applied in automotive interior parts. Particle foams are produced by micro-granulate, loaded with a blowing agent, which is foamed into the particles supplied by the manufacturer. Further processing to the final part is then conducted in five steps: foam pearls are introduced into the cavity of the tool under pressure (1.5–4 bar) and compressed. The amount of pressure influences the resulting density. The cavity is vented by means of steam. Subsequently, the steam causes softening of the polymers at the foam pearl surface which results in bonding between neighboring foam particles. The steam temperature during processing plays an important role, since the pearls are only meant to be softened at the surface. The center should stay cool, which is needed for the stabilization of the particles. Finally, cooling is conducted, and the part can be released.

In combination with fiber-reinforced PP skins, consisting of the same polymer used for the particle foam, an all-PP sandwich product can be realized in a single manufacturing

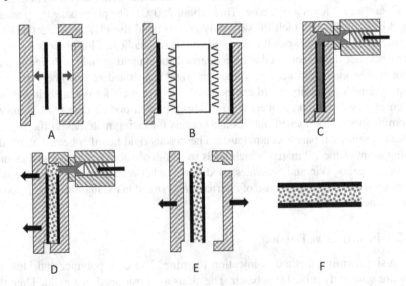

FIGURE 5.13 Manufacturing steps of injection foaming for making all-PP sandwich panel.[20]

step, called the in-situ bead foaming process. The beads are welded between the face sheets of the sandwich, and at the same time, the core is fused to the thermoplastic matrix of the facings. Bead foams such as expanded polypropylene (EPP) enables a low-core density of down to 30 kg/m^3. Furthermore, density gradients are also possible through the use of different beads. Bead foams allow for shaping very complex geometries and enable varying core thicknesses between app. 5 mm to several centimeters. In addition, these closed-cell foams also offer superior thermal and acoustic insulation.

For the in-situ production of thermoplastic sandwich structures with bead foam cores, it is better to use preconsolidated tape 0/90 laminate as the face sheets. These are prepared into designed shape and placed on both tool surfaces of the foaming mold as shown.[21] Beads are then filled-in between the face sheets. Due to the placement of the skin layers directly onto the mold surface, steam inlets are mainly covered. Steam can only penetrate in between the beads from the narrow sides of the cavity if no notches are placed in the face sheets.

A series of new products made by using foaming in-situ technology are called foamed in-situ thermoformable sandwich (FITS). These unique manufacturing processes produce an isotropic PEI, or PP, foam core bonded to fiber-reinforced PEI, or PP, face skins all in one step. The resulting panels are high-strength, high-stiffness, lightweight, and thermoformable. End-use products are quicker, and thus cheaper, to manufacture. Thermoforming revolutionizes edge close-outs, fastener inserts, and joining, eliminating the need for potting compounds, further reducing time, cost, and weight. FITS panel with PEI matrix skin and PEI also meet all flammability and smoke toxicity requirements for aircraft interiors.

In this process, a polymer film, for example, PEI, impregnated with a physical blowing agent (for some polymers, such as PP, the film can be made with chemical glowing agent) is placed between consolidated fiber-reinforced thermoplastic PEI facings (will use PP matric fiber-reinforced facings if PP film is used for making core), before the in-situ foaming takes place. In a press, a stack of the skins and the film is heated under high pressure until a homogenous temperature is obtained to thoroughly connect the PEI in facing to the core, as shown in Figure 5.14. Then the pressure is released, and the press is opened allowing the blowing agent to expand

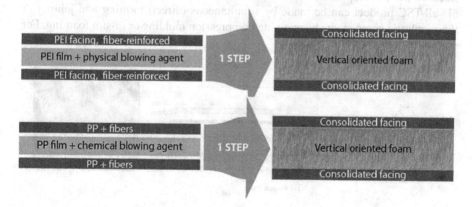

FIGURE 5.14 Foamed in-situ thermoformable sandwich process.[22]

the PEI film to create foam. The rate of press opening ensures the creation of vertical cells for high compression strength. At the end of the foaming curve, the sandwich panel is cooled to room temperature. After removal from the press, the blowing agent is removed by a special oven drying process.[22]

In the FITS process, the facings consist of consolidated fiber-reinforced thermoplastic layers. The fibers may be carbon, glass, or aramid in the formats of unidirectional or biaxial fabrics or as a combination thereof. The panels were manufactured in batch by a press of 1220 mm × 2440 mm in 7 minutes per panel, in a thickness of 3–25 mm, and at a density range of 80–250 kg/m^3 with a facing thickness of 0.1–0.6 mm.

5.4.4 Diaphragm Forming

The diaphragm forming (DF) process has gained considerable attention primarily in the forming of preconsolidated thermoplastics, due to its capability for producing structural parts with complex geometries at high-volume rates.[23] In the conventional form of DF, the preheating stage usually involves heating of the thermoplastic plate as well as the tool using conventional heating methods. This step consumes a substantial part of the overall process time, which makes the DF process cost-ineffective. Implementation of IR heating, which is also known as "cold" DF, includes two main steps. In the first step, the preconsolidated thermoplastic laminate is heated above melting point, while the mold is usually heated to the required temperature. In the second step, the laminate is fixed under vacuum between two thin plastically deformable diaphragm films, which are usually clamped around the edges; forming of the plate is performed over the heated tool, which achieves the desirable shape by introducing a suitable pressure gradient through the diaphragms. The formed part is removed from the tool after cooling under pressure to a suitable temperature at which material structural stability is achieved. A schematic representation of the process is shown in Figure 5.15.[24]

5.4.5 Manufacturing of 3D Thermoplastic Sandwiches by One-Step Forming

Because the thermoplastics can melt and be formed at the same time, a complex, 3D all-TSC product can be made by simultaneous (direct) forming and joining of thermoplastic components by applying compression molding or in-situ foaming. For

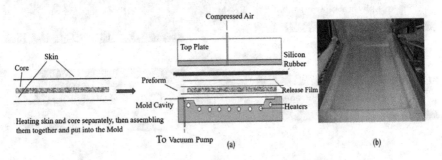

FIGURE 5.15 Diaphragm forming setup (a) and laminate after forming (b).[25]

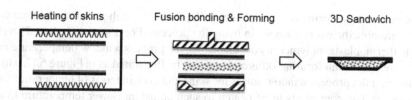

FIGURE 5.16 Direct forming and joining of skins and foam core by a curved mold.[1]

example, a PP-based composite sandwich laminate was manufactured by applying nonisothermal compression molding. The core and skins are stacked, heated in an IR-field, and then formed and joined in a single step. By carefully controlling temperature, with pressure as low as 10 bar, cycle times of about 1 minute are possible. One example is GF/PA12 and PA12 honeycomb core 3D sandwiches.

Vacuum assistance laminating may also be used for making 3D TSC products. The components are stacked and heated by contact and transferred into the press with a curved mold, and then by applying a vacuum in the negative mold to shape the sandwich structure. Figure 5.16 shows an example of direct forming and joining of skins and foam core by a curved mold.

The process window for fusion bonding of skins and core is narrow, since it is limited on the one hand by a weak interfacial bond at lower temperatures and on the other by core collapse and skin de-consolidation at higher temperatures, although this can be mitigated with short cycle times. Another point of complexity is the 3D shape for which the temperature and pressure distribution is more difficult to control. For manufacturing 3D sandwich structures, the mechanisms responsible for skin-to-core bonding are the same as for 2D sandwich structures. Lightly curved panels may even be manufactured in a similar way to flat panels with the use of a shaped mold. More complex geometries may require preforming the core to near net-shape prior to the application of the skins, which for example can be realized by machining, thermoforming, or expansion. The more complex the curvature, the greater the need for additional preforming of the skins. For highly complex parts, quality may be improved by bonding one skin and then the second in a two-step process.

The in-situ foaming methods mentioned early (film foaming, or injection melt, and bead foaming) can also be suitable for the production of 3D sandwich structures. Since the internal pressure during the foaming process can reach up to 10 bar for film and up to 20 bar for injection, the face sheets can be formed using this pressure and a 3D sandwich realized in one step. Nevertheless, the skins can be formed in a previous step as well. Thermoplastic bead foams can be easily introduced into a 3D mold with previously shaped skins.

5.5 POSTPROCESSING AND RECYCLING

Thermoforming of initially assembled flat sandwich structures is another method of realizing 3D sandwich parts. Although it is an additional step, it can be an effective way to obtain complex sandwich structures. The thermoplastic composite skins and the thermoplastic core are heated up to the softening temperature of the polymers, followed by a stamping step in a cold or tempered mold. Phenomena such as skin de-consolidation during heating or core collapse due to high pressure and temperatures are nonetheless challenges in using the process. For a sandwich, based on the same thermoplastic polymer in core and skins, the process window (temperature and pressure) for thermoforming skins and core can be illustrated as in Figure 5.17, which shows that the process window varies for skins and core in terms of temperature and pressure.[26] The core needs to be heated to such an extent (lower temperature limit) that it allows forming without destroying the cell structure (foam tearing) or causing internal stress after cooling. However, the core must not melt (upper temperature limit) in order to prevent the core from collapsing and to maintain a sufficient form- or compression-stability to enable precise shaping. A thermoforming example is shown in Figure 5.18.

Some thermoplastic cores offer suitable characteristics for thermoforming. Due to their amorphous structure, they feature a broad softening temperature range and only lose complete form stability at very high temperatures. One example is PEI-based sandwich by using glass fiber double warp knit-reinforced PEI as skin since they feature high drape ability, low forming energy, and quasi-isotropic behavior. PEI foam was used as the core. Skins and core were adhesively bonded by thermal fusion before being thermoformed in a secondary step.

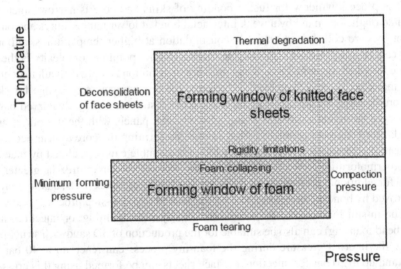

FIGURE 5.17 Schematic processing window for thermoforming a TPC sandwich based on single polymer in skins and core.[26]

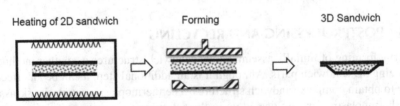

FIGURE 5.18 Example of thermoforming flat TSC structure into a shaped structure.[1]

All-Thermoplastic Sandwich Composites

The optimal process window was defined based on the mechanical behavior relative to the temperature of the PEI core and the PEI sandwiches separately. A processing temperature of 165°C–185°C seems to be optimal for the core, whereas the skins need to be heated up to above 280°C. A forming pressure range between 0.03 and 0.11 MPa is to be suitable for both parts. In order to fulfil the temperature process window, a strong thermal gradient must be followed between the skins and the core. A two-step heat conduction setup was conducted. In the first step, the temperature of the whole sandwich structure was elevated between two hot plates. In the following step, a fast heating of the skins was conducted before the sandwich was transported into the mold for shaping. The results showed that a PEI-based sandwich could be shaped successfully into a hemispherical-ellipsoidal shape in less than 7 minutes.[26] Stamp male mold with or without support cavity mold can be used for thermoforming TSC panel as shown in Figure 5.19.[22] Some PP honeycomb core TSC products made by the thermoforming process are shown in Figure 5.20.

The greatest driver for all thermoplastic sandwich composites is the ability to join components via fusion bonding/welding. This presents an attractive alternative to the conventional methods, such as mechanical fastening and adhesive bonding, used to join thermoset composite parts. The process of fusion-bonding involves heating and melting the polymer on the surfaces of the components to be joined and then pressing

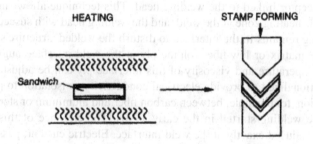

FIGURE 5.19 Stamp forming TSC panel with a mold cavity.[27]

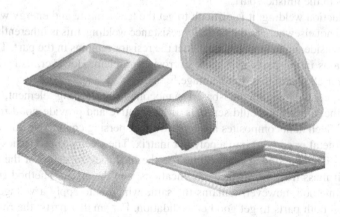

FIGURE 5.20 PP honeycomb core TSC products made by thermoforming process.[28]

these surfaces together for polymer solidification and consolidation. Resistance and induction welding are the two most established methods; others, including ultrasonic, laser, and conduction welding, are being advanced for use with composites.[29]

The induction welding technique involves moving an induction coil along the weld line. The coil induces eddy currents in the inherently conductive laminate with carbon fiber, which generate heat and melt the thermoplastic. The heat is generated in the weld line, where needed, but not the adjoining sections. Pressure should be maintained on the parts during welding. If welding fiber other than carbon, a susceptor or welding strip is required between two surfaces. In early iterations of induction welding, metal mesh susceptors were left in the weld, but this was seen as undesirable. More recent techniques have been able to eliminate susceptors because the carbon fiber in common aerospace laminates is electrically conductive. This has also enabled the use of carbon fiber materials as susceptors.[30]

A metallic wire mesh can be embedded in one part, be put in contact with another part, and induce current when two noncarbon fiber-reinforced TSC products are to be welded. Alternatively, a resistive patch of carbon fiber is used at the welding interface to generate the heat required.

A new development used a susceptor that is linked to and moves with the induction coil in the welding head, as well as an interface ply of unreinforced or low fiber-volume thermoplastic matrix, which can be tailored per application, which is a mobile susceptor linked to the welding head. This technique allows more perfect location of the heating zone at the weld, and the welding head with susceptor is moving so nothing remains in the interface to disturb the welded structure's properties. The pure TP matrix or low-fiber-volume ply at the weld interface augments resin flow. Melt temperature and viscosity of this interface ply can be adjusted and may also be functionalized to provide electrical conductivity or isolation to prevent galvanic corrosion, for example, between carbon fiber and aluminum or steel.

Resistance welding started in the early 1990s. The elegance of this method is that heat is produced exactly at the weld interface. Electric current, passed through a resistive element at the weld interface, creates heat and melts the thermoplastic polymer. However, this resistive element – a metal mesh or carbon fiber composite sheet stays in the finished part.

For induction welding, it is difficult to get the temperature and energy where you want it and not elsewhere in the part. For resistance welding, this is inherently solved, but the downside, up to now, has been that the resistor remains in the part. The metal mesh remains in the parts will cause corrosion problem. However, using a carbon fiber resistor alleviates this disadvantage.[31]

Also a conductive polymer-based nanocomposite heating element, compatible with the adherents, could serve as an alternative and provide good interfacial bonding.[32] Such nanocomposites are formed by dispersing conductive nanoparticles (typically metal or carbon) into a polymer matrix. The electrical properties of nanocomposites depend, among other things, on the intrinsic properties of the nanoparticles, their mass fraction, surface modifications, and the mixing method employed. The basic method, however, remains the same which is to apply a voltage and put pressure on both parts to get good consolidation. For smaller parts, the robotic end effector applies the pressure, but larger parts will need a jig to provide clamping.[32]

There is no limit to the part thickness being welded. It could be 3 or 30 mm, but care must be taken with thermal management at the welding line.

The third most common technique is ultrasonic welding. The process uses a sonotrode to generate high-frequency (20–40 kHz) vibrations that cause frictional heat and melting at the weld surfaces. The strain concentrates at the interface, in the energy directors, that melts because of the viscous dissipation. The triangles then progressively flow on the whole interface and perform welding of the two composite planes. Even if no fiber crosses the interface, the overlay is wide enough to transmit the stress which allows the process to be used for large-scale assemblies. It is also good for spot welds. It is very fast and highly automated. Ultrasonic welding has been used with plastics for several decades, typically with energy directors at the weld interface.[33] Now a research study has shown that 0.08 mm thick, unreinforced thermoplastic films may be used in their place.[34] A simple flat energy director shape, made of a loose film of neat resin, can be used to produce welds of high quality.[31] While ultrasonic welding is an efficient technique to join thermoplastic composite components, another potential application was shown to be the welding of thermoplastic and thermoset composites, enabled through the very short heating times in the ultrasonic welding process.

Current state of the art for continuous ultrasonic welding is robot-based. A robot manipulator with digital control is used to refine the head velocity and the energy that works best for each material and laminate thickness and to do very long welding.

Laser welding is also a technology to join all-TSC products. In this process, laser light is first passed through a part that is transparent or partially transparent in the near infrared spectral range (e.g., an unreinforced thermoplastic or glass fiber TPC skin). The light is then absorbed by carbon fiber or conductive additives in a second adjacent part, transforming the laser energy into heat, which creates the weld between the two materials.[35]

Many glass fiber-, synthetic fiber-, and natural fiber-reinforced thermoplastic skin sandwich products are laser transparent. There is great potential for using laser welding to achieve assembly of these products by adding a conductive layer on the surface of the adherents or welding to a carbon fiber-reinforced face product.

Recyclability has been a long-promised benefit of using all thermoplastic sandwich composites. Because the skin and the core can both be re-melted by heating, the scraps from production and product after service time can be reprocessed into different, new format products. An example by using the production scrap to make a new sandwich composite product is shown in Figure 5.21. The production scrap is chopped into small granulates, then mixed with a new matrix material and additive, and laminated with continuous fiber fabric as faces by heating and pressing into a new sandwich composite.

Basic processes for recycling thermoplastic sandwich composites are chopping or grinding down to chips or granulates, combining with raw materials, and applying heat and pressure to make new products.

Another example as shown in Figure 5.22 takes thermoplastic composite scrap from cutting and trimming (or from end-of-life parts), coarsely shreds it, and then uses a thermo-mechanical process to convert the shreds into organosheet, which becomes the facing for a composite structural panel. The term organosheet evolved

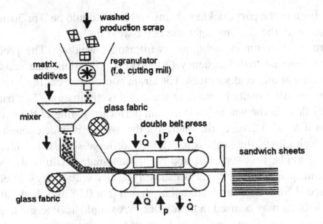

FIGURE 5.21 Process of using thermoplastic composite scape to make new sandwich panel.[36]

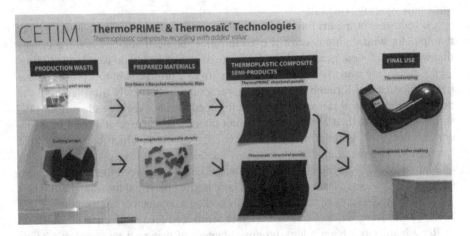

FIGURE 5.22 An example to make production waste into general-purpose panels.[37]

in Europe to describe fabric-reinforced thermoplastic prepreg, semi-preg, or preconsolidated blanks that could be thermoformed and over-molded into composite parts. The panels are quasi-isotropic, and fiber length is kept as long as possible in order to maximize mechanical properties. They are in the middle between short fiber sheet and classic organosheet.

REFERENCES

1. F. I. Von der Fakultät, "Thermoplastic composite sandwiches for structural helicopter applications," Ph.D. dissertation, Lehrstuhl für Polymere Werkstoffe Universität, Bayreuth, 2018.
2. "DuPont™ Styrofoam™ extruded polystyrene (XPS) foam," https://www.dupont.com/content/dam/dupont/amer/us/en.

3. "Dow STRANDFOAM™ EA 1000 Polypropylene Foam," http://www.lookpolymers.com/polymer_Dow-STRANDFOAM-EA-1000-Polypropylene-Foam.php.
4. Dow Chemicals, "Strandfoam Energy Absorbing Polypropylene Foams," Datasheet, 2001.
5. "Expanded Bead Foam," http://www.arplank.com/company.php.
6. "Light-Weight Reinforced Thermoplastic," https://www.azdel.com/products.html.
7. "SymaLITE®," https://www.mcam.com/na-en/products/composite-materials/symaliter/.
8. "Low density boards of natural fiber composites," https://www.flexformtech.com/Auto/Products/.
9. J.C. Velosa, J.P. Nunes, J.F. Silva, C.A. Bernardo and A.T. Marques, "Production of thermoplastic towpregs," *Materials Science Forum*, vols. 636–637, pp. 220–225, 2010.
10. "Fibrflex Product Matrix The fiber resin combinations," http://www.fibrtec.com/solutions.html.
11. "Melt Impregnation Process." https://www.azom.com/article.aspx?ArticleID=319.
12. "Thermoplastic Composite Tapes & Laminates," https://www.avient.com/products/advanced-composites/continuous-fiber-composite-tape-laminates-and-barstock/polystrand-continuous-fiber-tapes-laminates.
13. "T-UD Series CFRT UD Tapes," http://www.topolocfrt.com/t-ud-series-cfrt-ud-tapes/.
14. "Versitex Specifications," http://www.uslco.com/versitex.php.
15. N. Wiegand and E. Mäder, "Commingled yarn spinning for thermoplastic/glass fiber composites," *Fibers*, vol. 5, p. 26, July 2017.
16. "Fiberglass Products for Thermoplastics," https://www.jushi.com/en/product/product-introduction-146.html.
17. T. Gundberg, "Thermoply™ Commingled E-Glass/Polypropylene Woven & Stitch-Bonded Biaxial," https://www.compositesworld.com/cdn/cms/ThermoPly%20Paper%20(TG2).pdf.
18. L. McGarva, "Thermoplastic composite sandwich components: Experimental and numerical investigation of manufacturing issues," PhD Thesis, Royal Institute of Technology, Stockholm, Sweden, 2002.
19. "Continuous Fiber Reinforced Composites in Thermoplastic Sandwich Panels," http://www.topolocfrt.com/cfrt-in-sandwich-panels/.
20. A. Roch, T. Huber, F. Henning and P. Elsner, "LFT-Foams - Lightweight potential for structural components through the use of ling-glass-fiber-reinforced thermoplastic foams," *Proceedings of Polymer Processing Society 29th Annual Meeting (PPS29)*, Nuernberg, Germany, July 2013.
21. T. Neumeyer, T. Kroeger, J. Knoechel, P. Schreier, M. Muehlbacher and V. Altstaedt, "Thermoplastic sandwich structures processing, approaches towards automotive serial production," *21st International Conference on Composite Materials*, Xi'an, 20–25th August 2017.
22. "FITS PEI Thermoformable panels," http://www.fits-technology.com/FITS_PEI.html.
23. S.K. Mazumdar, *"Manufacturing Techniques, Composites Manufacturing,"* p. 222. Boca Raton, FL: CRC Press LLC, 2002.
24. G. Labeas, V. Watiti and V. Katsiropoulos, "Thermomechanical simulation of infrared heating diaphragm forming process for thermoplastic parts," *Journal of Thermoplastic Composite Materials*, vol. 21, pp. 353–370, July 2008.
25. H. Ning, G. Janowski, U. Vaidya and G. Husman, "Thermoplastic sandwich structure design and manufacturing for the body panel of mass transit vehicle," *Composite Structures*, vol. 80, pp. 82–91, 2007.
26. O. Rozant, P.-E. Bourban and J.-A.E. Manson, "Manufacturing of three-dimensional sandwich parts by direct thermoforming," *Composites Part A*, vol. 32, no. 11, pp. 1593–1601, Nov. 2001.

27. U. Breuer, M. Ostgathe, and M. Necitzel, "Manufacturing of all-thermoplastic sandwich systems by a one-step forming technique," *Polymer Composites*, vol. 19, no. 3, pp. 275–279, Apr. 2004.
28. T. Czarnecki, "Continuous production of thermoplastic honeycomb sandwich components for automotive interiors, low weight – Low cost technology," *SPE*, Novi, September 7–9, 2016.
29. G. Gardiner, "Multiple methods advance toward faster robotic welds using new technology for increased volumes and larger aero structures," *Composites World*, vol. 9, pp. 50–63, 2018.
30. T. Ahmed, et al., "Induction welding of thermoplastic composites—an overview," *Composites Part A: Applied Science and Manufacturing*, vol. 37, no.10, pp. 1638–1651, 2006.
31. G. Gardiner, "Multiple methods advance toward faster robotic welds using new technology for increased volumes and larger aero structures," *Composites World*, vol. 9, pp. 50–63, 2018.
32. J. R Tavares, M. Dubé, and D. Brassard, "Resistance welding of thermoplastic composites with a nanocomposite heating element," *Composites. Part B, Engineering*, vol. 165, pp. 779–784, Feb 2019.
33. A. Levy, S. Le Corre and A. Poitou, "Ultrasonic welding of thermoplastic composites: a numerical analysis at the mesoscopic scale relating processing parameters, flow of polymer and quality of adhesion," *International Journal of Material Forming*, vol. 7, no. 1, pp. 39–51, 2014.
34. G. Palardy, "Smart ultrasonic welding of thermoplastic composites," *Proceedings of the American Society for Composites, 31st Technical Conference*, ASC 2016.
35. I. Iwanowski, "Laser Transmission Welding Using Carbon Fiber Reinforced Polymers," https://www.inventionstore.de/en/offer/3728/.
36. "The GREEN alternative providing value and performance," http://www.uslco.com/ecotex.php.
37. G. Gardiner, "Gardiner G, Industrial production line for recycling thermoplastic polymers and composites into organosheet, Composites World, 3/12/2020," Composites World, Mar. 12, 2020.

6 Characterizations of Sandwich Structures

Wenguang Ma

Russell Elkin

CONTENTS

6.1 Face to Core Bonding Strength Tests 187
    6.1.1 Drum Peel Test 187
    6.1.2 Flatwise Tensile Test... 189
    6.1.3 Other Tests for Evaluating Strength of Skin and Core Bonding 191
6.2 Flexural Strength and Bending Stiffness Evaluations................................... 192
    6.2.1 Core Shear Properties of Sandwich Constructions by
          Beam Flexure Test 193
    6.2.2 Facing Properties of Sandwich Constructions by Long-Beam
          Flexural Test 198
    6.2.3 Test for Determining Sandwich Beam Flexural and Shear Stiffness ... 201
6.3 Flatwise and Edgewise Compressive Test 205
    6.3.1 Flatwise Compressive Test 205
    6.3.2 Edgewise Compressive Strength... 207
6.4 Concentrated Load and Wave Impact Tests... 211
    6.4.1 Concentrated Load Impact Tests 211
    6.4.2 Wave Impact Tests 218
    6.4.3 Air and Water Blast Tests by Shock Tubes.. 222
    6.4.4 Air and Water Blast Tests by Full-Scale Explosion........................... 225
6.5 Dynamic Fatigue Evaluation 227
    6.5.1 Flexural Fatigue Tests... 227
    6.5.2 Other Fatigue Tests 231
          6.5.2.1 Flatwise Compressive Fatigue Test..................................... 231
          6.5.2.2 Edgewise Compression – Compression Fatigue Test......... 233
          6.5.2.3 Two-Dimensional Simply Supported Distributed Load
                  Flexural Static and Fatigue Test.. 234
6.6 Fracture Toughness Test 236
6.7 Thermal Mechanical Tests 240
6.8 Nondestructive Tests.. 244
    6.8.1 Visual Inspection 246
    6.8.2 Tap Tests 246

DOI: 10.1201/9781003035374-6

6.8.3	Pitch-Catch Swept Method	246
6.8.4	Ultrasonic Tests	247
6.8.5	Shearography Test	248
6.8.6	Infrared Thermography Method	248
6.8.7	Industrial-Scale Inspection	249
References		249

Sandwich structures are defined here as those consisting of relatively thin face sheets that are strong and stiff in tension and compression compared to the low-density core material to which they are adhesively bonded. While the core material does resist shear loadings, its primary purpose is to keep the face sheets separated and thus maintain a high section modulus (a high "moment of inertia" or "second moment of the area" as defined by the stress analyst). The core material is typically of relatively low density, which results in high specific mechanical properties of the panel under favorable loadings (in particular, high flexural strength and stiffness properties relative to the overall panel density). That is, the sandwich structures are particularly efficient in carrying bending loads, although they have other important load application uses as well. For example, they provide increased buckling and crippling resistance to shear panels. Additionally, they provide buckling resistance to compression members.[1]

The testing of sandwich composites has existed for as long as these panels have been used as structural materials in high-performance applications, that is, for 70 years or more. The first thing we need to know is what properties or characteristics of the sandwich panel must to be evaluated. This is dictated by how the panel will be used and how it might fail. The panel can fail in any one of several different ways, depending upon the geometric and fabrication characteristics of the panel and how it is loaded. For example, a face sheet can fail in tension, compression, shear, or local buckling. Additionally, the core can fail in shear or by crushing. A face sheet can separate from the core due to excessive shear or normal tensile stress in the adhesive bond. Test methods have been developed to isolate and simulate each of these specific failure modes.

Because the use of sandwich construction continues to grow rapidly, particularly in structural applications, it is important that manufacturers and researchers become familiar with the various sandwich panel test methods. Although our general level of understanding continues to grow today, those involved in a testing program must have some assurance that the sandwich panels from which test coupons are extracted are representative of panels that will be used in the end application. Further, they must be sure that the properties obtained are reasonable and consistent with the current state of the art.

In this chapter, the tests for evaluating the adhesion strength between skins and core will be introduced (the sandwich panel will not deliver any function if the three layers are not connected together firmly). Then the second group of tests is the flexural tests or called the bending tests. They are a long or short sandwich beam to either three- or four-point loading tests. Sometimes, the multiple tests must be conducted to obtain a full characterization. Sandwich panel flexural stiffness, the tensile and compressive strengths of the face sheets, core shear strength and shear modulus, and even the face sheet-to-core bond can be evaluated by the flexural tests.

Characterizations of Sandwich Structures

Compressive properties are important also to the sandwich composites, which include flatwise and edgewise compressive tests. The edgewise test is used to determine the in-plane compressive strength of a sandwich panel. A rectangular panel is subjected to an in-plane compressive loading along two opposite sides, with no load on the other two sides. The failure models of the test are important because they show how a sandwich structure will fail when it bears a vertical or buckling load. The flatwise compression properties are critical due to they represent how the structure is stiff and strong when a force works on the panel in normal to panel surface.

Impact tests are performed to evaluate the property of the product against a dynamic load, which include the impact load in different velocity, with different indenter types, and in wave format. The specimen will be checked for damage degree and residual strength after testing. Other mechanical tests also will be presented in this chapter, including the tests for evaluating the properties of flatwise tensile, fracture toughness, dynamic fatigue, screw retention force, etc. Except for mechanical tests, physical and thermal tests for evaluating the properties of the sandwich composites, including material density and panel areal weight, water migration, water absorption, sound transmission, and thermal conductivity and thermal resistance, will be presented as well. Nondestructive test also will be unveiled because it is really useful for defect inspection and quality controlling.

6.1 FACE TO CORE BONDING STRENGTH TESTS

6.1.1 Drum Peel Test

The test is to determine the peeling force of an adhesive bonded, or a composite skin self-bonded, sandwich assembly by peeling a face skin around a circular drum. This is done by clamping the assembly in a universal test machine. The average peeling load is determined using a standard formula. This test method covers the determination of the peel resistance of adhesive bonds between a relatively flexible facing of a sandwich structure and its core, when tested under specified conditions.[2] The test setup is shown in Figure 6.1.

The thickness of the overall laminate must be high enough not to bend during the test. There is also a limit to the thickness of an individual face sheet. Too thick, and the skin will crack before peeling. Five specimens at least are needed for each evaluating test. The size of specimens is 305 mm × 76 mm, 25 mm overhang of one facing at each end.[3]

Test starts to clamp the specimens in the climbing drum peel apparatus. The grips of a Universal Test Machine are initiated at a separation significant enough to roll the drum upward.[4] The standard speed is 25.40 ± 2.54 mm/min. Apply an initial load, equal to that obtained in the calibration of the apparatus for load to overcome the resisting torque of the drum assembly, by loading the apparatus in tension. Determine the peel resistance over at least 152 mm of the bond by loading the apparatus in tension at a crosshead speed of 25.40 mm/min. The peel load vs. peeling distance is recorded in a computer or an autographic machine.

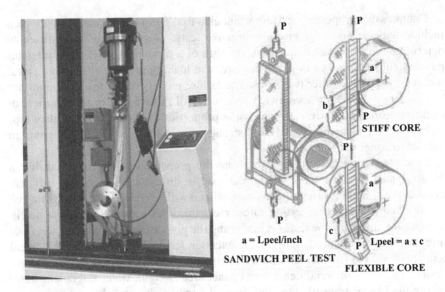

FIGURE 6.1 Sandwich structure skin peeling test setup.

To determine the average peeling load, the autographic curve is used for find the peel load for the 127 mm of peeling range between 25 and 152 mm of head travel. A curve of peeling load vs. propagation extension of drum peel test is shown Figure 6.2. From climbing drum peel results, the following data can be obtained: average peeling

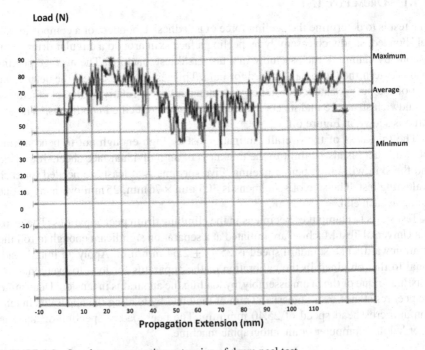

FIGURE 6.2 Load vs. propagation extension of drum peel test.

Characterizations of Sandwich Structures 189

load, maximum peeling load, and minimum peeling load. Then average peel torque T can be calculated by using the following equation. And failure type/mode can be found by inspecting each test specimen.

$$T = \left[(r_o - r_i)(F_p - F_o)\right] / W$$

where:
 T = average peel torque, mm·kg/mm of width,
 ro = radius of flange, including one half the thickness of the loading straps, mm,
 ri = radius of drum plus one half the thickness of the adherend being peeled, mm,
 Fp = average load required to bend and peel adherend plus the load required to overcome the resisting torque, kg,
 Fo = load required to overcome the resisting torque, in kg (reference the standard)
 W = width of specimen, mm.

The drum peel test method may be used to determine comparative rather than fundamental measurements of adhesion and is particularly suitable for process and quality control. As the skin thickness and/or the stiffness increases, the torque due to the skin and drum alone increases as a proportion of the total torque measured. The difference between these two values should not fall too low (say, less than 50% of the total torque value) if the test is to maintain sensitivity, particularly if the output trace oscillates. In addition, for a skin stiffness or thickness above a critical value, this test cannot be used for evaluating the bonding strength between skin and core. Also, the climbing drum peel test is not suited to high rate "peel" which might be experienced by marine sandwich constructions facing wave slam loading.

There are other test methods for determining the skin-to-core interfacial bond strength for sandwich structures consisting of relatively thick, stiff skins typical of those used in marine and other heavy-duty applications.

6.1.2 Flatwise Tensile Test

ASTM C297 "Standard Test Method for Flatwise Tensile Strength of Sandwich Constructions" was stated in Chapter 2 for evaluating the flatwise tensile strength of the core materials, but it is also considered that this method can be used to determine the bond strength between the skin and core. That is, the test will fail at the weakest link of core, skin, or core-skin interface. It should be noted that this weakest link also applies to the other interface tests. Depending on the system, the interface may be a wider zone than the original physical interface due to, for example, resin impregnation of the core.

The square specimens of 55 mm × 55 mm dimensions will be tested for the flatwise tension. The specimens should be bonded to the loading blocks using a room temperature two-part epoxy adhesive or other structural adhesives as shown in Figure 6.3. A crosshead speed of 2 mm/min is used for testing.

The test results will be recorded by computer software and will be shown as a load vs. deflection chart. Also, failure modes will be recorded and evaluated for a pass or failure decision. However, a stronger core will fail along the skin and core interface.[6]

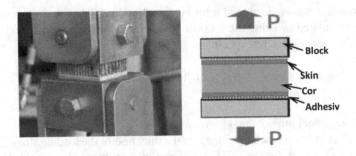

FIGURE 6.3 Flatwise tensile test setup.[5]

The failure could happen in different sections as shown in Figure 6.4. The core failure (Figure 6.4a) will be an ideal situation because the panel is strong as designed, in which the skin is strong and bonded to the core strongly. The failure at the core-facing bonding interface is not acceptable in most cases (Figure 6.4b) but depends on the failure load. Note that this type of "clean" failure does not necessarily mean the bond was poor. The strength at failure must be considered. If the maximum load passes the required value because the core is really strong as plastics and metal honeycombs, and end grain balsa wood core, the result is acceptable. The failure also could happen in the core-facing adhesive layer (Figure 6.4c), so a stronger adhesive should be used for achieving higher performance if this situation happens. The face section can split and fail (Figure 6.4d) because the thick face is made by multilayers of fiber fabrics, and bonding strength is not strong enough to hold together when a high pulling force acts on it. This case could lead to an early failure if a bending force works on the panel because the face layer separates before other section failure.

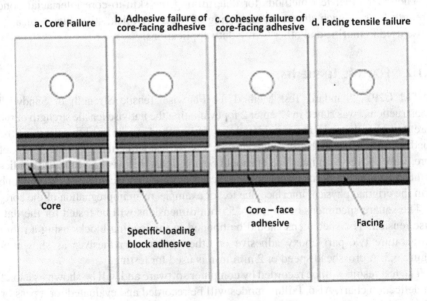

FIGURE 6.4 Different failure modes of flatwise tensile test.[7]

Characterizations of Sandwich Structures

Based on the analyses above, the flatwise tensile test is not just for evaluating the bonding strength between skin and core; we also can find some design and manufacturing problems from testing results.

6.1.3 Other Tests for Evaluating Strength of Skin and Core Bonding

A variety of test configurations for determining the skin-to-core bond strength of sandwich panels have been proposed by various literature studies.[8] Some of the most popular geometries are shown in Figure 6.5. It should be noted that many of the test configurations for sandwiches have evolved from adhesive test methods. In these test methods, the specimens for Test a, b, and c undergo a crack at the loading end. Then, the crack will extend when a testing load acts on the sample. A typical load vs. deflection curve of alternative tests is displayed in Figure 6.6. High bonding strength between skin and core resists crack propagation and gives a saw teeth curve.

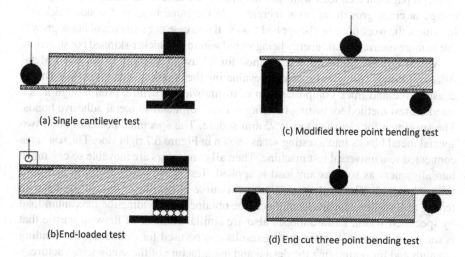

FIGURE 6.5 Alternative sandwich skin-to-core bond strength tests.[9]

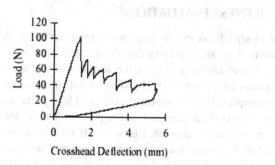

FIGURE 6.6 Typical load vs. deflection curve of alternative skin-to-core strength tests.[9]

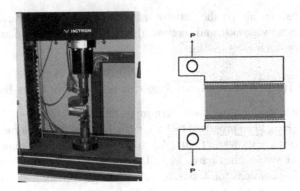

FIGURE 6.7 Sample bonded to block and test setup for cleavage test.

For sandwich structures with thick, heavy face sheets, the load required to begin crack propagation increases with skin thickness, and hence, the stored internal elastic energy at crack growth initiation increases. At the same time, as the skin thickness increases, the work done by the applied load will decrease per unit area of crack growth due to more internal elastic energy being stored within the thicker-skinned construction.

ASTM D1062 "Standard Test Method for Cleavage Strength of Metal-to-Metal Adhesive Bonds" can also be used for evaluating the bonding strength between high-density core and thick composite skin of a sandwich structure, even though it is a standard test method for testing bonding strength of metal-to-metal adhesive bonds. The specimen for cleavage test is 75 mm square. The specimen needs to bond two special metal blocks into a testing set as shown in Figure 6.7 right side. The test set is connected to a universal test machine. Then all connectors are movable so can move into alignment as soon as any load is applied. Test speed is 0.5 mm/min and until failure happens. The load vs. displacement curve is similar to that of the flatwise tensile test, so the cleavage strength can be obtained by dividing the maximum load by specimen width. Failure modes also are similar to those of flatwise tensile that is shown in Figure 6.4 so the testing results can be used for evaluating the bonding strength and for optimizing the design and manufacture of the sandwich structure.

6.2 FLEXURAL STRENGTH AND BENDING STIFFNESS EVALUATIONS

Because sandwich composite structures are subjected to flexural loading in typical applications, it is not surprising that flexure testing of composites is commonly performed. Flexure testing is among the simplest of the test types to perform, yet the state of stress and failure modes produced within a flexure test specimen is among the most complex. Under both three-point and four-point flexural loading, the tensile, compressive, and shear stresses that are produced vary along the length and through the thickness of the specimen. Flexure tests on flat sandwich construction may be conducted to determine the sandwich flexural stiffness, the core shear strength and shear modulus, or the facings' compressive and tensile strengths. Tests to evaluate core shear strength may also be used to evaluate core-to-facing bonds.

Characterizations of Sandwich Structures

Both long-beam flexure and short-beam flexure are used when addressing sandwich panel testing. The former is used to determine face sheet properties and the latter to determine core shear properties. Such a distinction is logical since we know that for a given applied loading, the flexural stresses (tensile and compressive) in the face sheets increase as beam length increases but the shear stresses in the core do not. That is, long beams produce high bending stresses, while short span lengths do not.[1] ASTM defines standard specimen and loading configurations but also permits nonstandard configurations. In fact, a nonstandard configuration could be required for flexure testing when the particular sandwich panel – short beam or long beam – does not meet certain criteria. These criteria include the required span length, core shear strength, and core compression strength. Each criterion is defined as a function of the face sheet's expected ultimate strength, face sheet thickness, core thickness, core shear allowable strength, core compression allowable strength, and even the width of the loading pads to be used. Appropriate formulas for evaluating these criteria are given in the standards.

6.2.1 Core Shear Properties of Sandwich Constructions by Beam Flexure Test

Determination of the core shear properties of flat sandwich constructions is subjected to flexure in such a manner that the applied moments produce curvature of the sandwich facing planes. All types of core material are permissible, both continuous bonding surfaces (such as balsawood and foams) as well as those with discontinuous bonding surfaces (such as honeycomb).

The test consists of subjecting a beam of sandwich construction to a bending moment normal to the plane of the sandwich. Force versus deflection measurements is recorded. The only acceptable failure modes are core shear or core-to-facing bond. Failure of the sandwich facing preceding failure of the core or core-to-facing bond is not an acceptable failure mode. Test method ASTM D 7249 is used to determine facing strength. The test method is limited to obtaining the core shear strength or core-to-facing shear strength and the stiffness of the sandwich beam and to obtaining load-deflection data for use in calculating sandwich beam flexural and shear stiffness using practice ASTM D 7250.

The beam flexural test can be used to produce core shear strength and core-to-facing shear strength data for structural design allowable, material specifications, and research and development applications; it may also be used as a quality control test for bonded sandwich panels.

Material selection and specimen preparation are important for performing the test. Poor material fabrication practices and damage induced by improper specimen machining are known causes of high data scatter in composites and sandwich structures in general. A specific material factor that affects sandwich cores is variability in core density. Important aspects of sandwich core specimen preparation that contribute to data scatter include the existence of joints, voids or other core discontinuities, out-of-plane curvature, and surface roughness. Specific geometric factors that affect core shear strength include core orthotropy (that is, ribbon versus transverse direction for honeycomb core materials) and core cell geometry.

The loading fixture shall consist of either a three- or four-point loading configuration with two support bars that span the specimen width located below the specimen and one or two loading bars that span the specimen width located on the top of the specimen (Figure 6.8). The force shall be applied vertically through the loading bar(s), with the support bars fixed in place in the test machine.[10]

Standard configuration test specimen shall be rectangular in cross section, with a width of 75 mm and a length of 200 mm. Proper design of the sandwich flexure test specimen for determining shear strength of the core or core-to-facing bond is required to avoid facing failures. The facings must be sufficiently thick and/or the support span sufficiently short such that transverse shear forces are produced at applied forces low enough so that the allowable facing stress will not be exceeded. However, if the facings are too thick, the transverse shear force will be carried to a considerable extent by the facings, thus leading to a high apparent core shear strength as computed by the equations given in the standard. The following equations can be used to size the test specimen (these equations assume that both facings have the same thickness and modulus and that the facing thickness is small relative to the core thickness [$t/c \leq \sim 0.10$]).

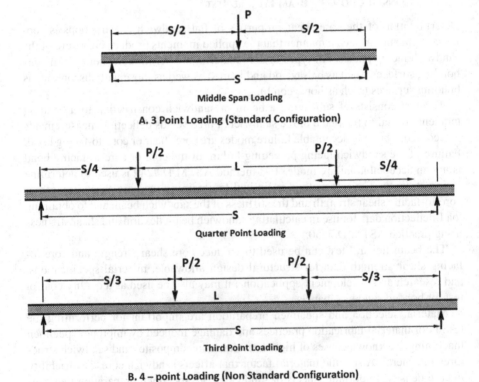

FIGURE 6.8 Sandwich flexural test setup configurations.

(three-point is standard configuration, use 150 mm support; four-point is nonstandard configuration, load span (L) is S/2 for quarter-span loading, S/3 for third-span loading.)

The support span length shall satisfy

$$s \leq \frac{2k\sigma t}{F_s} + L \tag{6.1}$$

or the core shear strength shall satisfy

$$F_s \leq \frac{2k\sigma t}{(S-L)} \tag{6.2}$$

The core compression strength shall satisfy

$$F_c \leq \frac{2(c+t)\sigma t}{(S-L)l_{\text{pad}}} \tag{6.3}$$

where:
   S = support span length, mm,
   L = loading span length, mm ($L = 0$ for three-point loading),
   σ = expected facing ultimate strength, MPa,
   t = facing thickness, mm,
   c = core thickness, mm,
   Fs = estimated core shear strength,
   k = facing strength factor to ensure core failure (recommend $k = 1.3$),
   $l$pad = dimension of loading pad in specimen lengthwise direction, mm, and
   Fc = core compression allowable strength, MPa.

For the standard test configuration, facings consisting of a laminated composite material shall be balanced and symmetric about the sandwich beam mid-plane. For the standard specimen, the facings shall be the same material, thickness, and layup. The calculations assume constant and equal upper and lower facing stiffness properties.

To measure facing thickness accurately is difficult after bonding or co-curing of the facings and core. For metallic or precured composite facings which are secondarily bonded to the core, the facing thickness should be measured prior to bonding. For co-cured composite facings, the thicknesses are generally calculated using nominal per ply thickness values, or cut from sandwich panel, and then the skin after cleaning the core material is measured.

The deflection of the specimen shall be measured in the center of the support span by a properly calibrated LVDT device having an accuracy of 61% or better. The use of crosshead or actuator displacement for the beam mid-span deflection produces inaccurate results, particularly for four-point loading configurations; the direct measurement of the deflection of the mid-span of the beam must be made by a suitable instrument. A four-point third-span loading setup with LVDT is shown in Figure 6.9. Setup and testing detail procedures can be found from the standard publication.

The test results will be recorded and displayed as force versus deflection detected by an LVDT device continuously. Also, how the specimen fails is an important phenomenon that should be recorded. When the initial failures are noted, record

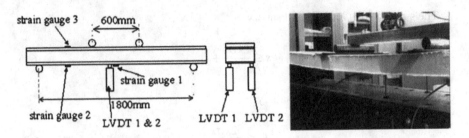

FIGURE 6.9 Four-point third-span loading setup with LVDTS and strain gauges.[11]

the force, displacement, and mode of damage at such points. The failure modes are listed in Table 6.1, including core crushing, skin to core delamination, facing failure, multimode combined, core transverse shear, and sample explosion. Failure area and location of each specimen should be recorded. The maximum force, the failure force, the head displacement, and the LVDT deflection at, or as near as possible to, the moment of ultimate failure also are important.

The different failure modes of the specimens after testing are displayed in Figure 6.10. For core shear strength calculation, only the shear failures of the sandwich core or failures of the core-to-facing bond are acceptable failure modes. Failure of one or both of the facings preceding failure of the core or core-to-facing bond is not an acceptable failure mode.

Plot and examine the force-deflection data to determine if there is any significant compliance change (change in slope of the force-displacement curve, sometimes referred to as a transition region) prior to ultimate failure (significant is defined as a 10% or more change in slope). Determine the slope of the force displacement curve above and below the transition point using chord values over linear regions of the curve. Intersect the linear slopes to find the transition point.[12]

The core shear ultimate stress can be calculated by using three-point mid-span loading test results and Equation 6.4:

$$F_s^{ult} = \frac{P_{max}}{(d+c)b} \qquad (6.4)$$

TABLE 6.1
Failure Modes of Flexure Tests of Sandwich Composite Beams[10]

Failure Type	Failure Area	Failure Location
Core crushing	At load bar	Core
Skin to core delamination	Gage	Core facing bond
Facing failure	Multiple areas	Bottom facing
Multimode	Outside gage	Top facing
Core transverse shear	Various	Both facings
Sample explosive	Unknown	Various
Others		Unknown

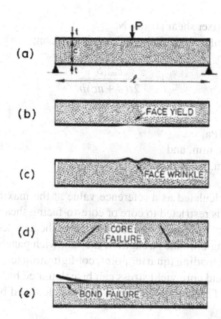

FIGURE 6.10 Different failure modes of sandwich beam after flexural test.

where

F_s^{ult} = core shear ultimate strength, MPa,
P_{max} = maximum force prior to failure, N,
t = nominal facing thickness, mm,
d = sandwich thickness, mm,
c = core thickness, mm ($c = d - 2t$) see Figure 6.11, and
b = sandwich width, mm.

The shear yield stress of core material that yields more than 2% strain can be obtained using the following equation[13]:

$$F_s^{yield} = \frac{P_{yield}}{(d+c)b} \tag{6.5}$$

where

F_s^{yield} = core shear yield strength, MPa, and

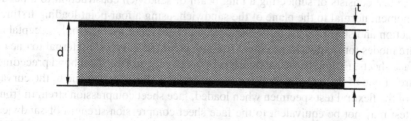

FIGURE 6.11 Sandwich panel thickness dimensions.

P_{yield} = force at 2% offset shear strain, N.
Facing stress is calculated using the following equation:

$$\sigma = \frac{P_{max}S}{2t(d+uc)b} \tag{6.6}$$

where
  σ = facing stress, MPa,
  t = facing thickness, mm, and
  S = span length, mm.

The facing stress is calculated as a reference value at the maximum applied force. Since this test method is restricted to core or core-to-facing shear failures, the facing stress does not represent the facing ultimate strength. The next section will introduce how to obtain the facing ultimate strength of the sandwich panels.

By using four-point loading (quarter point) configuration testing results, the core shear ultimate stress and core yield stress can be calculated by using Equations 6.4 and 6.5, respectively. The facing bending stress is calculated by using the following equation.

$$\sigma = \frac{P_{max}S}{4t(d+c)b} \tag{6.7}$$

The core shear ultimate stress and core yield stress can be calculated in case of four-point loading (third point) configuration and by using Equations 6.4 and 6.5, respectively, also. The facing bending stress is calculated by using the following equation.

$$\sigma = \frac{P_{max}S}{3t(d+c)b} \tag{6.8}$$

6.2.2 Facing Properties of Sandwich Constructions by Long-Beam Flexural Test

This test covers determination of facing properties of flat sandwich constructions subjected to flexure in such a manner that the applied moments produce curvature of the sandwich facing planes and result in compressive and tensile forces in the facings. Mostly, any type of sandwich laminate may be evaluated. Facing strength is best determined in accordance with this test.[13]

The test consists of subjecting a long beam of sandwich construction to a bending moment normal to the plane of the sandwich, using a four-point loading fixture. Deflection and strain versus force measurements are recorded. The only acceptable failure modes for sandwich face sheet strength are those which are internal to one of the face sheets. Failure of the sandwich core or the core-to face sheet bond preceding failure of one of the face sheets is not an acceptable failure mode. Due to the curvature of the flexural test specimen when loaded, face sheet compression strength from this test may not be equivalent to the face sheet compression strength of sandwich structures subjected to pure edgewise (in-plane) compression.

If the core material has insufficient shear or compressive strength, it is possible that the core may locally crush at or near the loading points, thereby resulting in face sheet failure due to local stresses. In other cases, facing failure can cause local core crushing. Pads may be added under the load points to prevent this. When there is both facing and core failure in the vicinity of one of the loading points, it can be difficult to determine the failure sequence in a postmortem inspection of the specimen as the failed specimens look very similar for both sequences.

Standard configuration as shown in Figure 6.12 needs the standard test specimen that shall be rectangular in cross section, with a width of 75 mm and a length of 600 mm. Nonstandard configurations need the specimen geometries with the width of not less than twice the total thickness, nor more than six times of the total thickness, and nor greater than one quarter the span length. The specimen length shall be equal to the support span length plus 50 mm, or plus one half the sandwich thickness, whichever is the greater.[13]

Proper design of the sandwich flexure test specimen for determining compressive or tensile strength of the facings is required to avoid core crushing, core shear, or core-to-facing failures. The facings must be sufficiently thin and the support span sufficiently long such that moments are produced at applied forces low enough so that the allowable core shear stress will not be exceeded. The core must be sufficiently thick to avoid excessive deflection. The following equations can be used to size the test specimen (these equations assume that both facings have the same thickness and modulus, and that the facing thickness is small relative to the core thickness ($t/c \leq {\sim}0.10$).

The support span length shall satisfy

$$s \geq \frac{2k\sigma t}{kF_s} + L \tag{6.9}$$

or the core shear strength shall satisfy

$$F_s \geq \frac{2k\sigma t}{k(S-L)} \tag{6.10}$$

The core compression strength shall satisfy

$$F_c \geq \frac{2(c+t)\sigma t}{(S-L)l_{pad}} \tag{6.11}$$

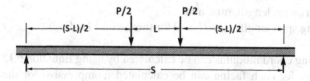

FIGURE 6.12 Loading confugurations of long-beam flexural test. (Standard: 4-poit with 560 mm support span (S), 100 mm load span (L); nonstandard 3 point mid-span, 560 mm support span; nonstantard 4-poit quarter span, 560 mm support span, S/2 load span; nonstandard four-point third span, 560 mm support span, S/3 load span.)

where
   S = support span length, mm,
   L = loading span length, mm ($L = 0$ for three-point loading),
   σ = expected facing ultimate strength, MPa,
   t = facing thickness, mm,
   c = core thickness, mm,
   F_s = core shear allowable strength, MPa,
   k = core shear strength factor to ensure facing failure (recommend $k = 0.75$),
   L_{pad} = dimension of loading pad in specimen lengthwise direction, mm and
   F_c = core compression allowable strength, Mpa.

The test data should be recorded as force versus head displacement, force versus strain, and force versus LVDT deflection continuously. If any initial failures are noted, record the force, displacement, and mode of damage at such points. Potential initial (noncatastrophic) failures that should be reported include face sheet delamination, core-to-face sheet disbond, partial core fracture, and local core crushing. The mode, area, and location of each initial failure should be recorded.

Facing ultimate stress is calculated by using Equation 6.12 that is valid for specimens with equal ($t_1 = t_2$) or unequal facing thicknesses with the condition that the facing thicknesses are small relative to the core thickness ($t/c \leq \sim 0.10$).

$$F_1^{ult} = \frac{P_{max}(S-L)}{2(d+c)bt_1} \tag{6.12}$$

$$F_2^{ult} = \frac{P_{max}(S-L)}{2(d+c)bt_2}$$

where
   F_1^{ult} = facing 1 ultimate stress, Mpa,
   F_2^{ult} = facing 2 ultimate stress, MPa,
   Pmax = maximum force prior to failure,
   $t1$ = nominal facing 1 thickness, mm,
   $t2$ = nominal facing 2 thickness, mm,
   d = measured sandwich total thickness, mm,
   c = calculated core thickness, mm (for specimens with equal facings, $c = d - 2t$;
       for specimens with unequal facings, $c = d - t1 - t2$),
   b = specimen width, mm,
   S = support span length, mm, and
   L = loading span length, mm ($L = 0$ for three-point loading)

Effective facing chord modulus can be calculated by using Equation 6.13. A separate modulus value for each facing can be calculated (compressive for the top facing, tensile for the bottom facing) if two separate LVDTs are used for recording stress vs. strain curves of top and bottom facing, respectively.

$$E_f = (\sigma_{3000} - \sigma_{1000})/(\varepsilon_{3000} - \varepsilon_{1000}) \tag{6.13}$$

Characterizations of Sandwich Structures

where
- E_f = effective facing chord modulus, Pa,
- σ_{3000} = facing stress calculated for applied force corresponding to ε_{3000}, Pa,
- σ_{1000} = facing stress calculated using for applied force corresponding to ε_{1000}, Pa,
- ε_{3000} = recorded facing strain value (magnitude) closest to 3000 micro-strain,
- ε_{1000} = recorded facing strain value (magnitude) closest to 1000 micro-strain.

Equation 6.14 is for calculating sandwich flexural stiffness. However, this equation is strictly valid only for cases where the shear flexibility of the sandwich beam is negligible. For procedures and equations for calculating the sandwich flexural and through-thickness shear stiffness in cases where the shear flexibility cannot be neglected, see the next section.

$$D^{F,\text{nom}} = \frac{d(S-L)}{4}\left(\frac{P^{3000}-P^{1000}}{\left(\varepsilon_{t\_3000}-\varepsilon_{t1000}\right)+\left(\varepsilon_{b\_3000}-\varepsilon_{b1000}\right)}\right) \qquad (6.14)$$

where
- $D^{F,\text{nom}}$ = effective sandwich flexural stiffness,
- ε_{t_3000} = facing of top surface recorded strain value (magnitude) closest to 3000 micro-strain,
- ε_{t_1000} = facing of top surface recorded strain value (magnitude) closest to 1000 micro-strain,
- P_{3000} = applied force corresponding to facing 1 strain ε_{t_3000}, N,
- P_{1000} = applied force corresponding to facing 1 strain ε_{t_1000}, N,
- ε_{b_3000} = facing of bottom surface recorded strain value (magnitude) corresponding to P_{3000}, and
- ε_{b_1000} = facing of bottom surface recorded strain value (magnitude) corresponding to P_{1000}.

6.2.3 Test for Determining Sandwich Beam Flexural and Shear Stiffness

This test covers determination of the flexural and transvers shear stiffness properties of flat sandwich constructions subjected to flexure in such a manner that the applied moments produce curvature of the sandwich facing planes. The calculation methods in this practice are limited to sandwich beams exhibiting linear force deflection response. The calculations in this section use test results obtained from the last two sections. This test covers the determination of sandwich flexural and shear stiffness and core shear modulus using calculations involving measured deflections of sandwich flexure specimens.[14]

The tests in this section can be conducted on short specimens and on long specimens (or on one material loaded in two ways), and the flexural stiffness, shear rigidity, and core shear modulus can be determined by simultaneous solution of the complete deflection equations for each span or each loading. Anyway, core shear strength and shear modulus are best determined in accordance with the test method introduced in Chapter 1 by testing pure core materials.

For cases where the facing modulus is not known, a minimum of two loading configurations must be selected. Refer to tests in the last two sections for the equations used to size the specimen lengths and loading configurations so that facing failure and core shear failures do not occur below the desired maximum applied force level. It is recommended that one loading configuration uses a short support span and specimen, and the other loading configuration uses a long support span and specimen. The purpose of this recommendation is to obtain force–deflection data for one test with relatively high shear deflection and one test with relatively high flexural deformation. If two short configurations or two long configurations are tested, measurement errors may be large relative to the difference in shear and flexural deflections between the two tests and may lead to significant errors in the calculated flexural and shear stiffness values.

If the facing modulus is known, a test of a single short support span loading configuration can be conducted using sandwich beam specimens specified in Section 6.2.1. Record force deflection curves for each test specimen using a transducer, deflectometer, or dial gage to measure the middle-span deflection. The use of crosshead or actuator displacement for the beam middle-span deflection would produce inaccurate results. All loading configurations used in this section have been presented in the last two sections that include three-point loading, four-point quarter-span loading, four-point third-span loading, and four-point standard-span loading configurations.

Calculation of flexural stiffness, transverse shear rigidity, and core shear modulus in the situation of the mid-span deflection of a beam with identical top and bottom facings in flexure is given by Equations 6.15 and 6.16. Given deflections and applied forces from results of testing the same sandwich beam with two different loading configurations, the flexural stiffness, D, and the transverse shear rigidity, U, may be determined from simultaneous solution of the deflection (Equation 6.15) for the two loading cases. The core shear modulus, G, may then be calculated using Equation 6.16.[14]

$$\Delta = \frac{P(2S^3 - 3SL^2 + L^3)}{96D} + \frac{P(S-L)}{4U} \qquad (6.15)$$

$$G = \frac{U(d-2t)}{(d-t)^2 b} \qquad (6.16)$$

where
Δ = beam middle-span deflection, mm,
P = total applied force, N,
d = sandwich thickness, mm,
b = sandwich specimen width, mm,
t = facing thickness, mm,
S = support span length, mm,
L = load span length, mm ($L = 0.0$ for three-point middle-span loading configuration),

Characterizations of Sandwich Structures

G = core shear modulus, Mpa,
D = flexural support span length, mm,
U = transverse shear rigidity, N.

Using the experimental values of the initial stiffness K ($K = P/\Delta$) of two span lengths of each group of specimens obtained from linear portion of each specimen's load deflection curve shown as Figure 6.13, Equation 6.15 is rearranged as Equation 6.17.[12]

$$1 = K_i \frac{(2S_i^3 - 3S_i L_i^2 + L_i^3)}{96D} + K_i \frac{(S_i - L_i)}{4U} \tag{6.17}$$

where K_i is initial stiffness in N/mm, S_i is the span length in mm, and L_i is the loading span in mm. Note that all occurrences of subscript i denotes that it is specific to each span length. For example, S_1 represents the span length of 150 mm, while S_2 represents the span length of 300 mm. Combining the equations for 2 span lengths, D and U are simplified to Equations 6.18 and 6.19. Then the core shear modulus, G, will be calculated using Equation 6.16.

$$D = \frac{m_2 n_1 - m_1 n_2}{96 \left(\frac{n_1}{K_2} - \frac{n_2}{K_1} \right)} \tag{6.18}$$

$$U = \frac{m_1 n_2 - m_2 n_1}{4 \left(\frac{m_1}{K_2} - \frac{m_2}{K_1} \right)} \tag{6.19}$$

where

$$m_i = 2S_i^3 - 3S_i L_i^2 + L_i^3 \tag{6.20}$$

$$n_i = S_i - L_i \tag{6.21}$$

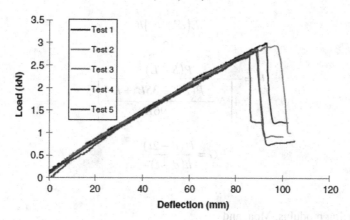

FIGURE 6.13 Load (P) vs. deflection (Δ) curves of four-point bending tests.[12]

If using two three-point middle-span loading configurations, the solution for this case ($L_1 = L_2 = 0$) is given by Equations 6.22 and 6.23. The flexural stiffness and shear rigidity for each selected value of P can be calculated by two equations. Then the core shear modulus will be obtained by using Equation 6.16.

$$D = \frac{P_1 S_1^3 \left(1 - S_2^2 / S_1^2\right)}{48 \Delta_1 \left(1 - P_1 S_1 \Delta_2 / P_2 S_2 \Delta_1\right)} \tag{6.22}$$

$$U = \frac{P_1 S_1 \left(S_1^2 / S_2^2 - 1\right)}{4 \Delta_1 \left(\left(P_1 S_1^3 \Delta_2 / P_2 S_2^3 \Delta_1\right) - 1\right)} \tag{6.23}$$

The flexural stiffness and shear rigidity can be calculated by using one three-point middle-span loading configuration and one four-point quarter-span loading configuration; the solution for this case $L_1 = 0$, $L_2 = S_2/2$ is given by Equations 6.24 and 6.25.

$$D = \frac{P_1 S_1^3 \left(1 - 11 S_2^2 / 8 S_1^2\right)}{48 \Delta_1 \left(1 - 2 P_1 S_1 \Delta_2 / P_2 S_2 \Delta_1\right)} \tag{6.24}$$

$$U = \frac{P_1 S_1 \left(8 S_1^2 / 11 S_2^2 - 1\right)}{4 \Delta_1 \left(\left(16 P_1 S_1^3 \Delta_2 / 11 P_2 S_2^3 \Delta_1\right) - 1\right)} \tag{6.25}$$

Calculations in the other configuration combinations, such as one three-point middle-span loading and one four-point third-span loading, one four-point quarter-span loading and one four-point third-span loading, one standard four-point loading and one four-point third-span loading, and one standard four-point loading and one four-point quarter-span loading, can be referenced to ASTM D7250/D7250M.

Calculation of core shear modulus using known facing modulus and the two facings is identical; the sandwich transverse shear rigidity (U) and core shear modulus (G) can be calculated from the results of a single loading configuration test on the sandwich using Equations 6.26–6.28.

$$D = \frac{E\left(d^3 - c^3\right) b}{12} \tag{6.26}$$

$$U = \frac{P(S - L)}{4 \left[\Delta - \dfrac{P\left(2 S^3 - 3 S L^2 + L^3\right)}{96 D}\right]} \tag{6.27}$$

$$G = \frac{U(d - 2t)}{b(d - t)^2} \tag{6.28}$$

where
E = facing modulus, Mpa, and
c = core thickness = $d - 2t$, mm.

6.3 FLATWISE AND EDGEWISE COMPRESSIVE TEST

6.3.1 FLATWISE COMPRESSIVE TEST

Even though the flatwise compressive test covers the determination of the compressive strength and modulus of sandwich cores which has been presented in Chapter 2, it also is important for determining the compressive strength of the sandwich products because the core usually changes after bonded with skins, especially laminated and bonded in one process by wet resin. The wet resin could invade the cores though open cells, cuts, grooves, and perforations and then increase the compressive properties of the cores. Some textile and specialty cores must be impregnated by resin to achieve their associated compressive properties.

Flatwise shall define compression normal to the plane of the flat panel. Deformation data can be obtained from a complete load-deformation curve. It is possible to compute the compressive stress at any load (such as compressive stress at proportional limit load or compressive strength at maximum load) and to compute the effective modulus of the panel. This section provides a method of obtaining the flatwise compressive strength and modulus for sandwich panels.

The equipment and fixtures for this test are same as those used for testing the core materials, which have been introduced in Chapter 2. Deflectometer or compressometer capable of measuring the displacement with a precision of at least ±1% should be used for measuring deformation of the core during compressive test.

Test specimens of sandwich construction shall be of square or circular cross section having areas not exceeding 10,000 mm², but the minimum cross-sectional area shall be 625 mm². Prepare the test specimens so that the loaded ends will be parallel to each other and perpendicular to the sides of the specimen. The laminate surface should be sanded smooth. Detailed testing procedures can be found from ASTM C365/365M. Load-deflection curves are recorded in testing process and will be used to determine the modulus of elasticity as shown in Figures 6.14 and 6.15. The flatwise compressive strength will be calculated by Equation 6.29 if the failure mode is as shown in Figure 6.15 with peak yield load.

$$\sigma^{fcu} = P_{max}/A \qquad (6.29)$$

where

σ^{fcu} = core compressive strength, MPa,
P_{max} = ultimate load, and
A = cross-sectional area, mm².

If the sandwich panel is strong and no clear yield point will be recorded as the curve shown in Figure 6.14, a deflection stress at 2% deflection or at any deflection point can be calculated by Equation 6.30.

$$\sigma^{fc0.02} = P_{0.02}/A \qquad (6.30)$$

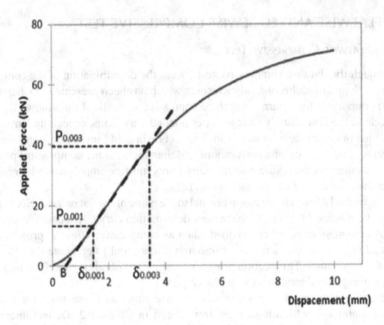

FIGURE 6.14 Description of determining flatwise compression modulus of sandwich panel.[15]

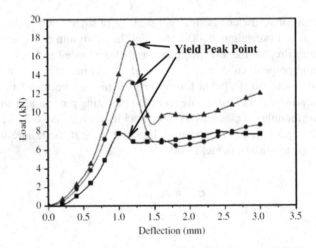

FIGURE 6.15 Load-deflection curves of the sandwich panel subjected to flatwise compression loading.[16]

where

$\sigma^{fc0.02}$ = flatwise compressive stress at 2% deflection, Mpa,
$P_{0.02}$ = applied force corresponding to 0.02 deflection, N.

Calculate the flatwise compressive chord modulus using Equation 6.31. The deflection values selected are intended to represent the lower half of the stress-stain curve.

Characterizations of Sandwich Structures

A deflection range of 25%–50% of ultimate is recommended. Anyway, an initial linear range of the curve should be used.

$$E^{fc} = t(P_{0.003} - P_{0.001}) / \left(A(\delta_{0.003} - \delta_{0.001})\right) \tag{6.31}$$

where
E^{fc} = core flatwise compressive chord modulus, Mpa,
$P_{0.003}$ = applied force corresponding to $\delta_{0.003}$, N,
$P_{0.001}$ = applied force corresponding to $\delta_{0.001}$, N,
$\delta_{0.003}$ = recorded deflection value such that δ/t is close to 0.003, and
$\delta_{0.001}$ = recorded deflection value such that δ/t is close to 0.001,
t = measured thickness of sandwich panel prior to loading, mm.

6.3.2 Edgewise Compressive Strength

Edgewise compression test consists of subjecting a sandwich panel to monotonically increasing compressive force parallel to the plane of its faces. The force is transmitted to the panel through either clamped or bonded end supports. Stress and strength are reported in terms of the nominal cross-sectional area of the two face sheets, rather than total sandwich panel thickness, although alternate stress calculation may be optionally specified. The test is dedicated mostly for sandwich panels, rather than for bare core materials. During that test, compressive loads are carried out by skins and the role of core is to keep them together and reduce bending effect.

Material and specimen preparation are important for obtaining designed properties of sandwich structures. Poor material fabrication, lack of control of fiber alignment, and damage induced by improper specimen machining are known causes of high data scatter in general, which include incomplete or no-uniform core bonding to facings, misalignment of core and facing elements, the existence of joints, voids, or other core and facing discontinuities, out-of-plane curvature, facing thickness variation, and surface roughness.

The test specimen shall be rectangular in cross section. The length (dimension parallel to the direction of applied load) of the specimen shall not be greater than eight times the total thickness of the specimen. The width of the specimen shall be at least 50 mm but not less than twice the total thickness, also not greater than the length, as listed in Table 6.2. The samples with varying aspect ratios across are

TABLE 6.2
Specimen Dimensions Recommended for Edgewise Compressive Test[17]

Dimension	Recommended Range
Length, L (mm)	$L \leq 8 \times t$
Width, W (mm)	50 mm $\leq W \leq L$; $W \geq 2 \times t$
t, total panel thickness (mm)	As required
t_{fs}, face sheet thickness (mm)	As required
t_c, core thickness (mm)	As required

FIGURE 6.16 Varying aspect ratios across testing specimens for edgewise compressive test.

shown in Figure 6.16. The yield stress generally reduces with an increase in aspect ratio. Care also shall be taken in preparing the test specimens to insure smooth end surfaces free of burs. The ends shall be parallel to each other and at right angles to the length of the specimens.

All specimens shall be laterally supported adjacent to the loaded ends on the facings of the sandwich by means of clamps, which shall be tightened to the extent that appreciable force shall be required to remove the specimen from the clamps. Each of the two clamps shall be made of two rectangular steel bars fastened together so as to clamp the specimen between them. The clamps will provide support for the facings and prevent an early buckling failure due to separation of the facing from the core at the point of contact with the loading plates. They should be placed on contact with the loading plates to approximate fixed end conditions of the specimen. The test fixture and setup are displayed in Figure 6.17 and ASTM C364/C364M – 07 Standard Test Method for Edgewise Compressive Strength of Sandwich Constructions.[17]

Data for load–deformation curves may be taken to determine the effective modules of elasticity, proportional-limit stress, yield stress, compressive strength, the approximated strain at failure, and the distribution of strain in each face. Deformations shall be read by means of a suitable gage – length compressometer attached over the length direction of the specimen. It is recommended that a strain-gaged specimen be used to determine the amount of bending inherent in the test configuration. A minimum of two axial strain gages, centrally located on opposite faces of the test specimen, is required for measuring the bending deformation. If more complete shear and bending information is desired, 4–12 gage configurations per test may be used.[19]

Except for recording load–length deformation and bending deformation curves during test, the failure modes also are important for evaluating the samples. The only acceptable failure modes for edgewise compressive strength of sandwich constructions are those occurring away from the supported ends. The sandwich column, no matter how short, usually is subjected to a buckling type of failure. The failure of the facings manifests itself by wrinkling of the facing, in which the core deforms the wavy shape of the facings; by dimpling of the facings into the honeycomb cells; by bending of the sandwich, resulting in crimping near the ends as a result of shear failure of the core; or by failure in the facing-to-core bond and associated face sheet buckling. ASTM C364/C364M[17] and Figure 6.18 shows different failure modes.

Characterizations of Sandwich Structures

FIGURE 6.17 Fixture and setup of edgewise compressive test.[18]

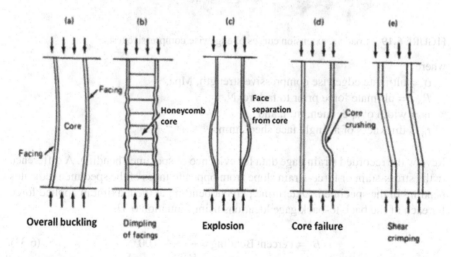

FIGURE 6.18 Commonly observed failure modes of edgewise compressive tests.[20]

The load – deformation curves are recorded in tests as presented in Figure 6.19, which is taken to determine the compressive modules of elasticity, compressive strength, and the approximated strain at failure. Ultimate strength can be calculated by Equation 6.32:

$$\sigma = P_{max} / \left[w(2t_{fs}) \right] \tag{6.32}$$

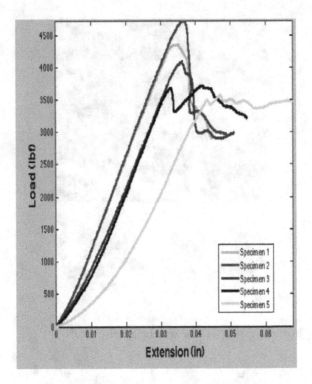

FIGURE 6.19 Load vs. extension curves of edgewise compressive tests.[21]

where
  σ = ultimate edgewise compressive strength, Mpa,
  P_{max} = ultimate force prior to failure, N,
  w = width of specimen, mm, and
  t_{fs} = thickness of a single face sheet, mm.

Review the recorded strain gage data for evidence of specimen bending. A difference in the stress-stain or force-strain slope from opposite faces of the specimen indicates bending in the specimen. Determine percent bending at the maximum applied force for each of the back-to-back gage locations using Equation 6.33.

$$B_y = \text{Percent Bending} = \frac{\epsilon_1 - \epsilon_2}{\epsilon_1 + \epsilon_2} \cdot 100 \qquad (6.33)$$

where
  ϵ_1 = indicated strain from gage on one face, and
  ϵ_2 = indicated strain from gage on opposite face.

The sign of the calculated percent bending indicates the direction in which the bending is occurring. This information is useful in determining if the bending is being introduced by a systematic error in the test specimen, testing apparatus, or test procedure, rather than by random effects from test to test.

Characterizations of Sandwich Structures

6.4 CONCENTRATED LOAD AND WAVE IMPACT TESTS

Concentrated load impact tests are test methods that use different shapes of indenters to act on the sandwich structures at different speeds, and then determine damage resistance properties of sandwich constructions. The shapes of the indenters could be hemispherical, conical, pyramid, and cylindrical. The impact speeds will be quasi-static, low velocity, and high velocity. Wave impact tests include water wave slam and air blast that typically generate high magnitude pressure pulses of very short duration moving across large area of the sandwich structure and resulting in high strain rates within the structure, both in the skin materials and the core. In this section, the concentrated load impact tests will be resented firstly to follow the standard test methods. Then the wave impact tests will be introduced by referring to the literature studies of investigation and application of the sandwich composite industries.

6.4.1 CONCENTRATED LOAD IMPACT TESTS

The concentrated load impact tests are performed by using different shape indenters at low to high velocity. First of all, it is important to define the difference between low-velocity impact and high-velocity or ballistic impact. In high-velocity impacts, damage is introduced in the plate before its motion is established, while in low-velocity, damage occurs after the establishment of plate's motion. The intermediate velocity impact is characterized by the existence of flexural waves, and the low velocity impact can be treated as a quasi-static indentation. In Figure 6.20, three impact regimes are shown.

In 2011, ASTM published the standard ASTM D7766, which brings together the concentrated load test methods found in the standards for solid laminates ASTM D6264 and D7136 to be applied to sandwich constructions. It consists of three main procedures called A, B and C, each having a different purpose and using different velocity and indenters with different fixture.

Procedure A involves an indentation test of a rigidly backed specimen at the low or quasi-static velocity. An out of plane concentrated force is applied by slowly pressing a displacement-controlled hemispherical indenter into the face of the specimen so that damage is imparted on the specimen surface. The size of the dent, location, and type of damage resulting from the applied indentation give an idea of the damage suffered by the panel.

Procedure B involves the low-speed indentation impact of a sandwich specimen, which is supported at the edges. This procedure is very similar to Procedure A, but the specimen is supported at the edges such that it is allowed to deflect. Damage on

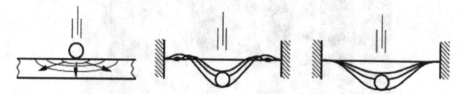

FIGURE 6.20 Impact regimes (from left to right): ballistic impact, intermediate velocity impact, and low-velocity impact.[22]

the specimen is again quantified by the depth and size of the indentation, the location and the type of failure observed at the skin and/or core.

Procedure C of the standard involves performing a drop weight impact test on a sandwich specimen supported at the edges. This test is conducted at intermediate velocity that is different from the previous two methods which are quasi-static, i.e. performed at low speeds. Damage is imparted through an out-of-plane, concentrated impact using a drop weight with a hemispherical striker tip. The damage resistance is quantified in terms of the resulting size, location, and type of damage in the sandwich specimen. For some structural applications, the use of a rigidly backed specimen in drop-weight impact testing may be appropriate.

The standard specifies that the most suitable procedure to evaluate and compare different damage characteristics for different sandwich panels is Procedure A, since the rigidly supported specimen prevents out-of-plane deflection of the specimen. Therefore, the sandwich flexural stiffness does not influence damage initiation in the specimen. However, the selection of a test procedure and associated support conditions should be done in consideration of the intended structural application, and as such Procedures B and C may be more appropriate for comparative purposes for some applications.[22] The damage response parameters can include dent depth, damage dimensions and location(s), indentation or impact force magnitudes, impact energy magnitudes, as well as the force versus time curve to compare quantitatively the relative values of the damage resistance parameters for sandwich constructions with different facing, core or adhesive materials.

These practices may be used desirably in conjunction with a subsequent damage tolerance test method to assess the residual strength of specimens. In this case, the tests should subject several specimens, or a large panel, to multiple indentations or impacts, or both, at various energy levels using these practices. A relationship between force or energy and the desired damage parameter can then be developed. Subsequent residual strength tests can then be performed using specimens damaged.

A hemispherical indenter is specified as having a geometry, which has, over the years, generated a larger amount of internal damage for a given amount of external damage, and used mainly in applications for plastics. However, in addition to the

FIGURE 6.21 Hemispherical, conical, pyramid, and cylindrical indenters.[24]

Characterizations of Sandwich Structures

standard hemisphere, other geometries, as shown in Figure 6.21, also produce damage of interest to sandwich manufacturers.

Attention should be paid to material and specimen preparation. High data scatter in test results in general could be caused by poor material fabrication practices, lack of control of fiber alignment, and damage induced by improper specimen machining. Specific material factors affecting test results include variability in core density and degree of cure of resin in both facing matrix material and core bonding adhesive.

For conducting test Procedure A, general apparatus shall be in accordance with Test Method D6264/D6264M with flat rigid support. The lower support fixture consisting of a flat rigid plate of dimensions 300 mm × 300 mm and thickness of about 30 mm.

A plate of dimensions of 200 mm × 200 mm of thickness about 40 mm (the thickness must be greater than the expected maximum indenter displacement), with an opening of 125.0 mm diameter, sits on top of the rigid plate to allow the simply supported configuration for Procedure B. The complete upper assembly is shown in Figure 6.22. Alternative opening geometries may be appropriate, depending upon the sandwich specimen geometry (especially thickness), flexural stiffness, through thickness shear stiffness, etc. It may be necessary to use alternative geometries to avoid core failure local to the edge support if the core has insufficient compression or shear strength. Tests conducted using alternative opening geometries must be designated as such, with the opening geometry reported with any test result.

A fixture for procedure C shall be in accordance with Test Method D7136/D7136M, with edge support utilizing a plate with a rectangular cut-out.[26] The cut-out in the plate shall be 75 mm by 125 mm. Clamps shall be used to restrain the specimen during impact. Alternative cut-out geometries and support conditions may be appropriate, depending upon the sandwich specimen geometry (especially thickness), flexural stiffness, through-thickness shear stiffness, etc. It may be necessary to use alternative geometries to avoid core failure local to the edge support if the core has insufficient compression or shear strength.

The indenter which is to strike, indent, and penetrate the sandwich panel is hemispherical with a diameter of 12.7±0.1 mm (default). In order to simulate the different indentation impact loads that can be experienced by the sandwich panels during

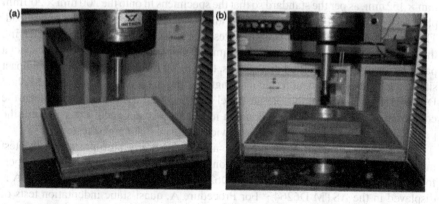

FIGURE 6.22 Assembled testing rig for impact test Procedure A (a) and Procedure B (b).[25]

operation, different indenters were designed to simulate varying and different damage requirements.[27]

A conical indenter simulates the rock type of a sharp cutting tip and bluff edges, which push and open through the composite laminate and core material. The apex angle is 30, with an indent depth of 50 mm. A pyramid-shaped indenter was designed, having a sharp point and four edges, leading to a square base having sides of 28 mm of diagonal 40 mm. A cylindrical indenter was also designed to simulate the rock impact. The cylinder is of diameter 13 mm and an overall length of 100 mm. The blunt flat front compresses the composite surface and core materials. The sharp circular edge cuts through the surface and foam. For procedure C following ASTM D7136, the indenter shall have a mass of 5.5 ± 0.25 kg and shall have a smooth hemispherical striker tip with a diameter of 16 mm and a hardness of 60 to 62 HRC.[28]

Impact devices for test procedure C, drop-weight impact testing, are shown in a Figure in ASTM D7136/D7136M.[26] At a minimum, the impact device shall include a rigid base, a drop-weight impactor, a rebound catcher, and a guide mechanism. The rebound catcher is typically an inertially activated latch that trips upon the initial impact, then catches the impactor on a stop during its second decent. The rebound catcher must not affect the motion of the impactor until after the impactor has lost contact with the specimen after the initial impact. If such equipment is unavailable, rebound hits may be prevented by sliding a piece of rigid material (wood, metal, and so forth) between the impactor and specimen, after the impactor rebounds from the specimen surface after impact. The guide mechanism includes single cylindrical tubes through which a cylindrical impactor travels, as well as double-column guides for a crosshead-mounted impactor. Other devices are force indicator that is capable of indicating the impact force imparted to the test specimen, and a velocity indicator to measure the velocity of the impactor at a given point before impact, such that the impact velocity may be calculated. The details of all devices can refer to ASTM D7136/D7136M.

By following the standards, the dimensions for Procedure A rigidly backed specimens should be a size of 75 mm × 75 mm. The size of the specimen chosen needs to ensure that the dent periphery was concentrated at the center and there was sufficient area around the dent periphery where no damage was visible.

For Procedure B with edge-supported specimens, the size is fixed at 152 mm × 152 mm, as per the standard so that the specimens fit onto the 200 mm × 200 mm plate containing the 127 mm hole, which allows the specimens to deflect under load. The size of the specimen for procedure C test is at least 100 mm × 150 mm by referring the standard for marching 75 mm × 125 mm cut out plate. It is permissible to impact a panel larger than the specified dimensions, then to cut out specimens (with the impact site centered) for subsequent residual strength testing in accordance with Test Method D7137/D7137M. Impacting a larger panel can help relieve interaction between the edge conditions and the damage creation mechanisms. Specimen size will limit the permissible overall thickness of the sandwich laminate that may be tested.

The test sequences of Procedures A and B are pretty similar, which the test machine preparation, specimen installation, loading and data recording shall be performed in accordance with test method D6264/D6264M. A setup for Procedure A is displayed in the ASTM D6264.[29] For Procedure A, quasi-static indentation tests of rigidly backed sandwich specimens, the suggested standard crosshead displacement

Characterizations of Sandwich Structures

rates are 0.25 mm/min for cores with high compression strength (e.g., balsa wood) and 1.25 mm/min for cores with low compression strength (e.g., foams, honeycombs). The test should be terminated before penetrating the back-side sandwich facing to avoid damaging the test apparatus. The unloading rate shall be the same as the loading rate. For Procedure B, quasi-static indentation tests of edge-supported sandwich specimens, the suggested standard crosshead displacement rate is 1.25 mm/min. The unloading rate shall be the same as the loading rate.[30]

After starting tests, the testing machine recorded values of force against crosshead displacement for Procedures A and B, which enabled a graph of indentation force versus indenter displacement to be plotted for each of the specimen as displayed in Figure 6.23. The curves represent that pressing force changed with the moving distance of the indenters after contacting top skin of the sandwich specimen. From the curves, the maximum force required to fully puncture the upper skin and

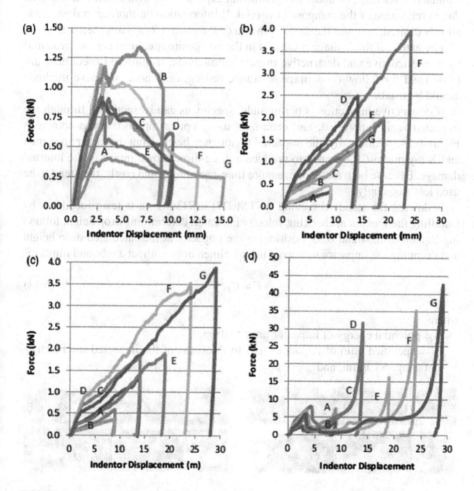

FIGURE 6.23 Impact force vs. indenter displacement of test Procedure A for hemispherical (a), conical (b), and pyramid (c).[25]

the core (except for the lower skin) was recorded so that the indenter caused the greatest amount of damage was noted. The damage incurred under each variation of indentation impact is described, in terms of force, absorbed energy and indentation displacement. Each force–indentation response can be analyzed and shown to be different for each structure with different indenter (however, the conical and pyramid show similar trends, but each provide essential damage information). By analyzing the curves, the relationship of the initial contact force and the experimental absorbed energy with the displacement can be found for each case.

At end of each test right after pulling out the indenter from the specimen, it is important to measure the depth and diameter of the damage or the dent accordance with test Procedures A and B as defined ASTM D7766.[23] The periphery of the dent shall be determined at eight points relative to the center of the specimen. Alternatively, automated algorithms may be used to define the dent periphery and to calculate the dent diameter. Over time, or under environmental exposure, the dent depth can decrease due to relaxation of the composite material. If information on short-term dent relaxation is desired, measure the dent depth and dent diameter 7 days after testing.

The extent of the damage produced in the composite specimen can be measured by nondestructive and destructive inspection methods. If impacted specimens are to be used for follow-on damage tolerance testing, only nondestructive methods should be used.

If destructive inspection is permissible, specimens can be sectioned through the impacted region, polished, and examined using optical microscopy as shown in Figure 6.24. Additionally, the impacted region may be cut from the larger specimen and X-ray micro-CT scanned to produce a three-dimensional image of the internal damage. Because both methods are more time-consuming and costly, they tend to be used less frequently.

Following the standard method ASTM D7136/7136M starts test procedures by installing the specimen, checking velocity indicator, determining drop height following Equations 6.34 and 6.35, positioning the impactor at the calculated drop height and dropping the impactor to impact the specimen once without a rebound impact.[26]

$$E = C_F t \qquad (6.34)$$

where

E = potential energy of impactor prior to drop, J,
C_F = specified ratio of impact energy to thickness of the impacted sandwich facing, 6.7 J/mm, and

FIGURE 6.24 Damage sustained on the sandwich panels with a rigidly backed configuration – Procedure A with different indenters.[25]

t = nominal thickness of impacted sandwich facing, mm.

$$H = E / (m_d g) \tag{6.35}$$

where
 H = drop-height of impactor, m,
 m_d = mass of impactor for drop height calculation, kg, and
 g = acceleration due to gravity, 9.81 m/s².

One thing should be known that a minimum drop height is 300 mm when selecting impactor mass for calculating the drop height.

Data of force versus time during contact should be recorded continuously if instrumentation is utilized. For this test method, a sampling rate of 100 kHz and a target minimum of 100 data points per test are recommended. Record the time at which the velocity indicator light beam is interrupted by each of the flag prongs.[26]

The onset of specimen impactor contact is noted by the detection of a nonzero contact force. As the impactor presses into the specimen, it will flex the specimen and form a local depression as the contact force increases. Sharp drops in recorded contact force indicate damage processes that result in a sudden loss of stiffness in the contact region.

Rapid increases and decreases in recorded contact force response with time can result from harmonic resonance of the impactor, load cell, or specimen during the impact event. For composite materials, the oscillations in the force response most often reflect actual forces applied to the specimen and should not be smoothed out. If smoothing is used in data interpretation, both the recorded data and the postprocessed "smoothed" data shall be paid attention. Parameters which can be determined from the contact force versus time curve(s) after the test include the F_1 force (recorded contact force at which the force versus time curve has a discontinuity in force or slope), the maximum contact force F_{max}, absorbed energy E_1 (at F_1 force), absorbed energy E_{max} (at maximum contact force), and contact duration t_T.

If the impact device is capable of detecting the velocity of the impactor, calculate the actual impact energy using Equation 6.36. This calculation is performed automatically by most systems with velocity detection capability but may be performed manually if necessary. The measured impact energy may differ from the nominal impact energy calculated by Equation 6.34 due to friction losses during the drop.[25]

$$E_i = \frac{m v_i^2}{2} \tag{6.36}$$

where
 E_i = measured impact energy, J,
 m = mass of impactor, kg, and
 v_i = impact velocity, m/s.

For procedure C, test as for Procedures A and B, the damage mode(s) observed for each specimen, and the surface(s) or through-thickness location(s), or both, at which the damage modes are observed, should be recorded in the same methods mentioned above. More than one damage mode may be present in a damaged specimen. A figure in ASTM D7766 illustrates commonly observed damage modes in sandwich construction damage resistance testing in procedure C.[23] There are nine potential damage locations in sandwich structure during procedure C impact test (two facings, two facing-to-adhesive interfaces, two adhesive layers, two adhesive-to-core interfaces, and core).

To measure the dent depth and diameter in accordance to procedure C specimens is the same as for Procedures A and B specimens. Common nondestructive measures of impact damage include surface indentation depth, measured with a depth gage micrometer. But for sandwich structures tested by procedure C, the damages under skin are not visible or barely visible, so some nondestructive instruments are useful for detecting real damages beyond the visible damages. Most commonly used instrument is ultrasonic C-scan inspection for producing the planar damage area. Using the pulse-echo method, an ultrasonic pulse is sent through the specimen thickness. The pulse reflects back to the probe when it reaches either the panel's back surface or internal damage, such as delamination. The planar damage area thus can be identified and measured, as the probe scans the region of impact.[31]

6.4.2 Wave Impact Tests

Sandwich composite materials used for the bottom panels of high-speed marine vessels can be subjected to dynamic pressure wave loads due to slamming with the water surface. Analysis of laboratory-based wave impact tests demonstrates that the transverse shear stress rates experienced by core materials in sandwich panels subjected to water wave impact can be significantly higher than those prescribed by an industry standard testing method. In particular, impact events typically generate high-magnitude pressure pulses of very short duration that move across the panel as the hull enters the water.

The water wave impact tests, also called slamming tests, include the tests by free dropping and by the servo-hydraulic slam testing system (SSTS). The free drop tests are to drop the specimen with frame from a defined height onto the water surface. The main drawback of drop tests is that there is no direct control of the specimen motion once it hits the water, so that the retardation rate primarily is dependent on the mass and geometry of the specimen and fixtures. In reality the velocity profile during a slamming event depends on the hydrodynamic behavior of the vessel and the position of the panel. It is difficult to reproduce particular conditions in real vessel testing due to the large number of variables that contribute to a slamming event including water entry velocity, wave height and frequency and dead rise angle during the slamming event. SSTS was developed in order to address these issues.[32]

There are two types of free drop tests, single slamming and repeated slamming tests. The specific free drop testing tasks are as follows: quantify the effect of slamming on the state of damage in sandwich composites as a function of drop height (energy of impact, E_n), dead rise angle (β), and then develop and implement a methodology to discern modes of failure and accumulated damage associated with the wave slamming on sandwich composites and a semi-empirical model to relate the effect of repeated slamming on the remaining strength and fatigue life of the material. Post slamming techniques, including nondestructive evaluation (NDE) and other

Characterizations of Sandwich Structures

test methods introduced in last sections, are generally employed to ascertain the state of damage in the material in order to determine cumulative damage and modes of failure in sandwich composites subject to single and repeated slamming scenario.

The free drop slamming system was designed and fabricated as shown Figure 6.25. Both single and repeated slamming can be performed from various heights corresponding to the desired energy level. The specimen sizes and dead rise angle (β) of the symmetric wedge can be varied. Various boundary conditions can be imposed. A gear system attached to a continuously rotating motor engages and disengages at preset time intervals in order to rise and release the wedge-shaped specimen assembly from predetermined heights. The interval between each slamming event can be set at 30 seconds (0.033 Hz) to allow the water to regain its initially calm state. A 1.80 m diameter and 1.25 m depth water tank is used to slam the symmetric wedge specimen holder. Baffles are used around the tank to minimize the wave reflection. The slamming specimens are instrumented with piezo-electric pressure, strain, transducers, and

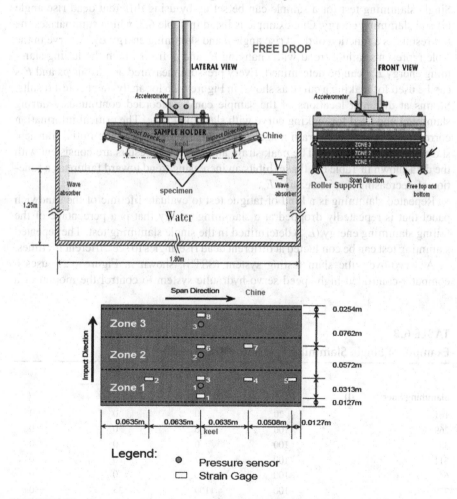

FIGURE 6.25 Free drop slamming test fixture and setup details.[33]

accelerometer. The pressure sensors are installed on the impact faces in the maximum pressure zone, whereas the strain gages are mounted on the back faces with the same configuration, see Figure 6.25. The accelerometer is mounted on the inside of the sample holder.[33]

The slamming energy can be determined by the following equation:

$$E_n = Hm_d g \qquad (6.37)$$

where

E_n = potential slamming energy prior to drop, J,
H = drop-height of fixture, m,
m_d = mass of all fixture including sample for drop height calculation, kg, and
g = acceleration due to gravity, 9.81 m/s^2.

Single slamming test for a sample can be set up by using different dead rise angle (δ) and slamming energy. One example is listed in Table 6.3, which summarizes the test results as a function of dead rise angle β and slamming energy E_n. A curve in the table represents failure trend with changing E_n and δ, from which the failing slamming energy E_{ult} can be determined. Every pressure measured at various βs and E_ns can be used for making a curve as shown in Figure 6.26 for analyzing the test results. Strains at different locations of the sample can be recorded continuously during slamming and used for making curve with slamming time. The critical information corresponds to the peak strain magnitude and the strain rate in the initial transient stage and recovery phase. The peak strain and strain rate results are consistent with the data shown in Table 6.3 that exhibits an increasing trend toward failure as a function of decreasing β or increasing E_n.

Repeated slamming is a kind of fatigue test to evaluate lifetime of the sandwich panel that is repeatedly dropped at a slamming energy that is a percentage of the failing slamming energy (E_{ult}) determined in the single slamming test. The repeated slamming test can be conducted at different dead rise angles to see different lifetimes.

A servo-hydraulic slam testing system (SSTS), shown in Figure 6.27, uses a computer-controlled high-speed servo-hydraulic system to control the motion of a

TABLE 6.3
Example of Single Slamming Test Setup and Results[33]

	Deadrise Angle (β)			
Slamming Energy, E_n (J)	0° (%)	15° (%)	30° (%)	45° (%)
161	20	0	0	0
269	100	0	0	0
386	100	0	0	0
511	100	67	0	0
642	100	100	0	0
779	100	100	33	0

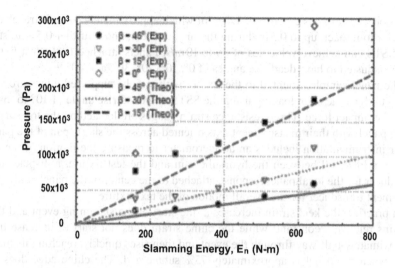

FIGURE 6.26 Pressure distribution profile vs. slamming energy with different dead rise angle of single slamming test.[33]

FIGURE 6.27 Panel sample mounted in SSTS and typical slamming event in SSTS.[32]

panel during water impact into a 3.5 m diameter, 1.4 m deep water tank. The specimen fixture slides on vertical rails and is attached to the hydraulic ram through a load cell. Hydraulic accumulators supply oil to the ram, and the velocity is determined by a servo-valve controlled by a closed-loop controller using position and acceleration feedback. Vertical panels on the sides and behind the panel constrain the flow to 2D behavior. The ram has a stroke of 1.4 m, of which approximately 0.4 m is traveled in air prior to impact, up to 0.5 m during the impact event, and a further 0.5 m to stop. The SSTS can achieve velocities of up to 10 m/s. Panels attached to the test fixture can be adjusted to have dead rise angles of 0°, 10°, 20°, 30°, and 40°.[32]

The panel sample mounted to the fixture with simply supported edges are subjected to dynamic slam loading using the SSTS at a dead rise angle of 10° to maximize the impact force. Three resistance strain gauges are located on the top surface of the panel with their measurement axis oriented across the short span of the panel. Other instrumentation includes an accelerometer to measure the motion of the test fixture, a load cell between the hydraulic ram and the test fixture, a displacement transducer for the local panel response attached to the center of the panel, and a displacement transducer for the overall motion of the text fixture.[31]

In practice, the keel strain increases at the start of the slamming event and then remains relatively constant, while the chine strain does not start to increase until approximately half-way through the event and then rises quickly, reaching its maximum when the panel is approximately 75% submerged. The chine edge does not experience significant transverse loads until relatively late in the slamming event, but it is then subjected to larger transverse shear loads and higher local loading rates than the keel edge. All failures of the panels should be by core fracture or permanent yield at the midpoint of the chine edge in the vicinity of the chine strain gauge.

From SSTS test, skin strain rate at each location of strain gauge vs. slamming time will be recorded, the maximum strain rate of each test can be determined. In the test, the skin strains measured at the chine edge do not directly measure the core shear strain but do provide an indication of the local bending moment and hence transverse force and its rate of increase at this position of high transverse shear loads. Then the corresponding core transverse shear stress can be predicted following modal transient finite element analysis.[32]

6.4.3 Air and Water Blast Tests by Shock Tubes

Sandwich structures are often used in civil and military structures, where air-blast or water-blast loading represents a serious threat. Materials testing under blast is now becoming more common in practice due to the unfortunate rise in use of explosive devices in civilian environments as well as in direct military combat. There are two types of blast test, full scale explosive test and shock tube test. Given the high costs and space requirements for explosive testing, shock tubes offer a cost-efficient and effective alternative for scaled experiments. Shock tubes enable a shock load to be produced in a controlled manner. Additionally, the experiment can focus solely on the shock wave incident rather than other factors inherent in blast situation, e.g., burning. An air shock tube consists of a long cylinder, divided into a high-pressure driver section and a low-pressure-driven section, which are separated by a diaphragm. To create a shock wave, the driver section is pressurized until the pressure difference across the diaphragm causes it to rupture. This rapid release of gas creates a shock wave, which travels down the tube to the test specimen. High-speed photography, with digital image correlation (DIC), and laser gauge systems were employed to monitor the deformation of these structures during the blasts.

A brief schematic of the shock tube facility is shown in Figure 2a of Reference 34. The driver and driven sections have a 0.15 m (6 in.) inner diameter, and the converging section begins as 0.15 m in diameter and ends as 0.07 m. The driver gas is helium for its lightweight, nonexplosive nature, low cost, and availability. The driven gas is the ambient air of the environment. The shock tube is instrumented with pressure and velocity measurements to provide real time data about the shock pressure and shock velocity. A dynamic pressure sensor is mounted at the muzzle section of the shock tube and graphite rods are used as break circuit initiators to measure the shock velocity. The driver pressure can also be measured for calibration purposes. Thin mylar sheets (10 mil) are used as diaphragm material in the shock blast experiments. The driver pressure and hence the shock pressure obtained from the burst of these diaphragms are controlled by a number of plies of sheets used.[34]

A typical loading history obtained from shock tube is plotted with two peaks as shown in Figure 3 of Reference 33. The first peak pressures obtained in such experiments are quoted as the "input shock pressure" or just the shock pressure to which the specimens are subjected. This remains constant for a given number of mylar diaphragms. The second peak is the "reflected shock pressure" from the specimen that the shock blast is impinged upon. This is dependent on the material and boundary condition of the specimen. Typically, this can vary between 1.5 and 3 times the shock pressure and were recorded for all the experiments in this study.

The first result of the air shock tube test is the damage progression in the sample plates as they are subjected to increasing shock pressures. Secondly, the samples will endure permanent deformation due to the shock. To measure the magnitude of this permanent deformation, a laser displacement sensor and an automated table will be used. The data obtained from the sensor can be plotted as deformation profiles. The maximum deflection is at the center and reduces gradually toward the fixed boundaries. The strain response of the sample panel subjected to explosive blast is third result from the air shock tube blast test, which shows a distinct initial compressive pulse, followed by a predominant tensile pulse. The strain response at two locations looks similar, due to axisymmetric loading.[34]

The shock tube also can be used for underwater blast test. An example is to use a cylinder-shaped underwater shock simulator, without using explosive. A water chamber made of a steel tube is incorporated into a gas gun apparatus as Figure 6.28. The specimen panel and a 22 mm thick piston are installed at the rear end and the front (right) end, respectively. The exponentially decaying pressure history is produced by impacting the piston with a 5 mm thick flyer plate launched by a gas gun. The pressure histories are measured by dynamic high-pressure transducers installed in the positions of A and B. The equipment is good for performing underwater shock experiments and accurately measure and analyze deformation and strain field in the response process of sandwich panel subjected to underwater impulse.[35]

Other two water shock tube setups are shown in Figure 6.29 with simply supported sample and back force transducer supported sample.

In an experimental procedure, a 3D digital image correlation (DIC) measurement system including two synchronized high-speed cameras in a stereoscopic setup, two halogen lamps, and specialized software was used to capture the transient response of the back (dry) face sheet of the panels, Figure 6.30a. The specimen with

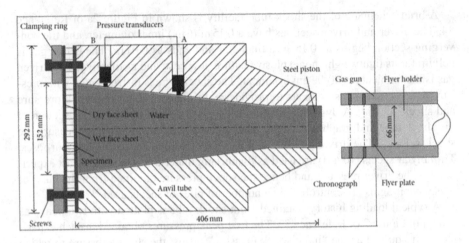

FIGURE 6.28 Schematic of water chamber with divergent diffuser as shock tuber.[35]

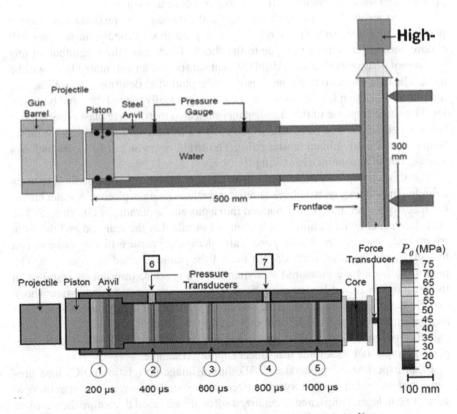

FIGURE 6.29 Schematic illustrations of air-backed loading shock tube.[36]

FIGURE 6.30 Photos of the experimental setup: (a) global view of 3D measurement system with two cameras and (b) area of interest of the clamped test specimen painted with black and white speckle pattern.[35]

a randomized speckle pattern was clamped using a steel ring with 12 screws at the rear of anvil tube, Figure 6.30b. Two high-speed cameras were positioned behind the anvil tube to record the speckles on the dry face sheet. A signal can be generated when the flyer plate impacts the piston. Thus, the pressure transducers and two high-speed cameras will be synchronized and triggered to record the pressure data and deformation images upon the arrival of shock wave. The post processing of deformation images was performed with a software to get the full-field shape and deformation measurements.

In the example test, the shock pressure histories, generated in a laboratory-scale water shock tube apparatus, are well captured for the velocity test. Sandwich panels with three thicknesses' core layer were chosen to be carried out in ten loading levels. A high-speed photography system with two cameras is utilized to capture real-time images of the panel during underwater blast loading. The full-field data of the deformation during the shock events are obtained utilizing 3D DIC techniques and analyzed compared with that of solid plate. Postmortem visual observations of the test specimens are carried out to identify the deformation modes. The results are analyzed in normalized forms to gain insight into underlying trends that can be explored in the design of blast resistance for sandwich panels.[35]

6.4.4 Air and Water Blast Tests by Full-Scale Explosion

When a designer considers using the sandwich structures against explosion threats, one has to consider the blast event (pressure wave), the surroundings (fluid medium and boundary conditions), and the component (material properties and construction). It is important to conduct full-scale blast loading tests on a sandwich composite product before using it in the civil and military structures to evaluate the properties of blast resistance. Two examples of full-scale air blast and water blast testing are presented below, respectively.

One air-blast test design is given in Figure 6.31 featuring all external instrumentation that was used during all air-blast tests. In each case, the spherical charge was raised of the floor to the mid-height of the target and positioned at the center of

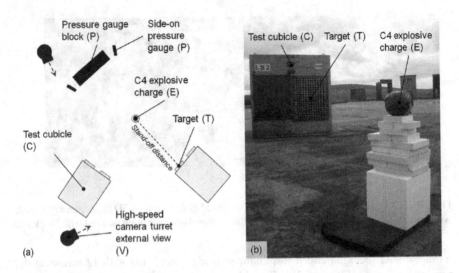

FIGURE 6.31 Air blast configuration: schematic diagram (a) and image of the test setup (b). (Features in each diagram are target to be tested (T), test cubic (C), high-speed camera and its relative location on the test pad (V) pressure sensor arrangements (P) and C4 explosive charge).[38]

the test pad at the given stand-off distance from the target. The test pad is approximately 100 m × 100 m in area and is made of concrete, the charge is positioned on a 150 mm thick steel plate to avoid cratering during blast. The test fixture was a steel test cubicle, which was used to test one sample at a time against a 38.4 kg TNT equivalent charge, although up to two test samples can be attached at one time in this fixture.[37]

Changing the medium used to carry the shock from a gas to a liquid (increasing the density) increases the speed of sound and generates a significant rise in pressures produced by a blast event. It is for these and related reasons that underwater shocks and their interaction with surrounding submerged structures are of particular interest to the naval industry. Figure 2 of Reference 38 shows a setup of underwater blast test. When an explosion occurs underwater, there is an intense release of energy, high pressure, and heat, similar to the air blast case. This is relieved by the formation of an intense (compressive) pressure wave, or shock wave, which radiates away from the source.[39]

In a test practice, the sandwich panels are subject to underwater-blast loading to observe the deformation of the targets during the blast and damage sustained. Surface strain measurements were taken during the blast event using strain gauges positioned at 12 different locations. Six gauges were positioned on front face, and the other six gauges were positioned behind these locations on the rear face as shown in Figure 3b of Reference 39. The strain gauges (and data acquisition hardware) chosen for this application was chosen specifically with the ability to monitor dynamic events (in terms of strain magnitude and strain rate).

Once the gauges were bonded, they were sealed in accordance to insulate from the environment and protect during impact, while maintaining a low profile and mass.

Characterizations of Sandwich Structures

The panels were first bonded into a steel frame (3 mm thick mild steel). They were then bolted into a substantial base frame, comprising 10 mm thick mild steel, prior to testing.[39]

Blast pressures experienced by two panels peaked at a shock pressure of 430 bar (1 kg of C4 explosive; 6 m depth; stand-off 1 m) and 300 bar (1 kg of C4 explosive; 6 m depth; stand-off 1.4 m) in the testing practices, respectively. The two pressure-time traces for the two blast scenarios, illustrating the ferocity of the blast event, note that the strain gauge data will be restricted to the initial response. The charts in Figure 11 of Reference 38 show pressure-time traces for 1 kg blasts at stand-offs of 1.0 and 1.4 m: the entire event including the 1st bubble pulse at ~200 ms (top left) and initial shock pressure including reflected shock at ~5 ms (top right). Sample strain gauge data of the panel response are given over these time periods for strain gauge position 1 on front and back face of the water backed panel when subjected to the 1 kg charge C4 at 1.4 m stand-off.[39]

The sandwich panel had its core crushed to reduce thickness by the shock. There were initial surface strains in the region of 3% and once the panel membrane response began, surface strains of around 1% remained causing severe cracks to form within the skins along the panel edges. This is evident in Figure 13 of Reference 39, in which the sandwich panel with shock 430 bar pressure (1 kg of C4; 6 m depth; stand-off 1.0 m) and diagrammatic representation of the panel showing signs of typical impulsive behavior with the top edge initially in compression while the remainder of the plate is in tension.

6.5 DYNAMIC FATIGUE EVALUATION

Sandwich structures are extensively used these days in structural applications such as components in space crafts, aircraft structures, marine vessels, transportation structures, tanks, refrigerator containers, bridge decks, and car body shells where minimum structural weight and maximum stiffness/strength are important. In such applications, composite sandwich structures are often subjected to repetitive loading, which may lead to fatigue failure (e.g., the repetitive loading of waves on the hull of a ship, the aero-acoustic excitation of a turbine engine housing by the rotating turbine blades, or the repeated loading of motor vehicle traffic over a bridge deck).

Firstly, we need to know different testing methods that can be used for obtaining data for predicting the fatigue life of the sandwich structures used in repeatedly loading applications. The testing results are either stress versus number of cycles (S–N), curves, strength degradation, stiffness reduction, cumulative damages, or some combination of these approaches. Then the different failure modes should be observed in dynamic fatigue loading on the sandwich structures. The fatigue life of the sandwich structures is also influenced by various factors such as loading frequency, environmental factors, and the nature of the repetitive loading, which will be discussed in each section.[40]

6.5.1 FLEXURAL FATIGUE TESTS

There are several fatigue testing methods, such as flexural, compressive, wave slamming, and hydromat (plate flexure) compressive tests. However, the most used testing method for evaluating fatigue resistance behavior of sandwich structures is the flexural test including three-point and four-point bending tests. ASTM D7774 gives the standard testing procedure for solid plastics but is also useful for sandwich composite structures.

There are two procedures in the test method: Procedure A, designed for materials that use three-point loading systems to determine flexural strength. Three-point loading system is used for this procedure as shown in Figure 6.32. Procedure B, designed for materials that use four-point loading systems to determine flexural strength. Four-point loading system is used for this procedure as shown in Figure 6.33. There are three different setup procedures for performing flexural fatigue tests: stress-controlled, displacement-controlled, and cycling number-controlled fatigue tests. For first two procedures, the tests will run until sample failing for obtaining the number of cycles (S–N). For the last procedure, after fatigue test, the sample will be evaluated for residue properties by deconstruction tests, such as the static compressive and flexural tests introduced in the previous sections, or by nondeconstruction tests that will be introduced later.

Before performing the flexural fatigue test, the material should be tested by quasi-static flexural test with the same point support and span loading procedure, such as three-point, four-point quarter-span, and four-point third-span testing setups, for obtaining maximum flexural displacement, maximum skin flexural strength, and maximum core shear strength by following the methods introduced in the previous sections. By using the equation for calculating the skin strength and the core shear

FIGURE 6.32 Setup for three-point loading system fatigue test.[41]

FIGURE 6.33 Setup for four-point loading system fatigue test.[42]

Characterizations of Sandwich Structures

strength, loading force can be found out for testing setup for stress-controlled fatigue test. Usually, 25%–75% of ultimate quasi-static face stress or core shear strength of the sample is used to set the maximum load applied for the fatigue test, depending on the loading force in the application.

For displacement-controlled fatigue test, a loading ratio r is defined by Equation 6.38:

$$r = D_{max}/D_f \qquad (6.38)$$

D_{max} and D_f represent the maximal displacement applied during the fatigue tests and the average displacement at failure measured during quasi-static tests respectively. Then the loading ratio r is used for determining the amplitude of displacement of the test. Due to the sample will lost stiffness in the test process, the amplitude of loading stroke will increase for the stress-controlled test, and the loading force will decrease for the displacement-controlled test.

Setting a loading striking frequency is important for the flexural fatigue test. In theory, a higher frequency test produces more heat in the sample which is especially problematic for a polymer-based sandwich structure. Based on ASTM D7774/D7774M, the material properties can vary with specimen test frequency. Test frequency can be 1–25 Hz, but it is recommended that a frequency of 5 Hz or less should be used. The adherence to the recommended frequency range should ensure that the results are independent of frequency and the laminate is not subjected to heating during the duration of the test.

Fatigue tests are most frequently carried out under stress-based constant-amplitude loading. Test samples can be subjected to a variety of waveform geometries. However, sinusoids are most prevalent. Figure 6.34 depicts a stress-based sinusoidal waveform showing fully reversed stress cycles. The maximum and minimum stresses are equal and opposite in a fully reversed cycle test. By convention, compressive stresses are negative. However, some examples exist where fully reversed loading is not performed either because it is not possible or during normal service a component is only subjected to forces in one direction. Examples of repeated stress-cycle compression-only fatigue tests on sandwich panel for truck floor.

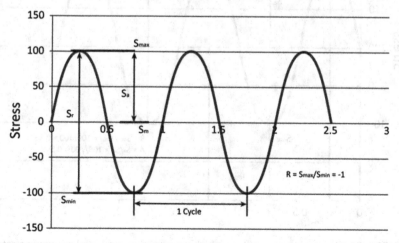

FIGURE 6.34 Fully reversed stress-based cycle.[43]

The following definitions and equations are used to express a stress-based waveform. (Refer to Figures 6.34 and 6.35 for further explanation.)

Stress Range, S_r – difference between the maximum and minimum stress.

$$S_r = S_{max} - S_{min} \qquad (6.39)$$

where

S_{max} = maximum stress,
S_{min} = minimum stress.

Stress amplitude, S_a – one half of the stress range, S_r.

$$S_a = S_r/2 = (S_{max} - S_{min})/2 \qquad (6.40)$$

Mean stress, S_m – average of the maximum and minimum stress.

$$S_m = (S_{max} + S_{min})/2 \qquad (6.41)$$

Stress ratio, R equals -1 for fully reversed loading

$$R = S_{min}/S_{max} \qquad (6.42)$$

Amplitude ratio, A – infinite for fully reversed loading.

$$A = S_a/S_m = (1-R)/(1+R) \qquad (6.43)$$

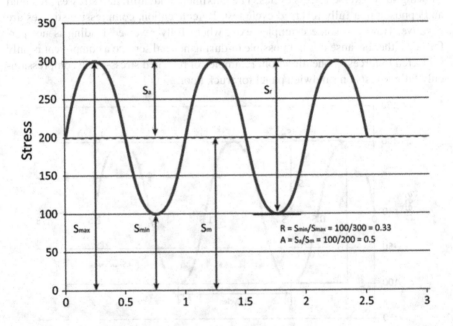

FIGURE 6.35 Repeated stress cycle in which the maximum stress and minimum stress are not equal in magnitude.[43]

Figure 6.35 depicts a repeated stress cycle in which the maximum stress and minimum stress are not equal in magnitude. Both stresses are in tension, but they could also be in tension and compression or compression only.

Most of the fatigue testing work in laboratories involves constant amplitude sinusoidal loading in three-point or four-point bend tests. But composite structures are rarely subjected to uniform constant amplitude loads in service. The load could fluctuate randomly according to a distribution, creating a load spectrum, or the load could vary in sequential steps (i.e., block amplitude loading) or the loading could consist of various combinations of these types of loading. An example could be the gust load spectrum applied to a composite aircraft wing during flight or the effects of ice motion or sea waves on composite sandwich ship hull materials.

An example to evaluate the fatigue behavior of sandwich beams under two-step and block loading regimes is shown below. A combination of low–high loads and then high–low loads was carried out to investigate the influence of load sequence. The beams were loaded for 50% of their average fatigue life at each respective load. Most of the beams did not fail and were tested for their residual static strength under static loading conditions. From the results of the residual strength of the specimens, it was concluded that load sequence affects the fatigue life, and a high/low load combination is more damaging than a low/high load combination.[44]

Sandwich composites can fail in several ways as mentioned in the previous sections. Failure modes in fatigue are often similar to those observed in static and impact loading. But under cyclic loading, sandwich beams are particularly prone to core shear failure. This is due to the fact that cyclic loading appears to reduce the residual shear strength of the foam core. The other failure modes include face-sheet yield and indentation. Face-sheet yield and indentation failure were limited only to specimens with very low face-sheet strength and thickness. And it was concluded that the predominant failure mode under cyclic fatigue is core shear.

6.5.2 OTHER FATIGUE TESTS

6.5.2.1 Flatwise Compressive Fatigue Test

The sandwich composite structures are being used as bridge decks, temporary bypass roadways, truck and bus floor, on which the compressive load repeatedly works every day. So it is also important to perform fatigue test by compressive load on the sandwich structure for determining life cycling time as well as its residual strength and rigidity after certain cycling time. A testing example of conducting the compressive fatigue test is below.

The purpose of the fatigue test was to determine the performance of the sandwich panel under repeatedly compressive loading. After testing specimens for 0.5, 1, 1.5, and 2 million cycles at certain load levels, the destructive tests were performed for studying the behavior of the tested specimens and comparing them to the samples observed in the static compressive test. Four different levels of load were imposed in this test. For the minimum level (Load Level 1) it was decided to take the value of 30% of the ultimate load determined from the static compressive test. The maximum level (Load Level 4) corresponds to 60% of the ultimate compressive capacity. The levels of load shown in Table 6.4 were calculated based on the ultimate static

TABLE 6.4
Loading Levels of Compressive Fatigue Tests

Load Values	Load Level 1 30% P_{ult} (kN)	Load Level 1 35% P_{ult} (kN)	Load Level 3 40% P_{ult} (kN)	Load Level 3 60% P_{ult} (kN)
Minimum	423	4.23	4.23	4.23
Maximum	25.27	29.5	33.73	50.55

compressive capacity of the sample. In the table, the ranges of the imposed loads for the different levels are presented.[45]

In the test, with a frequency of 5 Hz, the duration of the test for each imposed level of load was approximately 5 days, due to the fact that the four series of four specimens were stacked and it was removed one series at a time after each 0.5 million cycles until the last series accumulated 2 million cycles. For safety reasons, avoidance of horizontal displacement of the specimens, and to guarantee a uniform load distribution from the crosshead over each sample during conditioning, a steel frame was built and used for holding the specimens as shown in Figure 6.36.

FIGURE 6.36 Setup for a compressive fatigue test.[45]

6.5.2.2 Edgewise Compression – Compression Fatigue Test

The sandwich structures have a potential to be utilized in transport industries where structures are subjected to vibration or cyclic loading. Fatigue properties of sandwich materials under vertical compression are significant prior to using these materials in bulkheads, shear webs and side walls of truck, semitrailer, bus, and rail cars. The performance of sandwich structures under compression–compression fatigue is important for evaluating the fatigue life, the fatigue resistance, and the associated failure modes of the sandwich structures. An example of performing edgewise compression-compression fatigue test is presented below.

Compression–compression constant amplitude fatigue tests were performed to determine the influence of different loading parameters on balsa wood core sandwich beam.[46] The tests were conducted on a MTS testing machine with load capacity of 100 KN under ambient laboratory conditions at a frequency of 5 Hz. All compression fatigue tests were under stress-controlled mode, and an extensometer was used to obtain strain from the beam. During the compression-compression fatigue test, the value of R is determined from Equation 6.42:

$$R = S_{min}/S_{max}$$

where
  S_{min} is the maximum compressive stress and
  S_{max} is the minimum compressive stress, referring to Figure 6.37.

The maximum compressive stress (S_{min}) of the cyclic loading was chosen to be 40%–70% of the ultimate compressive static strength of the composite sandwich beams. Four different load levels in the range of 0.4–0.7 were chosen for the sandwich beams and three specimens were tested for each stress level. Load and displacement data collected using a PC-based acquisition system are obtained from the machine, which are converted into stress-strain curves. Typical compression cyclic load is shown in Figure 6.37.

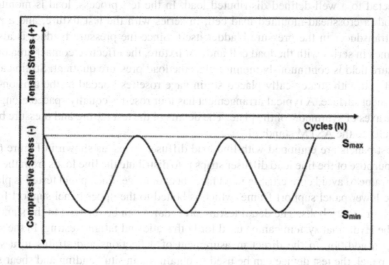

FIGURE 6.37 Typical compression-compression cyclic load.[45]

6.5.2.3 Two-Dimensional Simply Supported Distributed Load Flexural Static and Fatigue Test

To simulate more realistic test conditions, especially for marine structures where sandwich composites are used in hull materials, a test method was developed by using the hydromat test system (HTS), a new test system for applying more realistic distributed loading to two-dimensional sandwich panels, and which has been approved as ASTM standard D 6416/D6416M. This test method can be used to compare the two-dimensional flexural stiffness of a sandwich composite. Since it is based on distributed loading rather than concentrated loading, it may also provide more realistic information on the failure mechanisms of sandwich structures loaded in a similar manner. Properties that may be obtained from this test method include panel surface deflection at load, face-sheet strain at load, bending stiffness, core shear stiffness, strength, and failure modes. The test fixture uses a relatively large square panel sample which is simply supported all around and has the distributed load provided by a water-filled bladder as shown in a figure of ASTM D6416/D6416M.[47]

The test specimen shall be a uniform sandwich composite plate structure with a square perimeter and constant thickness. The average thickness as measured to the nearest 0.025 mm at the center of each edge. There should be no more than a ±2% variation in the thickness of each edge with respect to the average thickness. Specimen length and width should be 1.017 times the support span with a tolerance of ±0.0025a (a – support span). The difference between the length of the two opposite diagonals (measured from corner to corner) should be less than or equal to 0.005a.

Sample for the test is simply supported on all four edges and uniformly loaded over a portion of its surface by a water-filled bladder. Pressure on the panel is increased by moving the platens of the test frame. The test measures the two-dimensional flexural response of a sandwich composite plate in terms of deflections and strains when subjected to a well-defined distributed load. In the test process, load is monitored by both a crosshead-mounted load cell, in series with the test fixture, and a pressure transducer in the pressure bladder itself. Since the pressure bladder is also at all times in series with the load cell and test fixture, the effective contact area of the pressure field is continuously monitored as the load/pressure quotient. Strain can be monitored with strategically placed strain gage rosettes bonded to the tension-side face-sheet surface. A typical arrangement has four rosettes equally spaced along one of the axes of symmetry of the plate. The details of the test fixture and pressure blade can refer to the ASTM standard.

Test panels are equipped with line load diffuser strips, as shown in Figure 6.38. The purpose of the line load diffuser strips is to distribute the line loads from the support frame to avoid core crushing and face sheet damage. Each panel tested is placed in the lower panel support frame, which is bolted to the upper panel support frame with four corner bolts. The detail testing procedures can be referred to the standard.

The Hydromat system can be used for both static and fatigue testing. In the static tests, in addition to the direct measurement of deflections and strains in a sandwich panel, the test device can be used to obtain the in situ bending and shear stiffness of the sandwich panel using a combined analytical/experimental approach.

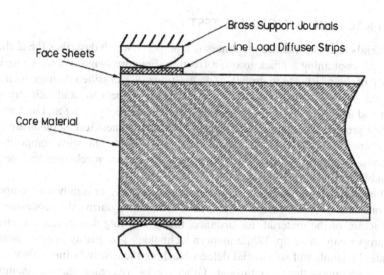

FIGURE 6.38 Sandwich sample plate section and edge support detail.[48]

Experimentally, the deflection is measured at a selected point x_d, y_d, in regions of low gradients, for example, near or at the center of the panel. The sum of the normal strain components ($\epsilon_x + \epsilon_y$) is measured at a second selected point x_s, y_s. Therefore,

$$\omega_e = \frac{C_1}{B} + \frac{C_2}{S} \qquad (6.44)$$

$$(\epsilon_x + \epsilon_y)_e = \left(\frac{c}{2} + f\right)\frac{C_2}{B} \qquad (6.45)$$

where
  ω_e = experimentally determined deflection at x_d, y_d,
  $(\epsilon_x + \epsilon_y)_e$ = experimentally determined sum of the normal strains at x_s, y_s,
  B = bending stiffness,
  S = shear stiffness,
  c = core thickness,
  f = face sheet thickness, and
  C_1, C_2 = constants resulting from the Navier solution.

Equations 6.44 and 6.45 can be solved sequentially for B and S. With $(\epsilon_x + \epsilon_y)_e$ known from experiment, Equation 6.45 can be solved for B and with ω_e, B known, Equation 6.64 can be solved for S.

Hydromat system can be used for a deflection-controlled compression-compression fatigue test. Setting up deflection ratio $r = D_{max}/D_f$ (as Equation 6.38) in different values, such as 0.30–0.70, and using typical compression-compression cyclic load shown in Figure 6.37, the hydromat system fatigue can be performed for evaluating rigid sandwich composite panels loaded with a uniform force.

6.6 FRACTURE TOUGHNESS TEST

In materials science, fracture toughness is a property which describes the ability of a material containing a crack to resist fracture. Fracture toughness tests provide a measure of the resistance to growth of an existing crack or called damage tolerance. The field of the damage tolerance is an imperative aspect to sandwich structural design and analysis. The damage tolerance focuses on recognizing and understanding crack growth. Although well-accepted fracture toughness test methods are available for composite materials, the same cannot be said for sandwich composites. In fact, considerably less attention has been given to fracture mechanics test methods for sandwich composites until recently.

All material systems, such as wood, metal, plastic, or sandwich composite, are prone to flaws and cracks. Some cracks are created during the processing and manufacture of the material, for instance, micro-cracking due to residual stresses or improper cure or layup. While modern technology and quality control measures mitigate the number of substantial defects reaching the assembly line, many microscopic cracks make their way through. Other cracks are created during assembly or service life; these include poor machining or impact. Regardless of how the crack is initiated, when one is detected, it must be either fixed, typically at high cost, or monitored to ensure that it doesn't grow and threaten the integrity of the structure.

The best way to describe the complexities associated with fracture toughness testing of sandwich composites is to begin with the existing test method for unidirectional composite materials – the double cantilever beam (DCB) test method, standardized as ASTM D5528. Figure 6.39 shows a conventional DCB specimen and the DCB test configuration adapted to a sandwich composite. For the sandwich DCB specimen, however, the initial crack is made by intention and placed at the location of interest, near the face sheet/core interface. Unfortunately, testing and analysis of this sandwich DCB configuration has identified several shortcomings. Although the starter crack is placed near the interface, crack growth is often into the core – an undesirable result. Additionally, the large specimen rotation introduces considerable tensile stress into the core and can cause core failure that is called core kinking as shown in Figure 6.40.

In view of the problems associated with adapting the DCB test configuration to sandwich composites, other test configurations have been looked at. One of the most

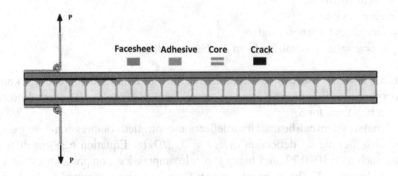

FIGURE 6.39 Schematic of a double cantilever beam sandwich specimen.[49]

FIGURE 6.40 Double cantilever beam sandwich test with large rotation and core kinking failure.[48]

promising is the single cantilever beam (SCB) configuration shown in Figure 6.41. To prevent bending deformation in the core and bottom face sheet, the bottom face sheet is affixed to a lower support plate, so that only the upper face sheet is considered a cantilever beam. An upward load is passed through a hinge bonded to the upper face sheet on the specimen's cracked end. Based on a detailed experimental and numerical investigation, the SCB test configuration has been selected for Mode I standard facture toughness test by the standard committee.

The static test procedure of SCB is based very closely on ASTM standard D5528 for Mode I Interlaminar Fracture Toughness of Unidirectional Fiber-Reinforced Polymer Matrix Composites. Excluding the test configuration (SCB versus DCB) and data reduction method, the test procedure is nearly identical. Three distinct values of strain energy release rate (GIC) were determined from the sandwich SCB test. The first and most conservative is the nonlinear (NL) GIC, based on when the load versus displacement curve deviates from linearity. The second and most subjective is the visual (VIS) GIC, which corresponds to when the crack first propagates. Lastly the 5% offset or max load (5%/Max) GIC which is determined by plotting the intersection of the load versus displacement curve and a 5% offset curve or taking the max load (whichever occurs first). These three values as well as all the propagation points are then plotted producing a resistance curve.

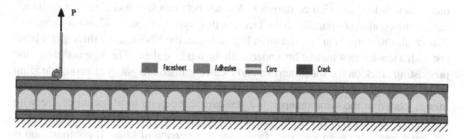

FIGURE 6.41 Schematic of a single cantilever beam sandwich specimen.[49]

The fracture toughness is measured using the strain energy release rate, as it is the most applicable with the specimen's test configuration and geometry, which is determined by Equation 6.46:

$$G_{IC} = \frac{3P\delta}{2ba} \qquad (6.46)$$

where
δ = End Displacement of the Cantilever Beam,
a = Crack Length,
b = Specimen Width,
G_{IC} = Mode I Fracture Toughness, and
P = Load.

Furthermore, the fracture toughness is often represented graphically by a resistance curve. A resistance curve will manifest in two primary forms. A level resistance curve, which represents brittle materials, has no stable crack growth and once the critical load is reach the crack propagates rapidly. A rising resistance curve, which represents more ductile materials, will experience slow stable crack growth until the critical load is reached. In either case the strain energy release rate that corresponds to the critical load is considered the materials fracture toughness.

Failure modes of fracture toughness Mode I tests include adhesive interface disbond and adhesive pullout failure, tensile core failure and tensile core pullout failure, and face sheet failure.

Mode I fracture toughness test is also can be used for performing a fatigue test for evaluating the crack tolerance and propagation at repeatedly force. An example of fatigue testing was performed at 4 Hz and at an R-ratio (P_{min}/P_{max}) of 0.1. The goal was to obtain da/dN versus ΔG curves for the parameters studied. Displacement control was used because the loads were relatively small for control feedback. The load was allowed to drop 10% before the test was stopped and restarted. At the beginning and end of each segment of cycles, peak loads and compliance were measured for use in analysis, and periodic visual measurements were taken.[50]

End Notched Sandwich (ENS) test configuration was selected for Mode II fracture mechanics testing of sandwich composites. This sandwich test configuration was motivated by the three-point end notch flexure test for monolithic composites, as well as modifications for sandwich composites. A crack between skin and core of a sandwich panel situated along or parallel to the faces, as displayed in Figure 6.42, is subjected to a state of almost pure shear. As shown in Figure 6.42a, the ENS test is a three-point bend test with a tensioned wire that fits underneath the top face sheet. The wire is slid into the preexisting crack on the specimen end and located immediately above the outer loading point. The height of the tensioned wire is positioned such that when the sandwich specimen is loaded in the same manner as a standard three-point bend specimen, the wire maintains separation between the face sheet and core and prevents the introduction of frictional forces between the crack faces. Another version of Mode II configuration is shown in Figure 6.42b, which is a modified cracked sandwich beam (CSB) with hinge.

Characterizations of Sandwich Structures

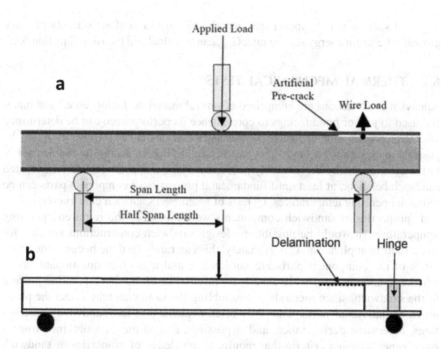

FIGURE 6.42 ENS with a wire (a) and CSB with hinge (b) test configurations.[51]

The procedure of the ENS test is similar to a three-point flexural test. Load vs. deflection curves can be recorded in the ENS test firstly as shown in Figure 6.42. Stable, semi-stable or unstable crack growth behavior can be determined by analyzing the curves as presented in Figure 6.43. Core type and properties, specimen dimensioning (i.e. specimen width W, specimen thickness t, and span length, L), and crack

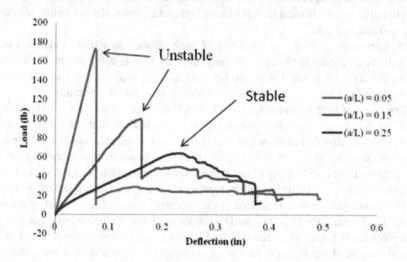

FIGURE 6.43 Load vs. deflection of ENS test and effect of precrack length on crack growth stability (a/L = precrack length/support span).[51]

length of specimen and support span length and their ratio affect stability of crack growth. The strain energy release rate, G_{IIC}, can be calculated by using Equation 6.46.

6.7 THERMAL MECHANICAL TESTS

Sandwich constructions are comprised of several materials, facing, cores, and material used to join or bond facings to cores; hence its performance can be determined after due consideration of structural properties of each material. However, it may be possible in assembling a sandwich construction to somewhat alter the characteristics of the components by the assembly processes. Thus, it appears logical to evaluated sandwich behavior, at least until fundamental properties of component parts can be found at operating temperatures, by tests of assembled sandwich construction.

If properties of sandwich components were known at the required operating temperatures, it would be possible to design sandwich constructions suitable for any structure application. Unfortunately, this can rarely be done because the properties of the component parts are not known, and it is often questionable as to whether component parts are exactly representative of the corresponding portion of the sandwich, since methods of assembling the sandwich may affect the properties of individual parts. Thus it is often necessary to test sandwich constructions, determine performance, and if possible, extend the data obtained toward more general design criteria that require a knowledge of properties of sandwich components.

What tests should be performed on structural sandwich constructions? From consideration of the primary use for sandwich construction, most of sandwich structures are used under flexural load scenarios. The flatwise flexural test can comprehensively evaluate the sandwich structure through skin tension and compression, core compression and shear, and adhesion strength between core and skins. Also the beam flexure tests appear to be the simplest types of tests, particularly at elevated temperatures, so the flexural tests at elevated temperatures on the sandwich structures will be discussed next.

In order to perform the flexural test of the sandwich structure at elevated temperature, two heating methods are usually adopted. One is using thermal laboratory ovens, and another method is to use infrared lamp to heat the sample. The testing examples using these two heating devices will be introduced separately.

Two types of heating ovens for testing at elevated temperatures are available: one is called thermcraft laboratory ovens for materials testing, and the other is called environmental chamber as shown in Figure 6.44. The latter can also be used for tests at below room temperature. The heating oven or environmental chambers enable testers to perform a wide array of mechanical tests in conditions that closely replicate the temperatures that materials and components experience in their actual operating environments, providing confidence in the operating range of a material in the real world. The heating device is mounted on the lower platen of a testing machine, which is an electrically heated or infrared heated furnace, and the tests are conducted in the central portion of the furnace. Loads are transmitted to the test specimens through stainless steel pipe extending from the testing machine platens through holes in the top and bottom of the furnace.

Characterizations of Sandwich Structures 241

FIGURE 6.44 Thermcraft laboratory ovens for materials testing (a)[52] and environmental chamber (b).[53]

To ensure accurate and consistent results, the heating chambers are designed to maintain a constant temperature within a few degrees of the desired setting and with very little temperature gradient across the specimen. Heating is performance with different temperature ranges. Forced convection heating ensures rapid heat transfer and overshoot protection. For environmental chamber, liquid nitrogen is used for providing efficient cooling. These chambers have precise temperature management and optional remote set-point control to regulate temperature from the system PC. The fan, its blades and a baffle help ensure uniform temperatures throughout the chamber while also shielding the specimen from direct exposure to radiant heat.

An example of performing a flexural test on sandwich panel in a heating oven is represented below. The composite sandwich panel used in this test was made up of glass fiber composite skins co-cured onto the modified phenolic core material using a toughened phenol formaldehyde resin. The top and the bottom skins of the sandwich panels was made up of 2 plies of stitched bi-axial (0/90) E-CR glass fabrics with a chopped strand mat and a toughened phenol formaldehyde resin was used as the laminate matrix.

In the testing practice, three point static bending tests were set up in a heating chamber as shown in Figure 6.45 at elevated temperature under nine different temperature ranges from 21°C to 180°C namely 21°C, 35°C, 50°C, 65°C, 80°C, 100°C, 120°C, 150°C, and 180°C. Test specimens with the dimensions of 220 mm × 50 mm × 20 mm were simply supported on two ends and the load was applied at the mid-point through a 100 KN servo-hydraulic loading machine. The effects of temperature are applied through an environmental chamber. The specimens were separated by a span of 180 mm for all tests. The rate of loading was 5 mm/min. A set of five specimens were tested for each temperature range.

Figure 6.46 shows the load – deflection diagram of the sandwich panel specimens tested for their flexure at elevated temperatures at left and chart of flexural modulus vs. testing temperatures at right from the testing practice.

The test revealed that the modulus of the sandwich panels steadily decreased until 120°C attributed to the effect of the temperature on the plastic foam core. A sudden decrease in the modulus is observed at 150°C and 180°C, which is attributed to the phenomenon of the decomposition of the resin in skins and core. The sandwich panel

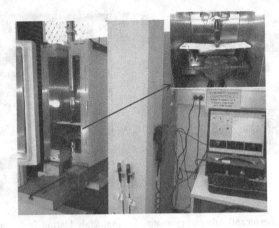

FIGURE 6.45 Flexure test with thermal chamber setup.[54]

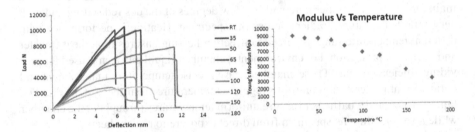

FIGURE 6.46 Deflection diagram at different temperatures (a) and modulus vs. temperatures of flexural tests in the heating chamber (b).[55]

specimens tested at room temperature and 35°C failed in a brittle manner with the failure dominated by the interlaminar shear of the core. The specimens tested at 50°C, 65°C and 80°C exhibited a plastic phase before failing in a brittle manner. The plastic phase exhibited by the sandwich panels is attributed to the effect of temperature on the core of the panel which is characterized by minor indentations on the core and on the top skin below the point of loading. Until 80°C, the skin de-bonds from the core due to the delamination of the individual plies of the skin. The specimens tested at 100°C and 120°C fail in a completely plastic manner which is attributed to the phenomenon of resin softening which is characterized by strong indentations on the top skin and the core at the point of loading. At 150°C and 180°C, the specimens fail in a completely plastic manner similar to the previous case.

At present, polymer foam cored sandwich structures are being used increasingly in applications such as marine structures and wind turbine blades. In service, sandwich structure components are frequently exposed to solar radiation which results in elevated surface temperature in the range of 50°C–100°C. Under these conditions the mechanical properties of the face sheet material (e.g. aluminum alloy or glass fiber-reinforced polymer, GFRP) are barely affected. However, the polymer core materials become much softer, losing both stiffness and strength. As the core material in sandwich structures subjected to bending and shear loading carries transverse shear stresses and stabilizes the face sheets, it is clear that the mechanical behavior of sandwich structures will be affected by the elevated temperature.[57]

When the sandwich structure is subjected to a uniformly elevated temperature, classical sandwich analyses are capable of predicting the mechanical behavior, if the face sheet and core material properties at elevated temperatures are known. However, defining the mechanical behavior becomes much more complex if a temperature gradient exists through the thickness of the material (e.g. solar radiation on only one surface), as the face sheet and particularly the core may possess inhomogeneous mechanical properties.

In one testing practice, the influence of elevated temperature on the stability of sandwich structures is investigated. The sandwich beam tests in the practice were loaded in a simply supported four-point bending configuration, where one of the face sheets was heated. The experimental approach utilized high-speed imaging where the strains were calculated from measured displacements obtained from digital image correlation (DIC). A shift of the failure mode from face sheet yielding to face sheet wrinkling was observed with increasing temperatures. The results from the new analytical method agree well with corresponding experimental results.

Sandwich panels were made by using 0.3 mm aluminum alloy sheet as the face sheet material because of its large heat conductivity and because its stiffness and strength are virtually independent of temperature over the range applicable in this work. The core material was chosen as crosslinked PVC foam. Sandwich panels with a size of 600 mm × 600 mm were manufactured by bonding two aluminum sheets, one 15 mm foam plank and epoxy films as adhesive. The specimens were cut into 50 mm wide, which were positioned on the four-point bending jig. The loading spans L_1 and L_2 (see Figure 6.47) were chosen to be 275 and 100 mm, respectively.

Figure 6.47 shows the top face sheet was heated by an infrared lamp, while the bottom face sheet was exposed to the ambient temperature. Thus a temperature gradient was introduced through the specimen thickness. During the heating process, the side surfaces of the specimen were insulated using polymer foam shields to minimize the heat conduction from the side surfaces to the ambient air. A linear temperature profile through the core thickness was observed after the specimen had reached the state of heat equilibrium. The temperature of the top face sheet (T_t) was controlled by adjusting the power of the infrared lamp and monitored by a thermocouple. Four temperature cases were experimentally studied in this work with the top face sheet heated to 25°C, 50°C, 70°C, and 90°C, respectively. The corresponding bottom face sheet temperatures (Tb) were measured by a thermocouple. It was observed that the bottom face sheet temperature also increased slightly when the top face sheet was heated.

In the practice tests, all the specimens failed by a sudden collapse, where the core near the mid-span was locally crushed with the face sheet folding into the core. The obtained load-deflection curves are shown in Figure 6.48, in which the horizontal axis represents the deflection of the top face sheet at the mid-span and the vertical axis represents the load P (see Figure 6.47). The legend shows the top and bottom face sheet temperatures for each test. It can be seen that the overall stiffness reduces with increasing temperature. This is as expected because the shear deflection is larger at elevated temperatures due to the reduction in the core shear modulus value. Similarly, the maximum load also decreases with rising temperatures, and the reduction is very significant from 70°C to 90°C.

It has been experimentally observed that the specimens failed by face sheet yielding at room temperature and by wrinkling (instability) when the top face sheet is heated to 70°C or higher. This failure mode shift corresponds well with the predictions by other tests.

6.8 NONDESTRUCTIVE TESTS

To ensure sound performance, structural integrity, and safety operation of the diverse group of sandwich structures, nondestructive inspection and testing methods will always be required. NDE results are needed to manage the accept/reject decisions and are used in the planning for repair, overhaul, and refurbishment of in-service structures.

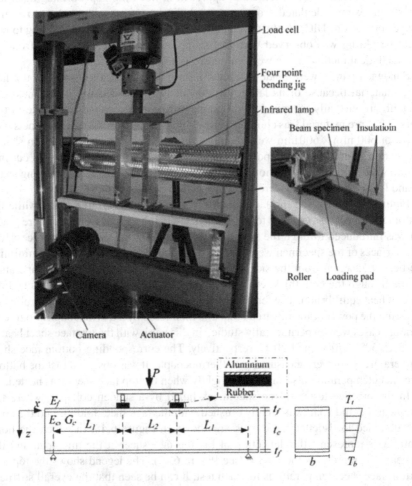

FIGURE 6.47 Schematic of four-point bending test of sandwich beam with infrared lamp and trough thickness temperature gradient.[55]

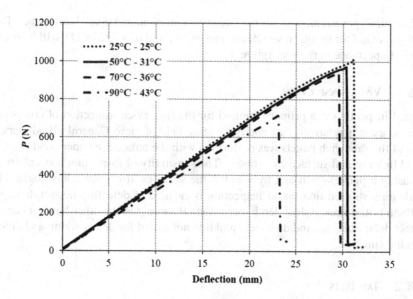

FIGURE 6.48 Load-deflection curves of sandwich beam specimens at different temperatures.[55]

The challenge for the NDE capability will continue to be the ability to not only detect the presence of the defects and damage, but the classification of the nature of the flaw and a quantitative evaluation of the severity of the damage and its effects on the performance of the structure. For sandwich structures, this would include the ability to differentiate between a skin-to-core disbond and a buckled core with the adhesive bond intact.

The targets of inspection include manufacture defects, in-service damage due to human and natural causes, and the inspection of repairs on sandwich structures. The emphases adopted in the development of NDE techniques and instruments for sandwich structures are imaging capability, quantitative inspection, and the ease of use in the field.

The nondestructive evaluation (NDE) of sandwich structures can be divided into two categories: inspection of manufacturing defects for quality assurance purposes and in-service inspection for damage originated from natural or human causes. Manufacturing defects of concern include delamination in the composite face sheet, disbond between the skin and core, foreign object inclusions, and porosity. While in service, sandwich structures can suffer damage or degradation that require periodic inspection or repair. The primary concern in composite sandwiches is damage caused by foreign object impact, especially those caused by low velocity foreign objects that defy visual detection; either invisible or barely visible (BVID). Other damage can be attributed to static overload, fatigue, and environmental factors such as water ingression, heat, chemical, ultraviolet radiation, extreme temperature excursion, and lightning strike. Like honeycomb sandwiches with composite face sheet, the foam-cored sandwich can also suffer substantial internal damage without leaving visible indication on the surface.

Basic methods of nondestructive testing of composites include visual, manual tap test and low-frequency bond test techniques, and ultrasonic, thermovision, radiological, penetrant, eddy current, acoustic emission, and optical techniques. Due to the demand for robots in world, one can expect that robotized NDT will be growing in importance in the near future.

6.8.1 VISUAL INSPECTION

Visual inspection is a primary method for the in-service inspection of composite structures. It is relatively fast and has a large field of view. General visual inspection of the sandwich panels was performed with the naked eye under conditions of good lighting and surface cleanliness. The sensitivity of inspection was enhanced by using a pocket-torch and by viewing the surfaces also from a low angle. The evaluation showed that visual inspection is capable of detecting impact damages with an initial impact dent depth larger than 0.5 mm. On the other hand, it cannot detect delaminations and disbonds, and it is not suited for defect sizing and defect depth estimation.

6.8.2 TAP TESTS

Sandwich structures have historically been inspected with simpler mechanical means of inspection, including manual tap test and low-frequency bond testing techniques. Bond testing techniques generally include mechanical impedance analysis, low-frequency resonance method, and pitch-catch mode.

The practice of tapping a sandwich structure with a mass to detect defect or damage based on the sound it makes is probably the most widely used nondestructive inspection in the field over the years. This inexpensive technique, which is most basic a coin or screw driver handle and one's own ear, is reasonably effective despite its qualitative and subjective nature for both mono-lithic and sandwich composites. For a more instrumented and data driven methods, handheld tap hammers that uses a battery-driven solenoid with a force sensor (accelerometer) built in the hammer tip.[56] Practically, the time during which the hammer is in contact with the surface of the test part is measured. This contact time will increase in areas with defects such as disbonds that lower the local contact stiffness of the part. For global inspection of large surface areas it is less suited because of its spot measurement performance. Impact damages are generally well detectable; only a few minor impacts were missed in the evaluation. The detectability for delaminations and disbonds, on the other hand, was varying and not always consistent. Furthermore, a varying skin thickness (e.g. lay-up differences) or the presence of back-up structure can influence the tap tester response. The capability for defect sizing is limited and the technique is not suited for defect depth estimation.

6.8.3 PITCH-CATCH SWEPT METHOD

Pitch-catch swept method was selected for the evaluation because of its relatively low cost and its multimode inspection capabilities (pitch-catch, mechanical impedance, resonance). The couplant-free pitch-catch technique proved to be the most promising inspection mode.[56] In the evaluation, however, this device showed a limited detection performance for in-service defects; the detectability was also quite varying. Impact damage is the defect type best detectable. Although couplant-free, this method is not well suited for global inspection of large surface areas.

6.8.4 Ultrasonic Tests

Handheld ultrasonic (UT) camera is a handheld ultrasonic imaging camera for fast and real-time UT inspection. It produces a high-resolution C-scan image over an area of the specimen utilizing an array of 120 × 120 piezoelectric sensing elements. The array is responsive over a wide array of ultrasound frequencies, although most imaging is done in the range of 1–7.5 MHz. A separate, "transparent" PVDF sensor in front of the main UT transducer is used to provide a conventional A-scan presentation of the test part with information about the depth of present defects. Both XY-Scanning (C-scan) image and amplitude scan (A-scan) presentation are displayed on the control unit of the camera head.[56] Inspection of using UT camera can indeed be used for relatively fast and real-time smaller areas where damage is suspected. For global inspection of large surface areas, however, the camera is less suited because of the limited field of view (about 1 square inch). A limitation, inherent to ultrasonic testing, is that a couplant is required between the camera and test part. The detectability for defects is generally good, especially when scanning the camera over the test specimen. For honeycomb structures, the camera was somewhat less successful in the evaluation: some disbonds and impacts were only detectable with limitation. The UT camera can further be used for defect sizing and for defect depth estimation when using the A-scan module of the camera.

Ultrasonic (UT) wheel probe is an ultrasonic transducer assembly that allows rolling contact of a transducer over a surface. The transducer assembly is "scanned" over the surface in a more or less straight line. An outer part of the assembly rotates, allowing the wheel to roll over the surface, while an inner part of the assembly holds the ultrasonic transducer at fixed angles relative to the surface. In a wheel probe housing a phased array transducer, the mechanical support for the transducer will be fixed, even though the angle may be scanned electronically. One type of wheel probe uses a water-filled polymer bladder to surround the transducer assembly. This bladder is usually shaped in the form of a small tire attached to the rotating part of the wheel and which rotates as the wheel is scanned over the surface. The tire may be made from polyurethane, silicone, or some more specialized material with improved ultrasonic and mechanical properties. This type of wheel probe can be called a "water-filled wheel probe".

The wheel probe can be used without couplant, but, generally, a fine water spray on the test part is used for optimum coupling. The larger probe is meant for flat surfaces, the smaller one can also be used on slightly curved parts. The transducer frequency can be selected as 2, 5, or 10 MHz. The handheld roller probe together with a 7-axis Faro scanning arm can be used for fast and real-time UT inspection of relatively large areas. The wheel probe is very suitable for the in-service inspection of sandwich composites. The detectability for defects such as delaminations, disbonds, and impact damages is excellent. A limitation of the

UT wheel probe can be the relatively high cost of a complete inspection system including scanning arm. The portability of the system may limit its use for inspection areas with limited access.

6.8.5 Shearography Test

Shearography is an optical method, based on speckle interferometry, for the noncontact measurement of out-of-plane deformations of a material surface. The method has been developed in particular to overcome the sensitivity to external vibrations that is common to standard interferometry techniques. This is achieved by not using a separate reference beam. Instead, the returning object beam is doubly imaged, with one of the coherent images slightly shifted or "sheared" relative to the other image. Then, a second similar recording is made with the object put under a slight strain. The two speckle patterns are superimposed resulting in a fringe pattern. The fringes do not show the contours of the displacement but of the derivative of the displacement (gradient of deformation). Digital image processing of the data is further done to enhance the defect presentation (e.g. filtering and fringe unwrapping).

The device of shearography inspection can be found as a mobile, portable shearography system suitable for in service inspections. Thermal load can be applied by heating lamps, and vacuum load can be applied by connecting a vacuum hood which sucks directly to the surface to be inspected. A practice evaluation showed that shearography is a relatively fast, noncontact technique that requires no coupling or complex scanning equipment. Because of the optical technique the specimen should not have a shiny surface (standard coating is acceptable). The inspection time is largely determined by the limitations of the field of view. The impact damages were readily detectable (including the nonvisible impacts). The detectable defect size decreases with increasing defect depth (defect diameter must exceed its depth). Shearography can be used with limitation for defect sizing, but the technique is not suited for defect depth estimation. The optimum loading technique depends on the specific inspection configuration but, in general, defects at larger depth are better detectable with vacuum loading. Shearography inspection seems most promising for the inspection of honeycomb sandwich structures.

6.8.6 Infrared Thermography Method

Infrared thermography is a noncontact NDI method that monitors the heat radiation pattern on the surface of a test part. The method employs light just above the visible part of the electromagnetic spectrum, in the range of about 2–14 µm. Passive and active IR techniques can be distinguished but for NDI purposes only the active technique is used: the object is here with excited either by an external heat source or by mechanical vibrations. Material defects are then detectable by the corresponding anomalies in the heat distribution pattern on the surface of the test part. Thermographic techniques are well applicable to composite materials because of their relatively low thermal conductivity which implies a slow lateral heat flow with closely spaced isotherms, resulting in a good defect resolution. The detectable flaw size is in general larger than the depth of the flaw. The evaluation showed that thermography is a fast, global, and noncontact method that requires no coupling or complex scanning equipment.[57]

This technique can be used with two experimental configurations. On the one hand, the transmission setup consists in having the heat source on the opposite side of the camera; the measure obtained is then the heat flow which has passed through the material and which is perturbed by internal defects. For the reflection device on the other hand, the heat source and the camera stand on the same side; the thermal mapping given by the camera corresponds therefore to the flow reflected by internal heterogeneities and the material surface which increases the measurement noise. In both cases, the analysis of thermal mapping enables to highlight defects within the composite material.

The thermography is a fast, global, and noncontact method that requires no coupling or complex scanning equipment. Panels with a surface area up to $1\,m^2$ can be inspected with a single exposure technique. The detectable defect size decreases, as for shearography, with increasing defect depth. Most impact damages were readily detectable, except for some smaller and nonvisible impacts. The detectability for disbonds was somewhat better: the larger skin-to-stiffener disbonds and skin-to-honeycomb core disbonds were readily detected in the evaluation. Thermography can be used for defect sizing, but the technique is not suited for defect depth estimation. Although not considered in this evaluation, thermography seems promising for the inspection of water ingress in sandwich composite structures.

6.8.7 INDUSTRIAL-SCALE INSPECTION

Critical sandwich composite structures must pass through nondestructive inspection process before assembled onto an application. The primary method used by the manufacturers is ultrasonic scanning. The inspection is typically carried out in a scan gantry with squirters in a through transmission ultrasonic (TTU) mode. Two water jets carry the ultrasonic beam from the transmitting transducer to the solid laminate or honeycomb sandwich structures, through the structure and then to the receiving transducer. The scan images produced by the ultrasonic squirter system can reveal the presence of foreign object inclusions, ply to ply delaminations, skin-to-core disbonds, and regions of excessive porosity. TTU scan images of solid laminates based on the amplitude of the transmitted ultrasonic signal (or "dB drop") are often used in the estimation of the porosity volume fraction.

Although the squirter-operated, water-coupled ultrasonic imaging using TTU scan gantries remains the prevalent method of quality assurance inspection for manufacturing of composite parts, air-coupled ultrasonic scan has become increasingly practical. Air-coupled ultrasound, being noncontact and noncontaminating, is particularly suited for inspecting honeycomb sandwiches with perforated face sheets and other water-sensitive composite structures.[58]

REFERENCES

1. D. Adams, "Sandwich panel test methods," *Composites World*, Sept. 1, 2006.
2. S. Khan, "Bonding Of Sandwich Structures - The Facesheet/Honeycomb Interface - A Phenomenological Study," E.I. DuPont de Nemours Co., Inc., Advanced Fibers System, http://www.foradenizcilik.com/kutuphane/wp.pdf.

3. P. Joyce, "Mechanical Testing of Composites," https://www.usna.edu/Users/mecheng/pjoyce/composites/Short_Course_2003/12_PAX_Short_Course_Mechanical-Testing.pdf.
4. http://www.gatewaymaterialstestcenter.com/.
5. A. Jedral, "Review of testing methods dedicated for sandwich structures with honeycomb core," *Transactions on Aerospace Research*, vol. 2, no. 255, pp. 7–20, 2019.
6. J.R. Correia, M. Garrido, J.A. Gonilha, F.A. Branco and L.G. Reis, "GFRP sandwich panels with PU foam and PP honeycomb cores for civil engineering structural applications," *International Journal of Structural Integrity*, vol. 3, no. 2, pp. 127–14725, May 2012.
7. http://www.materialstandard.com/.
8. W.J. Cantwell, and P. Davies, "A study of skin-core adhesion in glass fibre reinforced sandwich materials," *Applied Composite Materials*, vol. 3, no. 6, pp. 407–420, Nov 1996.
9. M. R. L. Gower and G. D. Sims, "Determining the Skin-to-Core Bond Strength of Sandwich Constructions," http://eprintpublications.npl.co.uk/3243/1/CMMT_MN44.pdf.
10. Subcommittee D30.09, "ASTM C393/C393M-06, Standard Test Method for Core Shear Properties of Sandwich Constructions by Beam Flexure," *ASTM International*, 2006.
11. H. Cuypers and J. Wastiels, "Analysis and verification of the performance of sandwich panels with textile reinforced concrete faces," *Journal of Sandwich Structures and Materials*, vol.13, no. 5, pp. 589–603, 2011.
12. A. McCracken and P. Sadeghian, "Partial-composite behavior of sandwich beams composed of fiberglass facesheets and woven fabric core," *Thin-walled structures*, vol. 131, pp. 805–815, Oct. 2018.
13. Subcommittee D30.09, "ASTM D7249/7249M-12, "Standard Test Method for Facing Properties of Sandwich Constructions by Long Beam Flexure," *ASTM International*, 2012.
14. Subcommittee D30.09, "ASTM D7250-06, Standard Practice for Determining Sandwich Beam Flexural and Shear Stiffness," *ASTM International*, 2006.
15. P. Sharafia, S. Nematia, B. Samalia, A. Mousavib, S. Khakpourc, and Y. Aliabadizadehd, "Edgewise and flatwise compressive behaviour of foam-filled sandwich panels with 3-D high density polyethylene skins," *Engineering Solid Mechanics*, vol. 6, no.3, pp. 285–298, 2018.
16. R.S. Jayaram, V.A. Nagarajan and K.P. Vinod Kumar, "Mechanical performance of polyester pin-reinforced foam filled honeycomb sandwich panels," *Science and Engineering of Composite Materials*, vol. 25, no 4, pp. 797–805, 2018.
17. Subcommittee D30.09, "ASTM C364/C364M-07 Standard Test Method for Edgewise Compressive Strength of Sandwich Constructions," *ASTM International*, 2005.
18. Sandwich Panel Edgewise Compression Test Fixture, https://wyomingtestfixtures.com/products/.
19. H. Lei, Y. Kai, W. Wen, H. Zhou and D. Fang, "Experimental and numerical investigation on the crushing behavior of sandwich composite under edgewise compression loading," *Composites Part B: Engineering*, vol. 94, no. 1, pp. 34–44, 2016.
20. N. Fathi, "Study of an axial loaded sandwich panel," Master degree thesis," Lulea University of Technology, 2017.
21. J. A. Tafoya, "Effect of Sustainable and Composite Materials on the Mechanical Behavior of Sandwich Panels under Edgewise Compressive Loading," M. S. thesis, California Polytechnic State University, San Luis Obispo, CA, 2015.
22. T. Castilho, "Impact Resistance of Marine Sandwich Structures Impact tests," M. S. Thesis, Instituto Superior Técnico, Universidade de Lisboa, 2014.
23. Subcommittee D30.09, "ASTM D7766/D7766M-11 Standard Practice for Damage Resistance Testing of Sandwich Constructions," ASTM International, 2011.
24. C. Muscat-Fenech, J. Cortis and C. Cassar, "Impact damage testing on composite marine sandwich panels. Part 2: Instrumented drop weight," *Journal of Sandwich Structures and Materials*, vol. 16, no. 5, pp. 443–480, 2014.

25. C. Muscat-Fenech, J. Cortis and C. Cassar, "Impact damage testing on composite marine sandwich panels, part 1: Quasi-static indentation," *Journal of Sandwich Structures and Materials*, vol. 16, no. 4, pp. 341–376, 2014.
26. Subcommittee D30.09, "ASTM D7136/D7136M-2012 Standard Test Method for Measuring the Damage Resistance of a Fiber-Reinforced Polymer Matrix Composite to a Drop-Weight Impact Event," ASTM International, 2012.
27. V. Dikshit, A. Nagalingam, Y. Yap, S. Sing, W. Yeong, and J. Wei, "Investigation of quasi-static indentation response of inkjet printed sandwich structures under various indenter geometries," *Materials*, vol. 10, no. 3, p. 290, Mar 2017.
28. S. Salman, Z. Leman, M. Ishak, M. Sultan and F. Cardona, "Quasi-static penetration behavior of plain woven kenaf/aramid reinforced polyvinyl butyl hybrid laminates," *Journal of Industrial Textiles*, vol. 47, no, 7, pp. 1427–1446, 2017.
29. Subcommittee D30.09, "ASTM D6264/D6264M-04, Standard Test Method for Measuring the Damage Resistance of a Fiber-Reinforced Polymer-Matrix Composite to a Concentrated Quasi-Static Indentation Force," ASTM International, 2004.
30. M. Bulut and A. Erklig, "The investigation of quasi-static indentation effect on laminated hybrid composite plates," *Mechanics of Materials: An International Journal*, vol. 117, pp. 225–234, Feb. 2018.
31. D. Adam, "Damage-resistance testing of composites," *Composites World*, Oct. 3, 2016.
32. M. Battley and S. Lake, "Dynamic Performance of Sandwich Core Materials," 16th *International Conference on Composite Materials*, 2007. http://www.iccm-central.org/Proceedings/ICCM16proceedings/contents/pdf/ThuJ/ThJA1-04ge_battleym223234p.pdf.
33. S. C. Mamani, "Experimental Study of Wave Slamming of Sandwich Composites Panels," Ph.D. thesis, University Of Puerto Rico Mayagüez Campus, 2009.
34. S. Tekalur, K. Shivakumar and A. Shukla, "Mechanical behavior and damage evolution in E-glass vinyl ester and carbon composites subjected to static and blast loads," *Composites. Part B, Engineering*, vol. 39, no. 1, pp. 57–65, Jan. 2008.
35. D. Xiang, J. Rong, and X. He, "Experimental investigation of dynamic response and deformation of aluminum honeycomb sandwich panels subjected to underwater impulsive loads," *Shock and Vibration*, vol. 2015, 2015. https://doi.org/10.1155/2015/650167.
36. S. Avachat, "Design of Composite Structures for Blast Mitigation," Ph. D. degree Dissertation, Georgia Tech, Atlanta, GA, 2015.
37. H. Arora, P. Del Linz and J. P. Dear, "Damage and deformation in composite sandwich panels exposed to multiple and single explosive blasts," *International Journal of Impact Engineering*, vol. 104, Jun., pp. 95–1060, 2017.
38. H. Arora, P. Hooper and J. P. Dear, "Dynamic response of full-scale sandwich composite structures subject to air-blast loading," *Composites Part A*, vol. 42, no. 11, pp. 1651–1662, 2011.
39. A. P. Hooper and J. P. Dear, "The effects of air and underwater blast on composite sandwich panels and tubular laminate structures," *Experimental Mechanics*, vol. 52, no. 1, pp. 59–81, 2011.
40. N. Sharma, R. Gibson and E. Ayorinde, "Fatigue of foam and honeycomb core composite sandwich structures: A tutorial," *Journal of Sandwich Structures and Materials*, vol. 8, no. 4, pp. 263–319, 2006.
41. K. Feichtinger, W. Ma and T. Touzot, "Performance Evaluation of Core Materials Used for Structural Sandwich Marine Transoms," *COMPOSITES* 2003 *Convention and Trade Show*, Composites Fabricators Association, Oct. 2003.
42. D. Roosen, "Fatigue Behaviour of Sandwich Foam Core Materials – Comparison of Different Core Materials," https://www.rohacell.com/product/.
43. R. Gedney, "Materials Characterization & Nondestructive Testing, Stress-Life Fatigue Testing Basics," https://www.industrialheating.com/articles/94569-stress-life-fatigue-testing-basics, Nov. 2018.

44. S. D. Clark, R. A. Shenoi and H. G. Allen, "Modelling the fatigue behaviour of sandwich beams under monotonic, 2-step and block-loading regimes," *Composites Science and Technology*, vol. 59, no. 4, pp. 471–486, 1999.
45. S.V. Rocca and A. Nanni, "Mechanical characterization of sandwich structure comprised of glass fiber reinforced core: Part 1," *Composites in Construction*, Third International Conference Lyon, France, July 11–13, 2005.
46. A. M. Al-Sharif, Effect of Impact Damage on Compression - Compression Fatigue Behavior of Sandwich Composites, Dissertation for the Degree of Doctor of Philosophy, Ph. D. Dissertation, Wayne State University, Detroit, MI, 2015.
47. Subcommittee D30.09, "ASTM D6416 / D6416M-16 Standard Test Method for Two-Dimensional Flexural Properties of Simply Supported Sandwich Composite Plates Subjected to a Distributed Load," ASTM International, 2016.
48. P. Melrose, "Elastic Properties of Sandwich Composite Panels Using 3-D Digital Image Correlation with the Hydromat Test System," M. S. Thesis, University of Maine, Orono, ME, 2004.
49. S. P. Denning, "Fracture Mechanics of Sandwich Structures," M. S. Thesis, Wichita State University, Wichita, KS, 2010.
50. C. Berkowitz, W. Johnson, "Fracture and fatigue tests and analysis of composite sandwich structure", *Journal of Composite Materials*, vol. 39, no. 16, pp.1417–1431, 2005.
51. D. O. Adams, J. Nelson, Z. Bluth, and C. Hansen, "Development and Evaluation of Fracture Mechanics Test Methods for Sandwich Composites," https://depts.washington.edu/amtas/events/jams_12/papers/paper-adams_sandwich.pdf.
52. https://thermcraftinc.com.
53. https://www.mts.com.
54. S. Surendar, "Flexural Behaviour of Sandwich Panels under Elevated Temperature," M. S. dissertation, University of Southern Queensland, Darling Heights, Australia, 2014.
55. S. Zhang, J.M. Dulieu-Bartonand and O.T. Thomsen, "The effect of temperature on the failure modes of polymer foam cored sandwich structures", *Composite Structures*, vol. 121, pp. 104–113, 2015.
56. J. Heida and D. Platenkamp, "Evaluation of Non-Destructive Inspection Methods for Composite Aerospace Structures," International Workshop of NDT Experts, Prague, 10–12 Oct. 2011.
57. E. Péronnet, M. Pastor, R. Huillery, O. Dalverny, S. Mistou and H. Welemane, "Nondestructive investigation of defects in composite structures by three infrared thermographic techniques," *International Conference on Experimental Mechanics ICEM 15*, 22–27, July 2012, Porto, Portugal.
58. D. K. Hsu, "nondestructive inspection of composite structures: Methods and practice," *17th World Conference on Nondestructive Testing*, 25–28 Oct. 2008, Shanghai, China.

7 Sandwich Structure Design and Mechanical Property Analysis

Wenguang Ma

Russell Elkin

CONTENTS

7.1 Structure and Load Distribution of Sandwich Composites 254
    7.1.1 Rigidity, Stress, and Deflection of a Sandwich Beam Subjected to Bending Moment 256
        7.1.1.1 Flexural Rigidity and Shear Rigidity 256
        7.1.1.2 Face Tensile/Compressive Stress 257
        7.1.1.3 Core Shear Stress 258
        7.1.1.4 Deflection of Sandwich Beam Subjected to Bending Moment and Shear Force Caused by Bending 258
    7.1.2 Stress, Strain, and Rigidity in a Sandwich Beam Subjected to Tension or Compression 259
7.2 Strength and Deflection of Simple Sandwich Elements at Different Supports and Loads 261
7.3 Edgewise Damage Prediction and Prevention 269
    7.3.1 General Buckling 269
    7.3.2 Local Buckling in the Sandwich 270
7.4 Design Principles of a Simple Sandwich Element 272
    7.4.1 Design for Strength of Facings and Core 272
    7.4.2 Design for Rigidity 273
    7.4.3 Design for Minimum Weight 274
        7.4.3.1 Minimum Weight for Given Stiffness 274
        7.4.3.2 Minimum Weight for Given Strength 277
    7.4.4 Other Design Considerations 277
7.5 Design Procedure from Simple Element to Large Complex Structure 278
    7.5.1 Determine Thicknesses of Simple Sandwich Element 278
    7.5.2 Design Routine of Simple Sandwich Element 279
    7.5.3 Scaling Up to Large Complex Structure 280
        7.5.3.1 Multilevel Scaling 280

DOI: 10.1201/9781003035374-7

7.5.3.2 Similarity Theory Scaling 281
7.6 Effects of Core Formats, Process Methods, and Application Conditions on Design Parameters 283
    7.6.1 Influence of the Core Formats on Design Parameters 283
    7.6.2 Influence of Laminating Methods on Design Parameters 285
    7.6.3 Influence of Product Application Conditions on Design Parameters ... 286
7.7 Principles and Examples of Using Hybrid Cores, Regional Reinforcements, Transitions, and Connection Elements 287
    7.7.1 Use Intelligent Combinations of Hybrid Core Materials for a Large Complex Product 287
    7.7.2 Regional (Local) Reinforcements 288
    7.7.3 Fasteners and Connections 289
        7.7.3.1 Connect to Solid Structure 289
        7.7.3.2 Straight Connect Sandwich Structures 289
        7.7.3.3 Right Angle Connection 289
        7.7.3.4 T-Joint Connection 291
        7.7.3.5 Fasten Sandwich Panel to Solid Structural Component 291
References 292

Predicting mechanical properties, such as rigidity, stress, displacement, etc., under flexural, tensile, compressive, or shear load, is an important step for designing a sandwich construction. The design concept is to optimize a sandwich structure with a high ratio of mechanical properties to weight and other best interesting properties. This chapter starts from brief examinations of stress and deflection for basic loading to forecasting damages and preventing failures of bending, shearing, and edgewise buckling by using the simplified mathematic equations provided in the article, the mechanical properties of the core and facing, geometric size of the product, and loading style. Then the design procedures are introduced from the basic sandwich element to large and complex structures. Strategy routines include multilevel and similarity theory scaling techniques. The chapter also highlights considerations for raw material format, lamination process, and application environmental conditions. Finally, a section on the use of hybrid solutions, multiple cores, and facing reinforcing materials for one optimized design, as well as concerns for the transitions, connections, and fasteners is provided.

7.1 STRUCTURE AND LOAD DISTRIBUTION OF SANDWICH COMPOSITES

We know that a sandwich structure, like the "I" beam, consists of strong skins (flanges) bonded to a core (web). The skins are subject to tension/compression and are largely responsible for the strength of the "sandwich". The function of the core is to support the face sheets so that they don't buckle (deform) and stay fixed relative to each other. The core experiences mostly shear stresses (sliding) as well as some degree of vertical tension and compression as shown in Figure 7.1. Its material properties and thickness determine the stiffness of such a panel. Unlike the simple

Sandwich Structure Design

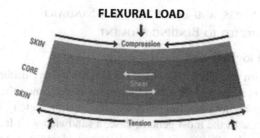

FIGURE 7.1 Sandwich structure under flexural load and deformation.[1]

I beam, which is designed to withstand stresses mostly along one axis and bending about another axis, the sandwich panel can be stressed along and about any axis laying in the plane. The implication is that such panel can extend "infinitely", forming a strong and continuous self-sustaining plate or shell. No reinforcing elements are needed because they are already built into the structure.

Sandwich is a type of stressed-skin construction in which the facings resist nearly all of the applied edgewise (in-plane) loads and flatwise bending moments. The thin-spaced facings provide nearly all of the bending rigidity to the construction. The rigid core spaces the facings and transmits shear between them so that they are effective about a common neutral axis. The core also provides most of the shear rigidity of the sandwich construction.

Figure 7.2 shows a sandwich beam to which a bending moment M, a shear force V, and a normal force N are applied. Obviously, the applied loads produce stress in each section of the sandwich, while straining it. The absolute value of the stresses produced by M, V, and N depends on the value of loads and moments; stress sign depends on the direction of the loads and moments. Therefore, to fully identify the stresses, a convention is necessary to define the direction of M, V, and N, based on the real situation. If the beam is a symmetric structure, which has the same skins on top and bottom and an uniform core, the bending moment M and the shear force V are not directional difference. Otherwise, we need to identify a positive direction. The normal force N is positive when it is a tensile force usually.

The external facings of a sandwich structure are not necessarily equal materials or thickness. In the next sections, mechanical analyses will be conducted by using the following basic hypotheses sometimes: the external facing thickness is negligible compared to the thickness of the core; the modulus of elasticity of the core is negligible compared to that of the facings; normal force and bending moments act on the external facings only; the shear forces act on the core only.

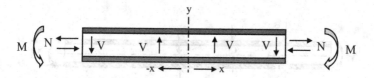

FIGURE 7.2 Sandwich beam subjected to a bending moment M, a shear force V, and a normal force N.

7.1.1 Rigidity, Stress, and Deflection of a Sandwich Beam Subjected to Bending Moment

7.1.1.1 Flexural Rigidity and Shear Rigidity

The sandwich beam in Figure 7.3, which we assume to have different external facings S_{f1} and S_{f2}, is subjected to a bending moment M. Since the core of a sandwich reacts negligibly to a bending moment, the flexural rigidity is entirely due to the external facings. In the most general case, a sandwich with facings of different thicknesses and made of different materials is considered now for calculating the flexural rigidity EI of a sandwich beam. The elastic moduli and the thickness of the two facings are identified as E_{F1}, E_{F2}, S_{F1}, and S_{F2}, respectively. The flexural rigidity is given by

$$EI = b\left(E_{F1}S_{F1}C_1^2 + E_{F2}S_{F2}C_2^2\right) \quad (7.1)$$

In Equation 7.1, b indicates the sandwich width and C_1 and C_2 the distances between the neutral axis of the sandwich and that of face 1 and face 2 as shown in Figure 7.3.[1]

After doing some formula transformations, Equation 7.1 can be represented as follows:

$$EI = bE_{F2}\frac{S_{F1}S_{F2}S_A}{S_{F1} + S_{F2}E_{F2}/E_{F1}}(C_1 + C_2)$$

$$C_1 + C_2 = S_A$$

$$EI = bE_{F2}\frac{S_{F1}S_{F2}S_A^2}{S_{F1} + S_{F2}E_{F2}/E_{F1}} \quad (7.2)$$

$$S_A = \frac{S_{F1} + S_{F2}}{2} + S_A^* \quad (7.3)$$

S_A^* is the core thickness. If the facings of the sandwich have the same elastic modulus $E = E_{F1} = E_{F2}$, Equation 7.2 is simplified:

$$EI = bE\frac{S_{F1}S_{F2}S_A^2}{S_{F1} + S_{F2}} \quad (7.4)$$

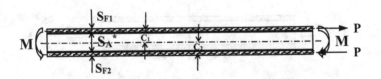

FIGURE 7.3 Sandwich beam subjected to a bending moment M.

Sandwich Structure Design

Finally, if the facings of the sandwich also have equal thickness $S_F = S_{F1} = S_{F2}$:

$$EI = bE \frac{S_F S_A^2}{2} \qquad (7.5)$$

Shear rigidity GA depends on the shape of the sandwich cross-section and on shear modulus of the core, which is given by Equation 7.6.

$$GA = bS_A G_A \qquad (7.6)$$

where G_A is shear modulus of the core, b is the beam width, and S_A is given by Formula 7.3.

7.1.1.2 Face Tensile/Compressive Stress

To calculate the face stresses under bending moment M, we have to find the relationship between them and applied moment. By referencing Figure 7.3, to multiply the stress by the facing area, we obtain the force P acting on the facing[2]:

$$P = bS_{F1,2} \sigma_{F1,2} \qquad (7.7)$$

where b is the beam width. Of course, the two forces acting on the two facings must be equal in absolute value, since their sum must be zero. Moreover, the moment generated by these forces must be equal to the applied moment. Since each P is applied to the center of a facing, their distance is S_A obtained by Equation 7.3:

$$M = PS_A = bS_{F1,2} \sigma_{F1,2} S_A$$

$$\sigma_{F1,2} = \frac{M}{bS_{F1,2} S_A} \qquad (7.8)$$

Formula 7.7 allows for the calculation of the absolute value of stress σ_F in a facing of a sandwich structure subjected to a bending moment M; according to the convention adopted for the moment sign in Figure 7.3, σ_F is positive (tensile) in the upper facing and negative (compression) in the lower facing when the moment is negative. The symbol $\pm$ can be added to Formula 7.7.

Example
  Calculate the stress in the facings of the sandwich with $S_{F1} = 2\,\text{mm}$, $S_{F2} = 1\,\text{mm}$, core thickness $S_A^* = 20\,\text{mm}$, subjected to a moment $M = 4 \times 10^6\,\text{N mm}$ and positive direction. The beam width is $b = 1{,}000\,\text{mm}$.
  From Equation 7.3:

$$S_A = \frac{2+1}{2} + 20 = 21.5\,\text{mm}$$

The moment M is positive; therefore, σ_{F1} is a compression stress, and σ_{F2} is a tensile stress: using Equation 7.8 for the facings

$$\sigma_{F1} = \frac{4 \times 10^6}{1{,}000 \times 2 \times 21.5} = -93\,\text{N/mm}^2$$

$$\sigma_{F2} = \frac{4 \times 10^6}{1{,}000 \times 1 \times 21.5} = +186\,\text{N/mm}^2$$

7.1.1.3 Core Shear Stress

Referencing Figure 7.2, shear force V in sandwich generated by bending moment M is

$$V = dM/dx \qquad (7.9)$$

Maximum shear force is at the interface between facing and core. So shear stress τ_A is given by

$$\tau_A = \frac{V}{bS_A} \qquad (7.10)$$

where b is the beam width, and S_A is given by Equation 7.3.[2]

Example:
A sandwich, made of two facings 1.5 mm thick and a core 20 mm thick, is subjected to a shear force $V = 20 \times 10^3\,\text{N}$. Calculate the shear stress in the core, knowing that the beam is 1,000 mm wide.

From Equation 7.3, $S_A = \dfrac{1.5 + 1.5}{2} + 20 = 21.5\,\text{mm}$

using Equation 7.10, the shear stress τ_A is

$$\tau_A = \frac{20 \times 1000^3}{1{,}000 \times 21.5} = 0.93\,\text{N/mm}^2$$

7.1.1.4 Deflection of Sandwich Beam Subjected to Bending Moment and Shear Force Caused by Bending

Considering the calculation of the deflection y_m of the beam due to the bending moment M as shown in Figure 7.2, it must be remembered that y is related to curvature c through the following relation:

$$\frac{d^2 y_m}{dx} = c$$

The curvature c is related to facing stress σ_F (Equation 7.8), facing elastic modulus E, and the natural distance of facings S_A (Equation 7.3). So, the last equation can be rewritten into

$$\frac{d^2 y_m}{dx} = \frac{2M}{bES_F S_A^2}$$

Sandwich Structure Design

From Equation 7.5, using EI replaces $Eb\dfrac{S_F S_A^2}{2}$ in the last equation. Then, the deflection y_m of the beam can be calculated by integrating the last equation[2] twice:

$$y_m = \iint \frac{M}{EI} dx\, dx \qquad (7.11)$$

According to the definition of shear modulus, the angle γ of rotation of the horizontal surfaces of the beam in relation to the vertical surface is given by

$$\gamma = \tau_A / G_A$$

where τ_A is shear stress under shear force V (Equation 7.10), and G_A is the shear modulus of the core. Based on the definition, in the case of small strain, dy_v can be calculated by

$$dy_v = \gamma\, dx = \frac{\tau_A}{G_A} dx$$

using Equation 7.10, replace τ_A,

$$dy_v = \frac{V}{b S_A G_A} dx$$

Deflection y_v generated by shear force V and be calculated by integrating the equation below:

$$y_v = \int \frac{V}{b S_A G_A} dx \qquad (7.12)$$

By analogy with EI, the quantity GA (Equation 7.6) is defined as shear rigidity of the sandwich[2]:

$$y_v = \int \frac{V}{GA} dx \qquad (7.13)$$

The overall deflection of a beam is of course given by the sum of the components due to bending moment and shear force. In conventional beams, the shear component of deflection is usually very small and negligible, whereas in sandwich beams, it is usually significant and becomes even more relevant as the ratio between core thickness and beam length increases.

As shear rigidity depends on G_A, the choice of a core with a higher shear modulus can considerably reduce sandwich deflection.

7.1.2 Stress, Strain, and Rigidity in a Sandwich Beam Subjected to Tension or Compression

It has been often repeated that the tensile or compression strength of the core is negligible if compared to that of the facings, because of its low elastic modulus. The load

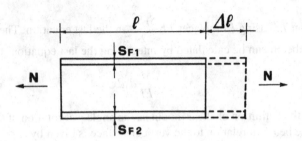

FIGURE 7.4 A sandwich structure beam subjected to a tension load N.[2]

N is therefore withstood mainly by the two facings as shown in Figure 7.4. Assuming for the sake of simplicity that the facings are of the same material in most cases, since their strain is equal so is their stress:

$$N_{F1} = \sigma_{F1} b S_{F1}, \quad N_{F2} = \sigma_{F2} b S_{F2},$$

where N_{F1} and N_{F2} represent the components of load N withstood by the two facings, respectively, and b is the beam width. Because N is a normal force along x-direction straightly, so

$$N_{F1} = N_{F2} = N/2.$$

$$\sigma_{F1,2} = N_{F1,2} / bS_{F1,2} = N / 2bS_{F1,2} \quad (7.14)$$

If $S_{F1} = S_{F2} = S_F$, $\sigma_{F1} = \sigma_{F2} = \sigma_F$,

$$\sigma_F = \frac{N}{2bS_F} \quad (7.15)$$

We shall now deal with the calculation of the total extension Δl that the sandwich undergoes when force N is applied to it, assuming that the facings are made of the same material of elastic modulus E_F. By definition, strain ε is given by the ratio between extension Δl and initial length l:

$$\varepsilon = \Delta l / l \quad \text{then} \quad \Delta l = \varepsilon l$$

ε is correlated with stress through $\varepsilon = \sigma_F / E_F$, from which the expression for σ_F given in Equation 7.15:

$$\varepsilon = N / 2bS_F E_F$$

and substituting in the last equation $\Delta l = \varepsilon l$

Sandwich Structure Design

$$\Delta l = \frac{Nl}{2bS_F E_F} \tag{7.16}$$

Analogously, for a sandwich made of facings of different thicknesses, we have[2]:

$$\Delta l = \frac{Nl}{b(S_{F1} + S_{F2})E_F} \tag{7.17}$$

With reference to the extension Δl of a sandwich subjected to tension or compression, Formula 7.17 shows that in this case, as well as in the case of strains caused by a bending moment and a shear force, a rigidity quantity can be found, depending on the shape of the cross-section and on the materials constituting the structure, which express the capacity of the sandwich to withstand strain. Such quantity, indicated by EA, is given by

$$EA = 2bS_F E_F \tag{7.18}$$

Equation 7.18 is for the two facings having the same elastic modulus and thickness:

$$EA = bE_F(S_{F1} + S_{F2}) \tag{7.19}$$

Equation 7.19 if for the facings having different thicknesses.

By analogy with EA and GA, the quantity EA is called tensile rigidity. If the normal force is a compressive force in negative direction, all Equations 7.14–7.19 are useful after adding a negative sign.

7.2 STRENGTH AND DEFLECTION OF SIMPLE SANDWICH ELEMENTS AT DIFFERENT SUPPORTS AND LOADS

It is a normal application of using flat sandwich element, such as beam and plate in many industries. The elements subjected to different pressure loads at different supporting types will suffer facing tension or compression stress, core shear stress, and deflection in certain range. It is important to predict the stresses and deflections by using some theoretical formulae during the design processes. The formulae listed in Table 7.1 can be used for calculating all parameters mentioned above at maximum situation in static state. The designer can consider about a safety factor with the result as a design criterion. In Table 7.1, b is the width of a sandwich beam; S_A is the distance between centers of facings from Equation 7.3; EI is the flexural rigidity of a sandwich beam from Equation 7.3 and 7.4; G_A is the core shear modulus. A maximum deflection f_{max} of a sandwich beam at a flexural load is the sum of maximum flexural deflection and maximum shear deflection ($f_{max} = f_f + f_t$).

Several examples of calculation by using the formulae in Table 7.1 are presented below for understanding how to use them.

TABLE 7.1
Strength and Deflection of Simple Sandwich Beam at Different Supports and Loads[2]

Support Type	Load Style	Diagram	Maximum Facing Normal Stress	Maximum Core Shear Strength	Maximum Flexural Deflection	Maximum Shear Deflection
Simply supports	Load at center		$\sigma_F = \dfrac{Fl}{4bS_A S_F}$	$\tau_A = \dfrac{F}{2bS_A}$	$f_f = \dfrac{Fl^3}{48EI}$	$f_t = \dfrac{Fl}{4G_A S_A b}$
	Two symmetrical loads		$\sigma_F = \dfrac{Fa}{bS_A S_F}$	$\tau_A = \dfrac{F}{bS_A}$	$f_f = \dfrac{F}{24EI}\cdot\left[l^3-(l-2a)^3-3a(l-2a)^2\right]$	$f_t = \dfrac{Fa}{G_A S_A b}$
	Uniform load		$\sigma_F = \dfrac{ql^2}{8S_A S_F}$	$\tau_A = \dfrac{ql}{2S_A}$	$f_f = \dfrac{5ql^4 b}{384EI}$	$f_t = \dfrac{ql^2}{8G_A S_A}$
Clamped ends	Load at center		$\sigma_F = \dfrac{Fl}{8bS_A S_F}$	$\tau_A = \dfrac{F}{2bS_A}$	$f_f = \dfrac{Fl^3}{192EI}$	$f_t = \dfrac{Fl}{4G_A S_A b}$

(Continued)

Sandwich Structure Design

TABLE 7.1 (Continued)
Strength and Deflection of Simple Sandwich Beam at Different Supports and Loads[2]

Support Type	Load Style	Diagram	Maximum Facing Normal Stress	Maximum Core Shear Strength	Maximum Flexural Deflection	Maximum Shear Deflection
	Two symmetrical loads		$\sigma_F = \dfrac{Fa}{blS_AS_F}$	$\tau_A = \dfrac{F}{bS_A}$	$f_f = \dfrac{Fla^2}{2EI}\left(\dfrac{1}{4} - \dfrac{1}{3}\dfrac{a}{l}\right)$	$f_t = \dfrac{Fa}{bG_AS_A}$
	Uniform load		$\sigma_F = \dfrac{ql^2}{12S_AS_F}$	$\tau_A = \dfrac{ql}{2S_A}$	$f_f = \dfrac{ql^4b}{384EI}$	$f_t = \dfrac{ql^2}{8G_AS_A}$
Cantilever beam	End load		$\sigma_F = \dfrac{Fl}{bS_AS_F}$	$\tau_A = \dfrac{F}{bS_A}$	$f_f = \dfrac{Fl^3}{3EI}$	$f_t = \dfrac{Fl}{bG_AS_A}$
	Uniform load		$\sigma_F = \dfrac{ql^2}{2S_AS_F}$	$\tau_A = \dfrac{ql}{S_A}$	$f_f = \dfrac{ql^4b}{8EI}$	$f_t = \dfrac{ql^2}{2G_AS_A}$

Example 7.1

Calculate the stresses and maximum deflection of a simply supported sandwich beam subjected to a central load $F = 5,000\,N$, knowing that:
$b = 1,000\,mm$, $l = 2,000\,mm$, $S_{F1} = 5\,mm$; $S_{F2} = 1\,mm$, $S_A^* = 47\,mm$, $El = 5 \times 10^{10}\,N\,mm^2$, and $G_A = 10\,N/mm^2$.

Solution:

From Equation 7.3: $S_A = 47 + (5+1)/2 = 50\,mm$
By using formulae in the first row of Table 7.1, obtain:
$\sigma_{F1} = 10\,N/mm^2$, $\sigma_{F2} = 50\,N/mm^2$, $\tau_A = 0.05\,N/mm^2$, $f_f = 16.7\,mm$, $f_t = 5\,mm$
Maximum deflection $f_{max} = f_f + f_t = 16.7 + 5 = 21.7\,mm$

Example 7.2

Calculate the stress and the maximum deflection of a simple supported sandwich beam loaded by two forces $F = 1,000\,N$ symmetrical with reference to the center, knowing that:
$b = 1,000\,mm$, $a = 1,000\,mm$, $l = 5,000\,mm$, $S_{F1} = 5\,mm$, $S_{F2} = 1\,mm$, $S_A^* = 47\,mm$, $El = 5 \times 10^{11}\,N/mm$, and $G_A = 10\,N/mm^2$.

Solution:

From Equation 7.3: $S_A = 47 + (5+1)/2 = 50\,mm$
By using formulae in the second row of Table 7.1, obtain:
$\sigma_{F1} = 4\,N/mm^2$, $\sigma_{F2} = 20\,N/mm^2$, $\tau_A = 0.05\,N/mm^2$, $f_f = 5.9\,mm$, $f_t = 2\,mm$
Maximum deflection $f_{max} = f_f + f_t = 5.9 + 2 = 7.9\,mm$

Example 7.3

Calculate the stress and the maximum deflection of a simple supported sandwich beam subjected to a uniformly distributed load $q = 1,000\,N/m^2$, knowing that:
$b = 1,000\,mm$, $l = 2,000\,mm$, $S_{F1} = 5\,mm$, $S_{F2} = 1\,mm$, $S_A^* = 47\,mm$, $El = 5 \times 10^{10}\,N/mm$, and $G_A = 10\,N/mm^2$.

Solution:

From Equation 7.3: $S_A = 47 + (5+1)/2 = 50\,mm$
By using formulae in the third row of Table 7.1, obtain:
$\sigma_{F1} = 2\,N/mm^2$, $\sigma_{F2} = 10\,N/mm^2$, $\tau_A = 0.02\,N/mm^2$, $f_f = 4.2\,mm$, $f_t = 1\,mm$
Maximum deflection $f_{max} = f_f + f_t = 4.2 + 1 = 5.2\,mm$

Example 7.4

Calculate the stresses and maximum deflection of a clamped sandwich beam subjected to a central load $F = 5,000\,N$, knowing that:
$b = 1,000\,mm$, $l = 2,000\,mm$, $S_{F1} = 5\,mm$; $S_{F2} = 1\,mm$, $S_A^* = 47\,mm$, $El = 5 \times 10^{10}\,N\,mm^2$, and $G_A = 10\,N/mm^2$.

Solution:

From Equation 7.3: $S_A = 47 + (5+1)/2 = 50\,mm$
By using formulae in the fourth row of Table 7.1, obtain:
$\sigma_{F1} = 5\,N/mm^2$, $\sigma_{F2} = 25\,N/mm^2$, $\tau_A = 0.05\,N/mm^2$, $f_f = 4.2\,mm$, $f_t = 5\,mm$
Maximum deflection $f_{max} = f_f + f_t = 4.2 + 5 = 9.2\,mm$

Example 7.5

Calculate the stress and the maximum deflection of a clamped sandwich beam loaded by two forces $F = 2,000\,N$ symmetrical with respect to the center, knowing that:
$b = 1,000\,mm$, $a = 1,000\,mm$, $l = 5,000\,mm$, $S_{F1} = 5\,mm$, $S_{F2} = 1\,mm$, $S_A^* = 47\,mm$, $EI = 5 \times 10^{10}\,N/mm$, and $G_A = 10\,N/mm^2$.

Solution:

From Equation 7.3: $S_A = 47 + (5+1)/2 = 50\,mm$
By using formulae in the fifth row of Table 7.1, obtain:
$\sigma_{F1} = 1.6\,N/mm^2$, $\sigma_{F2} = 8\,N/mm^2$, $\tau_A = 0.04\,N/mm^2$, $f_f = 18.3\,mm$, $f_t = 4\,mm$
Maximum deflection $f_{max} = f_f + f_t = 18.3 + 2 = 22.3\,mm$

Example 7.6

Calculate the stress and the maximum deflection of a clamped sandwich beam subjected to a uniformly distributed load $q = 2,000\,N/m^2$, knowing that:
$b = 1,000\,mm$, $l = 2,000\,mm$, $S_{F1} = 5\,mm$, $S_{F2} = 1\,mm$, $S_A^* = 47\,mm$, $EI = 5 \times 10^{10}\,N/mm$, and $G_A = 10\,N/mm^2$.

Solution:

From Equation 7.3: $S_A = 47 + (5+1)/2 = 50\,mm$
By using formulae in the sixth row of Table 7.1, obtain:
$\sigma_{F1} = 2.7\,N/mm^2$, $\sigma_{F2} = 13.3\,N/mm^2$, $\tau_A = 0.04\,N/mm^2$, $f_f = 1.7\,mm$, $f_t = 2\,mm$
Maximum deflection $f_{max} = f_f + f_t = 1.7 + 2 = 3.7\,mm$

Example 7.7

Calculate the stresses and maximum deflection of a cantilever sandwich beam subjected to an end load $F = 5,000\,N$, knowing that:
$b = 1,000\,mm$, $l = 2,000\,mm$, $S_{F1} = 5\,mm$; $S_{F2} = 1\,mm$, $S_A^* = 47\,mm$, $EI = 1 \times 10^{12}\,N\,mm^2$, and $G_A = 10\,N/mm^2$.

Solution:

From Equation 7.3: $S_A = 47 + (5+1)/2 = 50\,mm$
By using formulae in the seventh row of Table 7.1, obtain:
$\sigma_{F1} = 40\,N/mm^2$, $\sigma_{F2} = 200\,N/mm^2$, $\tau_A = 0.1\,N/mm^2$, $f_f = 13.3\,mm$, $f_t = 20\,mm$
Maximum deflection $f_{max} = f_f + f_t = 13.3 + 20 = 33.3\,mm$

Example 7.8

Calculate the stress and the maximum deflection of a cantilever sandwich beam subjected to a uniformly distributed load $q = 1,000 \text{ N/m}^2$, knowing that:
$b = 1,000 \text{ mm}$, $l = 2,000 \text{ mm}$, $S_{F1} = 5 \text{ mm}$, $S_{F2} = 1 \text{ mm}$, $S_A^* = 47 \text{ mm}$, $EI = 5 \times 10^{10}$ N/mm, and $G_A = 10 \text{ N/mm}^2$.

Solution:

From Equation 7.3: $S_A = 47 + (5+1)/2 = 50 \text{ mm}$
By using formulae in the eighth row of Table 7.1, obtain:
$\sigma_{F1} = 8 \text{ N/mm}^2$, $\sigma_{F2} = 40 \text{ N/mm}^2$, $\tau_A = 0.04 \text{ N/mm}^2$, $f_f = 40 \text{ mm}$, $f_t = 4 \text{ mm}$
Maximum deflection $f_{max} = f_f + f_t = 40 + 4 = 44 \text{ mm}$

The formulae introduced below allow for the calculation of the maximum stress in the facings, maximum shear stress in the core, and maximum deflection of a simply supported rectangular sandwich plate under uniformly distributed normal load as shown in Figure 7.5. It is supposed that the sandwich facings have equal rigidity in the direction a and b and that the core is isotropic.

The calculation of maximum stresses and deflection is carried out in two steps. In the first step, the plate is considered as a simply supported beam of length equal to the minor in-plane dimension (b in Figure 7.5) and width equal to the major in-plane dimension (a in Figure 7.5) of the plate, subjected to the same distributed load. Through the formulae and diagram listed in the third row in Table 7.1, the maximum normal stress in the facings, σ_{Fb}, shear stress in the core, τ_{Fb}, flexural component of maximum deflection, f_{fb}, and shear component of maximum deflection, f_{tb} are obtained.

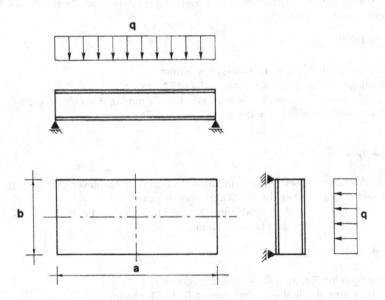

FIGURE 7.5 Simply supported sandwich plate subjected to a uniformly distributed normal load.[2]

Sandwich Structure Design

In the second step, the corresponding values for the sandwich plated are calculated by the following formulae:

$$\sigma_F = m_1 \cdot \sigma_{Fb} \tag{7.20}$$

$$\tau_A = m_2 \cdot \tau_{Fb} \tag{7.21}$$

$$f_f = m_3\left(1 - v_F^2\right)f_{fb} \tag{7.22}$$

$$f_t = m_4\left(1 - v_F^2\right)f_{tb} \tag{7.23}$$

The maximum deflection is given by

$$f_{max} = f_f + f_t \tag{7.24}$$

In Equations 7.20–24:

σ_F = maximum normal stress in the facings (N/mm²)
τ_A = maximum shear stress in the core (N/mm²)
f_f = flexural component of maximum deflection (mm)
f_t = shear component of maximum deflection (mm)
f_{max} = maximum deflection (mm)

$m_1, m_2, m_3,$ and m_4 = correction factors to be determined from Figure 7.6a–d
v_F = facing Poisson's ratio

It can be seen from Figure 7.6 that in order to determine m_4, it is necessary to know the quantity S, called "shear deformation factor", given by

$$S = \frac{\pi^2 EI}{G_A S_A b^2 a \left(1 - v_F^2\right)} \tag{7.25}$$

Example 7.9

Calculate the stress and the maximum deflection of a simply supported rectangular sandwich plate under uniformly distributed normal load $q = 8{,}000\,\text{N/m}^2$ as shown in Figure 7.5, knowing that:

$a = 1{,}000$ mm, $b = 600$ mm, $S_{F1} = 5$ mm, $S_{F2} = 1$ mm, $S_A^* = 47$ mm, $EI = 5 \times 10^9$ N/mm, $G_A = 10$ N/mm², and $v_F = 0.33$.

Solution:

From Equation 7.3: $S_A = 47 + (5+1)/2 = 50$ mm
By using formulae in the third row of Table 7.1, obtain:
$\sigma_{Fb1} = 1.44$ N/mm², $\sigma_{Fb2} = 7.2$ N/mm², $\tau_{Ab} = 0.048$ N/mm², $f_{fb} = 2.7$ mm, $f_{tb} = 0.72$ mm
$b/a = 0.6$, refer to Figure 7.6, then have $m_1 = 0.7$, $m_2 = 0.87$, and $m_3 = 0.65$

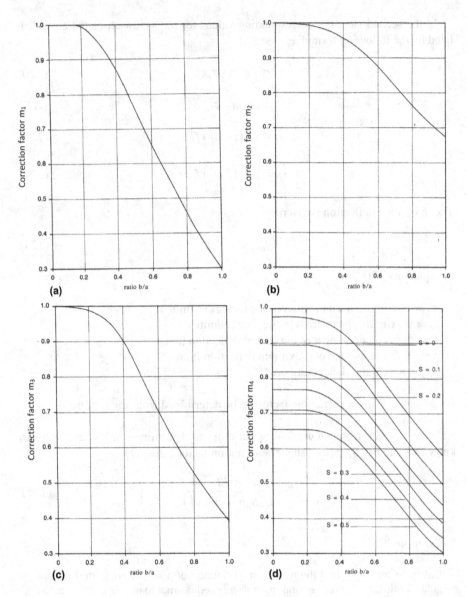

FIGURE 7.6 Correction factors m_1, m_2, m_3, and m_4 for calculating maximum normal stress σ_F in the facings of the sandwich plate.[2]

Using Equation 7.25 obtains $S = 0.3$; then from Figure 7.6d, have $m_4 = 0.62$
By using Equations 7.20–7.23, have:

$$\sigma_{F1} = 1.0\,\text{N/mm}^2, \sigma_{F2} = 5.0\,\text{N/mm}^2, \tau_A = 0.04\,\text{N/mm}^2, f_f = 1.8\,\text{mm},$$

$$f_t = 0.45\,\text{mm}$$

Maximum deflection $f_{\max} = f_f + f_t = 1.8 + 0.45 = 2.25\,\text{mm}$

Sandwich Structure Design

7.3 EDGEWISE DAMAGE PREDICTION AND PREVENTION

7.3.1 GENERAL BUCKLING

A sandwich bam subjected to compression may fail because of a condition of instability that interests the whole beam, for this reason called "general buckling (Figure 7.7a). General buckling may also occur when the stress in the facings and in the core is lower than the allowable stress. The load that determines sandwich instability depends on such parameters as the beam in-plane size and the constrain conditions, which can only be partially modified at design stage.

Other quantities, which are equally important for the definition of buckling load, depend directly on the type of sandwich. They are flexural rigidity of the sandwich, thickness of the facings, elastic properties of the facings, core thickness, and shear modulus of the core.[2] To avoid this type of failure, it is necessary to ensure that the general buckling load is higher, according to a suitable factor of safety, than the predicted edgewise compression stress. A good design should therefore result in a critical buckling stress (σ_{cr}) at least equal to or higher than the strength of the material selected for facings (σ_f).

The critical stress causing the general buckling of a sandwich beam under compressive load is given by

$$\sigma_{cr} = \frac{1}{2bS_{Ft}\left(1+\frac{\pi^2 EI}{l_{cr}^2 GA}\right)} \cdot \frac{\pi^2 EI}{l_{cr}^2} \tag{7.26}$$

where
$S_{Ft} = S_{F1} + S_{F2}$
σ_{cr} = critical buckling stress (N/mm^2)
EI = sandwich flexural rigidity (N mm^2)
b = beam width (mm)

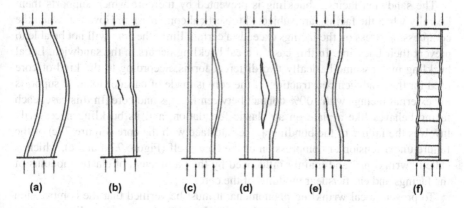

FIGURE 7.7 Edgewise compression failure modes. (a. general global buckling, b. shear crimping, c. facing buckling, d. core splitting, e. core crushing, and f. honeycomb sandwich dimpling.)

$S_{F1,2}$ = thickness of facings 1, 2 (mm)
l_{cr} = free length of the sandwich beam (see Figure 7.8 for determining the value) (mm)
GA = sandwich shear rigidity (Equation 7.6) (N)

It is easily verified that when $l_{cr}^2 \ll EI$, the formula 7.20 reduces to

$$\sigma_{cr} = \frac{G_A S_A}{2 S_{Ft}} \qquad (7.27)$$

where G_A is core shear modulus, and S_A is given by Equation 7.3.

If general buckling is feared, one may use facings with a higher elastic modulus material, increase facings' thickness, increase core thickness, or use a core material with higher shear modulus.

In the case of sandwich with relatively low length/thickness ratio and where shear rigidity is small in comparison with flexural rigidity, sandwich general buckling will assure typical configuration, shown in Figure 7.7b; this failure mode, typical of sandwich structures, is called "shear crimping". The total load per unit length capable of producing crimping is practically independent of facing properties; on the other hand, it increases linearly with the following parameters of the thickness and shear modulus of the core. The total critical crimping load can be increased by increasing the core thickness, and a higher shear modulus for the core can be used.

7.3.2 Local Buckling in the Sandwich

The local modes of failure may occur in sandwich panels under edgewise loads or normal loads. In addition to overall buckling and local modes of failure, sandwich is designed so that facings do not fail in tension, compression, shear, or combined stresses due to edgewise loads or normal loads, and cores and bonds do not fail in shear, flatwise tension, or flatwise compression due to normal loads at design load.

The sandwich facings' buckling is prevented by the core which supports them laterally when the facings are subjected to vertical compression. However, when the compression stress on the facings exceeds a certain limit, the core will not be able to prevent their buckling. In this case, a local buckling occurs in the sandwich. Local buckling may assume basically two different forms, according to the kind of core used for the sandwich construction. If the core is made of balsa or foam, it supports the external facings with 100% contact between facings and core. In this case, each facing behaves like a plate on an elastic foundation, and its buckling necessarily involves the failure of the bonding at the interface with the core (Figure 7.7c) or the failure under tension or compression of the core itself (Figure 7.7 d and e), which is called "wrinkling" and mainly influenced by the parameters of elastic modulus of the facings and elastic/shear modulus of the core.

To prevent local wrinkling phenomena, it must be verified that the compression stress in the sandwich is lower than the critical buckling stress, according to a suitable factor of safety. If local wrinkling is feared, the method for preventing is to use a facing material with a higher elastic modulus and/or to use a core material with

Sandwich Structure Design

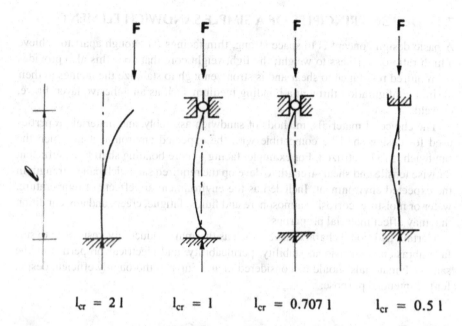

FIGURE 7.8 Free length of the sandwich beam for different cases of constraints.[2]

higher elastic properties. A foam core usually presents both of higher elastic and shear moduli with higher density. Since the two moduli both affect the critical stress for local instability, a higher density of the core is twice as efficient in the prevention of the phenomenon.

When the core is made of honeycomb, the bonding between facings and core is obtained only on honeycomb cell external borders. When facings are subjected to compression, they may therefore undergo buckling in the free spaces within the single cells, generating the "dimpling" phenomenon (Figure 7.7f). Dimpling depends mainly on the factors of facing's elastic modulus, facing's thickness, and core cell size.

The methods that can increase the critical dimpling stress are to use the facings with higher elastic modulus, to use thicker facings, and/or to use a core with smaller size cells.

It is important to note that different from the case of wrinkling, the phenomenon of dimpling does not determine the failure of the structure. This type of buckling, therefore, is not necessarily catastrophic. However, it is advisable to take dimpling into account at the design stage both because it may prelude to wrinkling failures and because it may produce irreversible deformation in the structure.[1]

Design procedures involving buckling are based on theoretical buckling coefficients. These coefficients are in fair agreement with average test results, but allowance can be made in the final design to account for the scatter characteristic of buckling test results, perhaps by choosing a slightly thicker core, so that buckling of the sandwich component does not occur at design load.

7.4 DESIGN PRINCIPLES OF A SIMPLE SANDWICH ELEMENT

A basic design concept is to space strong, thin facings far enough apart to achieve a high ratio of stiffness to weight; the lightweight core that does this also provides the required resistance to shear and is strong enough to stabilize the facings to their desired configuration through a bonding medium such as an adhesive layer, braze, or weld.

The choice of materials, methods of sandwich assembly, and material properties used for design shall be compatible with the expected environment in which the sandwich is to be utilized. For example, facing to core bonding shall have sufficient flatwise tensile and shear strength to develop the required sandwich panel strength in the expected environment. Included as the environment are effects of temperature, water or moisture, corrosive atmosphere and fluids, fatigue, creep, and any condition that may affect material properties.

Certain additional characteristics, such as thermal conductivity, resistance to surface abrasion, dimensional stability, permeability, and electrical properties of the sandwich materials should be considered in arriving at a thoroughly efficient design for the intended purpose.

7.4.1 DESIGN FOR STRENGTH OF FACINGS AND CORE

When dimensioning a sandwich structure, the designer's main task is to ensure that the structure be safe under the assigned loads. It is therefore necessary to know the possible failure modes the structure can undergo in order to take the necessary measures. This matter is extreme importance for a sandwich structure, besides the common failure modes characteristic of its peculiar construction.

A design that takes into account firstly the static safety of the structural part is generally defined as "strength design". However, in some cases, it is not enough to guarantee that the designed structure does not yield under the assigned loads. In such cases, in order to ensure its functionality, it is necessary for the strains caused by the applied loads not to exceed the limits specified in the design: the design is then called "rigidity design". It is obvious that a rigidity design will be followed by a strength verification. Then it is also important to know the most common types of failure in a sandwich and the means available to the designer to avoid them in a rigidity design. Finally, the possible criteria to keep the strains of a sandwich within acceptable limits in a rigidity design will be discussed.

The main causes for sandwich structure failure are tensile or compression failure of the facings, shear failure of the core, failure under general instability, and failure under local instability. These types of failure must be adequately taken into account and avoided in a strength design.

The results of the last section allow us to say that the bending moment M generates a tensile stress in one facing and a compression stress in the opposite facing; the normal force N causes stresses of the same sign in the two facings and tensile (compression) stresses if N is a tensile (compression) force; the shear force V does not affect the stress in the facings.

Sandwich Structure Design

The actual stress in a sandwich is of course given by the algebraic sum of the components due to M and N. If the stresses exceed the corresponding ultimate stresses of the constitutive materials of the facings, they will fail in a catastrophic way.

In design calculations, the strength verification of the facings is usually carried out by comparing the stresses caused by external loads with allowable stresses for the constitutive materials of the facings. The allowable stresses are obtained dividing the strengths by suitable factors of safety with taking into account the variable properties of the materials, the approximations in the structure schematic design, the accidental loads, fatigue phenomena, etc.

When the calculated stressed exceed the allowable stresses, a change in the sandwich sizing is required. In such case, one may use a material with higher allowable stresses for the facings and increase facings' thickness, thus reducing the applied stresses, and increase core thickness to decrease the stresses in the facings.

From the last section, we also obtain that bending moment M and normal force N do not cause stresses in the core; shear force V is withstood by the core, generating evenly distributed shear stresses. The sandwich core is then subjected to shear stress only. If the value of the shear stress is greater than the shear strength of the core material, the core fails cause the failure of the structure.

The shear verification is carried out by comparing the calculated stress generated by the design loads with the allowable stress, which is calculated dividing the shear strength of the core material by a suitable factor of safety. If the calculated stress exceeds the allowable stress, one may use a core material with higher allowable shear stress or increase the core thickness, thus reducing the shear stress. By the way, using a different material for the facings or increasing facings' thickness does not affect the shear stress in the core.

If a sandwich structure is used as a floor subjected to a uniform or a concentrated compression load, the force will act on the core only. In this case, the required compression stress will be calculated and compared to the flatwise compression strength by following the same processes as that for the facing strength and the core shear strength, and then, verification and adjustment are done in the same way.

7.4.2 DESIGN FOR RIGIDITY

In the case of design for rigidity, the designer must not only fulfill the strength requirements of the structure but also limit deflections within specifications. The process of dimensioning is therefore initially based on rigidity requirements. Once these are met, a strength verification must be carried out following the procedures stated in the last section.

When a sandwich is subjected to a set of loads made of a bending moment M and a shear force V, as seen in Section 7.1.1 in detail, it undergoes a deflection due to the bending moment M and a deflection due to the shear force V. Obviously, the total value of the deflection is the sum of the two components generated by M and V. A possible normal force N does not contribute to the deflection as it causes only a variation in the length of the sandwich.

The deflection component due to the bending moment depends, through flexural rigidity, on the flowing typical sandwich parameters: elastic modulus of the facings, facings' thickness, and core thickness. To limit this component, the changes can be to use a high elastic modulus material for the facings, to use thick facings, and to use a thick core as presented in the last section.

Since flexural rigidity changes according to the square of the central distance between the facings, increasing it is the most efficient way to stiffen the structure with respect to the action of a bending moment. Using a stiffer core also can increase the rigidity but produces no significant effects.

The deflection component due to the shear force depends, through shear rigidity, on the following typical sandwich parameters: shear modulus of the core and thickness of the core. To limit this component, it is effective to use a high shear modulus material for the core or use a thicker core. As it can be seen, a thick-enough core means a limitation of both the bending and shear components of the deflection. The elastic moduli of the facings and their thickness do not affect the shear deflection, while the shear modulus of the core has no significant effect on the bending deflection.

7.4.3 Design for Minimum Weight

Since the main feature of the sandwich structure is low weight with high stiffness and strength, the objective is essentially one that minimizes the total weight of the structure. Producing a light structure with relatively affordable cost without sacrificing strength has always been a challenging task for designers. The design is subjected to equality and inequality constraints related to allowable bending stress, shear stress, and deflection, where the face sheet thickness and the core thickness are considered as design variables. Many sandwich structures are designed to resist bending load; hence, considerable effort has been devoted to optimization through minimizing the weight to achieve bending stiffness requirement. Some new optimal designs aimed to the sandwich structures, especially the core structures, are of current academic and industrial interest due to their superior properties like high specific strength, high specific stiffness, high energy absorption ability, etc.

7.4.3.1 Minimum Weight for Given Stiffness

For a sandwich structure with a given bending stiffness, a core with certain density, and a facing with known elastic modulus, the minimum weight can be determined by the following process.[3]

A sandwich bending stiffness per unit width D can be derived by elementary mechanics and is given by the following formula for sandwich with dissimilar thin faces and a weak (negligible stiffness) core,

$$D = \frac{E_1 t_1^3}{12} + \frac{E_2 t_2^3}{12} + \frac{E_1 t_1 E_2 t_2 d^2}{E_1 t_1 + E_2 t_2}$$

If the faces are thin, we have $t_f \ll t_c$ and $d \approx t_c$, the two first terms are negligible relative to the third term, and sandwich bending stiffness per unit width can be written as follows:

$$D = \frac{E_1 t_1 E_2 t_2 t_c^2}{E_1 t_1 + E_2 t_2}$$

When the sandwich has identical facings, i.e., $t_1 = t_2 = t_f$, $E_1 = E_2 = E_f$, we have

$$D = \frac{E_f t_f t_c^2}{2} \tag{7.28}$$

where
t_1, t_2 – thickness of two facing skins
t_f – thickness of facing skins (assuming that both facing skins are the same)
d – sandwich panel thickness
E_1, E_2 – elastic (Young's) modulus of two facing material
E_f – elastic (Young's) modulus of facing material (assuming that both facing skins are the same)
D – sandwich plate bending stiffness per unit width
t_c – core thickness

The weight of a sandwich with thin facings neglecting the weight of bond can be expressed by the formula

$$W = W_1 + W_2 + W_C = \rho_1 t_1 + \rho_2 t_2 + \rho_c t_c$$

When the sandwich has identical facings, $\rho_1 = \rho_2 = \rho_f$, $W_1 = W_2 = W_f$, then we have

$$W = 2\rho_f t_f + \rho_c t_c \tag{7.29}$$

where
W_1, W_2 – weight of two facing skins
W_f – weight of facing skins (assuming that both facing skins are the same)
W_c – weight of the core
ρ_1, ρ_2 – density of two facing material
ρ_f – density of facing skins (assuming that both facing skins are the same)
ρ_c – density of the core material

Equation (7.28) can be written as $t_f = 2D/(E_f t_c^2)$, and substitution of this into (7.29) results in

$$W = \frac{4D\rho_f}{E_f t_c^2} + \rho_c t_c \tag{7.30}$$

The functional relationship following Equation 7.30 between the weight of a sandwich structure, having a certain bending rigidity, and the core thickness is shown as Figure 7.9, in which W_{opt}^B is optimal sandwich weight with bending stiffness constraint; $t_{c,opt}^B$ is optimal sandwich core thickness with bending stiffness constraint.

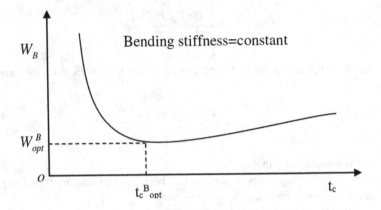

FIGURE 7.9 Weight vs. core thickness curve with a constant bending stiffness.[3]

The optimum solution for the core thickness can be obtained by minimizing the total weight, given by Equation 7.30, with respect to the core thickness t_c, yielding

$$t_c^3 = \frac{8D}{E_f}\frac{\rho_f}{\rho_c} \qquad (7.31)$$

Substitution of expressions of D, given by Equation 7.28, into Equation 7.31 yields the following equality:

$$\rho_c t_c = 4\rho_f t_f$$

$$W_c = 4W_f \qquad (7.32)$$

For sandwich with dissimilar facings, following the same way above, obtain

$$W_c = 2(W_1 + W_2)$$

So the weight of the core is four times that of a single face skin for a minimum weight sandwich structure with identical facings. The exact weight depends, of course, on the required bending stiffness. The optimum thickness of the core and the facing c_{opt} and $t_{f,opt}$ can be expressed as

$$t_{c,opt}^B = 2\left(\frac{\rho_f D}{\rho_c E_f}\right)^{\frac{1}{3}} \qquad (7.33)$$

$$t_{f,opt}^B = \frac{2D}{E_f \left(t_{c,opt}^B\right)^2} \qquad (7.34)$$

In this case of core properties varying with density, there are three variables, two thicknesses, and the core density. Considering the definition of the core material properties, and the thickness of facing and core, yields no optimum solution for minimum weight. However, the weight decreases asymptotically to a finite value when the core thickness is increased while the face thickness and core density simultaneously are decreased. This is however, in reality, not optimum. Letting the face thickness be predetermined leads to the same result as above, and infinitely thick and infinitely light core yields a minimum weight. Hence, for these cases, the optimum design is by choosing as thick and as light core as possible in conjunction with other constraints on the structure. This leads automatically to a given minimum face thickness. By instead predetermining the core thickness, which often is stipulated in practical cases due to requirements on the space or thermal insulation, it is possible to find an optimum face thickness and to that a corresponding core density by plotting weight versus facing thickness and graphically determine the optimum face thickness.[4]

7.4.3.2 Minimum Weight for Given Strength

If one now instead is aiming at finding a minimum weight design for a given strength, it is convenient to use the failure mode maps as shown in Figure 7.10. One approach is to use each of the failure modes as constraints and optimize the structure with respect to this constraint alone, using the methodology described above for the stiffness problem. One may then compare the results for the different "optima" and on the basis thereof choose a minimum-weight design. The chosen design should then be plotted on the failure mode map to ensure that it will fail in the given mode.

Using failure mode maps is a technique to design sandwich structures in way that will improve the performance of the sandwich from the view that no single component is over-designed with respect to the other components. The technique also allows the designer to choose the anticipated failure mode or making two different failure modes equally likely to occur. This can be advantageous in some cases where certain failure modes should be avoided.

For making a failure mode map, first is to define the principal failure modes: face yield, face wrinkle, or core shear is independent mode and of great importance to a structure. The variables used are, apart from face and core thicknesses, also the core density. One may use core density as a variable providing the core properties can be related to its density. Figure 7.10 shows a failure mode map with core density ρ_c as x-axis and a ratio of facing thickness t_f to beam length L as y-axis. The map shows that a sandwich will fail in core shear mode when a low-density core is used and in a face wrinkle mode when a thin skin is used.

7.4.4 OTHER DESIGN CONSIDERATIONS

There are various options for designing a sandwich structure. Any variation in face sheet, core type, and overall geometry will have a great impact on the general structural properties. The above constitute the design constraints on a structure but often there are other constraints that appear in an application that are not directly of a mechanical nature but rather of a practical one.

Depending on the application, they may of course differ, but some common constraints are as follows:

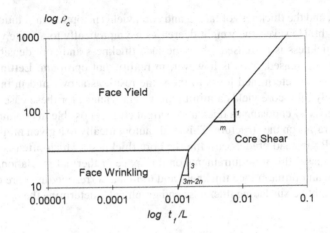

FIGURE 7.10 Failure mode map for a sandwich beam in three-point bending for which the properties vary linear with core density and facing thickness.[5]

1. Use right core thickness and select specific core materials for thermal insulation purposes. Most foam core materials have good thermal insulation property that changes with the raw material, cell structure, and density of the core. The core type, density and thickness should be determined firstly if the thermal insulation is required in a design and is significant. The design should then be followed for the stiffness, strength, and weight to determine facing type and thickness;
2. Minimize face thickness (based on available fabric weights) and consider about a combination of face and core materials for given impact resistance.
3. Consider specific face material and face thickness for surface wear resistance.
4. Choose specific face material for surface finish.
5. Specify core and face material for environmental resistance, such as temperature resistance, UV-resistance, sound insulation, water resistance, and corrosive resistance.
6. Use specific core and facing materials for fire resistance requirement.
7. Material selection should also take account of available manufacturing facilities, especially cure temperature capability.

Furthermore, there are always some other requirements on the materials depending on the manufacturing or assembly method, working environment, waste, and recycling, to only mention some.

7.5 DESIGN PROCEDURE FROM SIMPLE ELEMENT TO LARGE COMPLEX STRUCTURE

7.5.1 Determine Thicknesses of Simple Sandwich Element

Consider a beam or a panel subjected to a transverse load and find the core thickness required to carry this load. The structural member must fulfil all constraint given to it. The ultimate face stress is not only given by the yield or fracture strength of materials, but also may be limited by the wrinkling or, if a honeycomb core is used, dimpling strength of the face in compression. The local buckling stress may be substituted for an expression valid for biaxial loading, if that is the case. The maximum shear stress in the core must be lower than the ultimate shear strength. In the case of a panel, the same thing must be performed in both x- and y-directions. Hence, there are three constraints to be considered when sizing the sandwich.

If the face thickness is known, then use the equations presented in the last sections to calculate three values of the thickness. If the core thickness was known beforehand, the face thickness can be calculated in a similar manner. For a sandwich panel, the sizes can be found in the same way, it is all a matter of calculating the deflection as a function of the geometry and using formulae valid for sandwich plates for the strength constraints. If the sandwich is subjected to in-plane compressive loads, then the constraints will be buckling constraints.

A way of improving a design is to recognize that a design that allows for simultaneous failure of the face and the core has a potential of being closer to the optimum design, that is, to find a design that stresses all components to their limits at maximum load, for example, to design for simultaneous face yield and core shear fracture or face wrinkling and core shear failure. It is merely a matter of recognizing the important failure modes that may appear and finding the sizes and even materials that allow for simultaneous failure in these modes.

In the simplest possible way, the design of a sandwich can be described as follows:

1. The core thickness is determined by the fact that it should not fail under the applied shear stress. Beware, the shear strength of the core may vary in different directions if the core is orthotropic.
2. The face thickness is then determined by that the faces should be able to carry the applied bending moment, the facing tensile strength also may vary in different directions if the faces are orthotropic.
3. The calculated component thicknesses should then be checked so that maximum deflection must be less than the maximum allowed deflection.
4. Other constraints, such as local buckling, dimpling, etc. must subsequently be checked, and, if violated, sizes or materials must be modified until all are satisfied.

If any constraint is violated, then either the facing thickness or the core thickness, or both, must be increased until the necessary stiffness is obtained. A design of this kind is obviously simple and may in most cases serve as an initial design. This design could then be changed, allowing for dissimilar faces, orthotropic material components and a full set of constraints, in the pursuit of improved performance.

7.5.2 Design Routine of Simple Sandwich Element

The routine is to start with a preliminary design, something feasible achieved by some simple sizing rules and using materials that the designer assumes appropriate. Bending and shear stiffness properties are then calculated using formulae given early. Strength properties can be achieved in several ways, from material manufactures, data sheet, tests, or calculated by some proposed formulae, for example, for fiber composites. The structural analysis is then performed, e.g., using the proposed formulae in previous sections, finite element analysis or any other tool that would give the deflection, transverse force, and bending moment values.

Once that is done, the constraints must be checked, that is, after consideration about safety factors, the maximum deflection of the structure is less than the allowable, the stresses nowhere in the structure exceed the strength, that buckling does not occur, no wrinkling, no dimpling, or that any other constraint is violated. Even if all constraints are satisfied, there might be room for improvements, some components may be over-sized, and weight savings might be possible. As long as the designer believes that the structure needs to be improved, modifications can be implemented, properties updated, and the calculations performed again.

Afterwards, the structure still can be improved by considering about raw material availability, manufacturing method, and cost. Then the prototype will be made for testing to confirm the design parameters including destructive and nondestructive tests. The modification should be followed if the tested values are not equivalent to the theoretically calculated data. Finally, the design scheme including the structures, raw materials, processing method, and processing parameters can be decided.

7.5.3 Scaling Up to Large Complex Structure

7.5.3.1 Multilevel Scaling

A large complex sandwich structure consists of a few of simple elements. For example, a composite fuselage is made of a large area of curve panel with uniform thickness and numerous stringers that are made with most of sandwich structures. For designing the large-size complex product, we need to start from a simple element, for example, from a flat sandwich panel and a trapezoid shape sandwich column. By setting up the constrains and criteria and using the design principles and the formulae presented in previous sections, the structures of the elements can be designed, prototyped, and tested in a procedure as shown in Figure 7.11.

After all test results meet the requirements, the first scaling level will start to design and make subcomponent parts for testing. In the example as shown in Figure 7.11, the subcomponent panels constructed by combining a curve panel and stringers are made for conducting all tests, including bending, compression, buckling, weight checking, and other tests. If the testing results do not meet the requirements, the designer should go back to the first step to change the basic design of the elements. Then the designing, prototyping, and testing works should be repeated to find right results for meeting the new requirements of the first scaling level part. Afterwards, a few of new first scaled panels should be made and tested again. The works between the basic design and the first scaling level could be repeated for a couple of times until the requirements are met.

A second scaling level prototype will be made by using the designing and testing results of the first two steps for mechanical and application tests. The tests could be destructive or nondestructive because it is expensive to make. After testing, if

Sandwich Structure Design   281

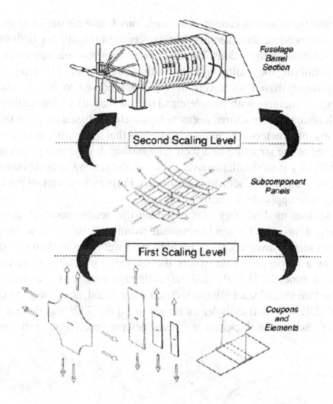

FIGURE 7.11 Design scaling-up for a fuselage made of composites.

anything doesn't meet the designing requirements because of the wrong choice of raw materials, inaccurate design, and inappropriate manufacturing methods, the designer should return to the basic step and then the first scaling level to adjust the design, make the prototype, and do the tests. The final product will be made and tested again until meeting all the requirements. All the works mentioned above can be simulated by using the FEA process before starting all steps or before starting each step. By referencing the FEA results, a designer can save time and cost for making a right product.

7.5.3.2 Similarity Theory Scaling

For designing a large sandwich composite product, laboratory-scale structures are commonly used for the experimental study of mechanical behavior to reduce the cost of experimental validation and certification of large structures. It has been shown that such a laboratory-scale structure can be effective for developing design guidelines for designing many products. However, scaling can be difficult for some products, such as a sandwich shell structure due to the small thickness, manufacturing considerations, and the fact that the buckling response is closely related to the relative stiffness properties of the structure.

Historically, small-scaled models have been built through the use of dimensional analysis to obtain similarity conditions. This dimensional analysis is employed to deduce a form of the system of characteristic equations. Complete similarity is obtained when all the independent dimensionless parameters are the same for both the scaled and real size baseline configurations. The main disadvantage of this methodology is the difficulty in identifying the scaling laws, due to the large number of design parameters.

Similarity theory based on governing equations proved to be effective in the design of scaled structures with complete and partial similarity. The scaling laws for the sandwich composite structures needs to be developed based on plate theory and nondimensional parameters. The main advantage is that the scaling laws are deduced from properties of the structure and their relationship through the governing equations. The difficulty is to simultaneously fulfill all the scaling laws while remaining within the design and manufacturing constraints. Furthermore, lack of perfect similarity can limit the applicability of the results.

Another scaling methodology was based on the nondimensionalization of the theoretical equations and the nondimensional parameters. In order to simplify the number of parameters involved, stacking sequences were chosen such that they were determined by a single ply angle, where the sandwich structures were considered balanced and symmetric. This allowed the scaling to be reduced to a three-step process: first, the face sheets' stacking angle θ was determined, next, the geometry ratio, for example, R/L of a shell cylinder as shown in Figure 7.12, was determined, and finally, the sandwich core thickness, t_{core}, was determined. Through this process, it

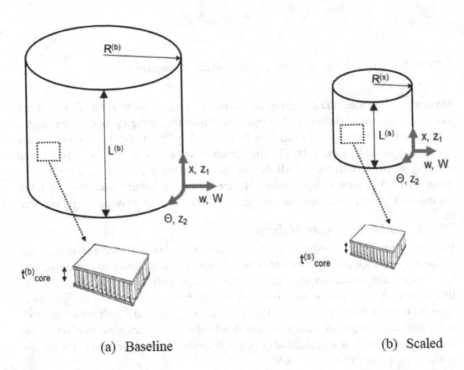

(a) Baseline   (b) Scaled

FIGURE 7.12 Geometric variable and coordinate system of the real-size baseline and the scaled.[6]

Sandwich Structure Design

was possible to find scaled configurations with the same nondimensional parameters as those from the real-size baseline configurations, which reproduce the loading response.

The applicability of the nondimensional methodology was limited by two initial simplifications: ignoring transverse-shear deformations and the flexural anisotropy parameters. The fact that these were neglected can explain some of the differences with the results of the finite-element analyses. Extending the methodology to include the transverse-shear and flexural anisotropy would likely extend the applicable range.[6]

7.6 EFFECTS OF CORE FORMATS, PROCESS METHODS, AND APPLICATION CONDITIONS ON DESIGN PARAMETERS

To design a sandwich structure, it is important to choose raw materials, processing method, and conditions and consider about application conditions, because all these will affect the product properties. The properties of the sandwich structure are the combination of the properties of the facing, adhesive, and core, which also depend on the manufacturing process of making the product. The property values of raw materials from the technical data sheet are generated from a standard format and a typical condition. They will change in the specific design and application conditions. As a designer, one should understand the relationship of the design parameters and all those mentioned above.

7.6.1 Influence of the Core Formats on Design Parameters

When choosing a core material, a designer will check the mechanical properties from its technical data sheet (TDS). However, the property values in the TDS are obtained by testing the core material at standard style. For example, the properties of the foam and balsa wood cores are acquired from the rigid board format as shown in Figure 7.13a, which represent the behavior of the pure core material. Actually, the pure core format is not used in all applications. Different format options as shown in Figure 7.13 are available for meeting the requirement of diverse lamination methods. The mechanical properties of each core material after laminated with facings should be estimated or tested as the designing parameter, and over-designing by using the properties at standard style should be avoided.

By referencing the figure, the application and the property changes of each core option are explained below:

A. Rigid board is an original format of the core that is used for press, compression molding, and pultrusion process. Hand lay-up may be achieved only by bonding/laminating one skin over the core (top side) and then flipping after cure and laminating the second face. The core properties after lamination should be similar to the value in the TDS. The bonding strength between core and facings may not be high enough if the core surface is too smooth, or foam cell size is too small so the resin or adhesive cannot anchor into the cavities of the foam surface.

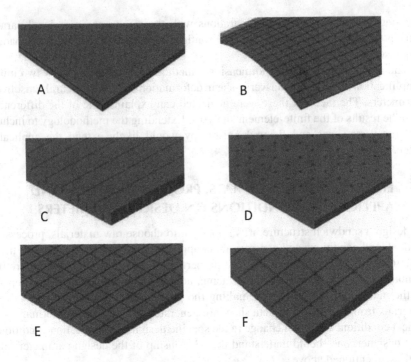

FIGURE 7.13 Foam and wood core formats: (a) rigid board, (b) contoured sheet, (c) double contoured board, (d) perforated board, (e) grooved board, and (f) grooved and perforated board.

B. Contoured sheet is cutting on one side along the direction of length and width to form small squares. Glass fiber scrim is laminated on another side, which is used for laminating on the curved mold surface in the processes of hand lay-up, vacuum bag, VARTM, and other resin infusions. After laminated with facings in the co-curing process, the resin filled the cutting slots, so that the density and weight of the core increase, then the shear, compression, and tensile properties increase. The bonding strength is high because the resin becomes the part of the core, which holds the facing fiber and the core together. Balsa, foam, and honeycomb are all supplied in this format

C. If a flexible sheet is desired without a scrim, single or double contoured board is cut on both sides and both directions. The cuts on two sides missing each other are for avoiding breakage. The double-cut sheet is contourable to both surfaces and can be used for hand layup and resin infusion processes. The density and mechanical properties will change with the same tendency as format B with lower increase. These formats are generally available for foam core only

D. Perforated board is used for vacuum bag, prepreg, RTM, VARTM, and resin infusion process in that the liquid resin can flow from the top surface through the perforated holes to the bottom surface. The core density will increase after laminated with the facings. Also, the cured resin in the

Sandwich Structure Design

holes will connect two facings together so the compressive properties, facing bonding strength, bending stiffness, and strength all rise compared with the structure with the standard style of the core.

E. Surface grooves provide flow channels in the resin infusion lamination and enhance the bonding strength between core and facing. The grooves can be on one side or both sides and in both the length and width directions. The core density after laminating also increases in a certain degree depending on the width and depth of the grooves.

F. Combinations of features B, D, and/or E are also available. These are mainly used in the vacuum infusion process. The weight of the core will increase dramatically after laminated with the facings. All mechanical properties also increase.

7.6.2 Influence of Laminating Methods on Design Parameters

The laminating method affects the properties of the sandwich structures, especially in the co-curing process by using the liquid resin and the fiber materials for skin reinforcement. The effects are mostly on the mechanical and thermal properties of facings. A designer should test the facing materials made by different laminating methods to acquire their mechanical, thermal, and other properties for designing calculations. The principles of the effects of processing methods on properties of the sandwich structures are described blow.

1. Hand layup and spray layup with open mold curing: The fiber content in the skins is low because air voids exist in the cured facings so that the mechanical properties of the facing are relatively low. The resin cannot completely impregnate the core surface so maximum bond strength between facings and core is not achieved.
2. Hand layup and spray layup with vacuum bag pressure curing or compression curing: All properties are better than those of the laminates made by the method 1.
3. RTM-Lite: This process fills the part under ½ atmosphere vacuum pressure. Most voids are eliminated, but mechanical properties are limited with generally higher resin content than vacuum bag or infusion.
4. Resin infusions: All infusion (VARTM) processes have a bag with vacuum or a mold with pressure assistances to draw close to 100% of air out from the laminate before introducing the liquid resin. The resin will fully fill the voids in the facing and core so that the fiber content can achieve high percentage, and the bonding strength between the facing and core also is high. A certain level of porosity is still present depending on type of resin and cure profile. Most porosity can be eliminated by autoclave cure.
5. Prepreg laminating by heating and pressure curing: The facing properties are similar to that of the sandwich made by resin infusions, but the bonding strength between the skin and core may not be as good as that of the sandwich made by the resin infusions because the resin in the prepreg sheet may not have a low viscosity even at high temperature, and the air in the

cavities on the core surface cannot be completely extracted out, unless cure is achieved in an autoclave.
6. Pultrusion: The resin is cured at a high temperature and high pressure in the pultrusion process so that the mechanical properties and the bonding strength between the facings and core of the sandwich laminate is the highest in all process methods if using the same resin system. Also, the temperature resistance of the facings is high due to the resin cured at the high temperature so no post curing and resin shrinkage will happen after laminating.

7.6.3 Influence of Product Application Conditions on Design Parameters

The application conditions are important for designing a sandwich structure. These conditions could include a hot or cold environment, a static or dynamic load, a dry or wet circumstance, a corrosive medium, and vibration state. A designer should conduct tests on the raw materials and the laminated elements in the application condition for obtaining the performances of the sandwich, then set up safety factors, and make the decision of choosing the right materials and processing method. Some attentions that should be paid in each application condition are presented below:

1. High-temperature application environment: The shear and compression moduli of the core and the tensile modulus of the facing materials will go down when a sandwich structure is used at a temperature that is higher than room temperature; especially, the core's properties will be lost remarkably. Also, the creeping property of the product at a high load and at a high temperature will increase.
2. Low-temperature application environment: The stiffness of a sandwich structure will increase when it is used at a temperature lower than the room temperature; especially, the shear stiffness has a big rise. However, the structure will become more brittle and less impact resistant. A designer should have the relationships of the raw materials' properties versus the temperature changing or should test the raw materials for abstaining the relationship for the best design if the product will be used at a high or low temperature.
3. Dynamic load: If a dynamic load is applied to a sandwich structure, the stiffness and strength of the core and facings will weaken with time (fatigue). Safety factors should be adjusted for the design requirements in the dynamic loading condition.
4. High humidity and under water application environment: The water and moisture will invade a structure through the facing into the interface of the core and facings if the facings are not waterproof materials. The water at the interface will expand at a high temperature close to the boiling point of water and at a low temperature below the freezing point, which reduces the bond strength between the skin and the core. In this case, a water resistance material must be considered to use as a facing material, or part of the facing materials, such as the adhesive and the liquid resin for the co-curing process.

Sandwich Structure Design

5. Instant impact on the structure: An instant impact could cause a facing crack, or a delamination between the facings and core, so that the product will lose function. The safety factor should be high enough for designing a structure for preventing the impact damage that could happen in handling, transportation, and application times.

7.7 PRINCIPLES AND EXAMPLES OF USING HYBRID CORES, REGIONAL REINFORCEMENTS, TRANSITIONS, AND CONNECTION ELEMENTS

Sandwich structures are usually not used alone; they need to be connected to a solid composite or metal components. Sometimes, the sandwich structure is not composed of uniform core and skin materials. For products with large sizes and complex shapes, different sections have different performance requirements, so the core materials could have different densities and thicknesses, and skin materials have different properties. It is an important task for designers to design the connections between the sandwich structure and the solid component. It is also an important task to select the core material with suitable density and thickness and the skin material with suitable strength according to the design requirements.

7.7.1 USE INTELLIGENT COMBINATIONS OF HYBRID CORE MATERIALS FOR A LARGE COMPLEX PRODUCT

In many applications, the requirements vary throughout the entire part. Some areas may be more weight-sensitive, while maximum stiffness is required in others. Using a single material configuration for the entire part may be simple, but in most cases, intelligent combinations of different materials will offer a better solution that optimally meets customer-specific target criteria.

Wind turbine blade is a very large product partially made by the sandwich structures as shown in Figure 7.14. Typically, a wind blade is made by the solid composite shear

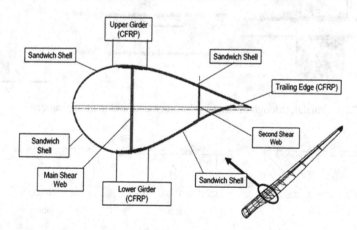

FIGURE 7.14 Wind turbine blade structure.

webs and girders, and sandwich structure shells. A large wind blade is about 60 m long with different sections from the root, middle to tip. Each section has requirements for the stiffness and antibulking. A designer should use the core materials with the different mechanical properties and thickness in different section. An end-grain balsa core usually is used in the section close to root, while a medium density foam core is used in the middle section, and a low-density foam core is used in the tip. The thickness of the cores also changes from thick to thin in the front to middle to tip section.

7.7.2 Regional (Local) Reinforcements

Sandwich structures are generally poorly suited for carrying localized loads, for the simple reason that the core does not have the stiffness necessary to distribute the forces effectively. Any localized load directly applied to a sandwich panel will cause deformations rather unlike those in a similar steel or aluminum structure. In a real application, a sandwich structure will be connected or fastened to the other sandwich or to a solid component by screw, bolt, joint, or adhesive. All joint and fastening areas need to be strengthened by a reginal reinforcement or called insert. A reginal reinforcement is a local change in stiffness and strength of the sandwich panel, the purpose of which is to distribute a localized load in an appropriate manner to the sandwich panel.

There are different types of reinforcements as shown in Figures 7.15–7.17 for different applications. The materials for reinforcements could be high density foam, solid wood, solid composite, and metal. A designer should test the properties of the reinforcements to make sure of the good bonding strength to the facings, also meeting the requirements of further applications.

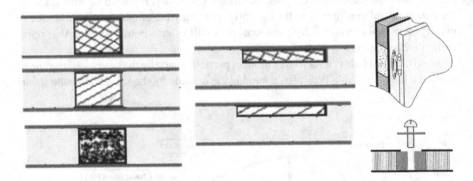

FIGURE 7.15 Partial, through thickness reinforcements and their applcations.

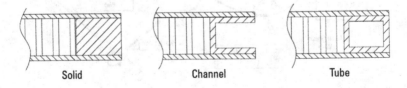

FIGURE 7.16 Edge reinforcements.

Sandwich Structure Design

7.7.3 Fasteners and Connections

Sandwich composite material is a structural material with good rigidity in terms of large size, but it is weak locally. Therefore, when connecting to each other or fastening with other rigid structural parts, surface contact should be achieved as much as possible, so that the bonding force is evenly transferred, and stress concentration is avoided. Also, a structural adhesive, such as epoxy, polyurethane, or acrylic adhesive, should be used along the interface to create face bonding strength. In addition, the edges and corners of the material should be reinforced and protected as much as possible during connection to avoid direct exposure. Any connection method should be tested or be confirmed by FEA modeling simulation.

7.7.3.1 Connect to Solid Structure

If to join a sandwich structure to a solid metal or composite structure as shown in Figure 7.18, one needs to reduce the thickness of the core gradually to zero, so the panel becomes a solid laminate at the edge, and then connect two components by facing joining with structural adhesive and bolt fastening. More examples are shown in Figure 7.19.

7.7.3.2 Straight Connect Sandwich Structures

To connect two sandwich structures straightly, a joint is necessary to transfer the force from one structure to another as shown in Figure 7.20. The joint could be an I-beam, a rectangular tube, rectangular solid high-density foam, or other structural materials. A structural adhesive will be used to bond the joint and panels together.

7.7.3.3 Right Angle Connection

Many joints can be used for connecting two sandwich panels in right angle as shown in Figure 7.21. The legs of the joints should be long enough to protect the edge for the panel and transfer the force from one panel to the other. Structural adhesives should be used for bonding the panel with the joints.

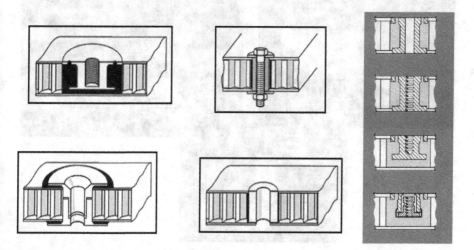

FIGURE 7.17 Reinforcements with fasteners.[7]

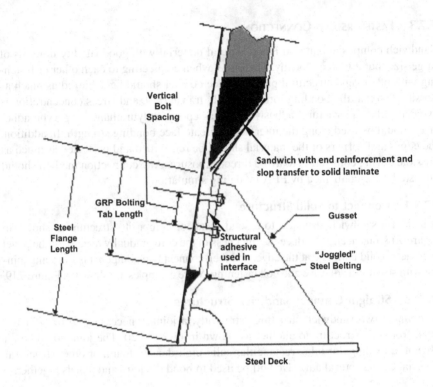

FIGURE 7.18 Example of joining a sandwich to a solid structure.[8]

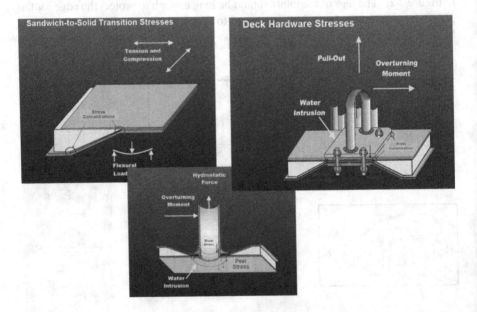

FIGURE 7.19 Example of joining a sandwich to a solid structure.[8]

Sandwich Structure Design

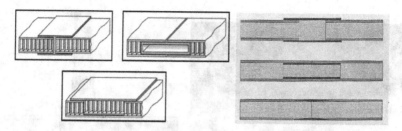

FIGURE 7.20 Connect two sandwich structures straightly by using a joint.

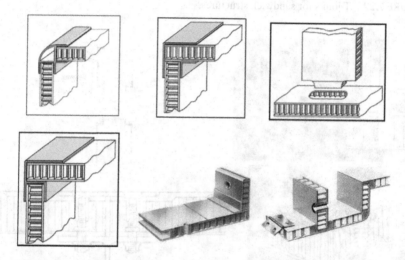

FIGURE 7.21 Examples of connection of sandwich structures in right angle.

7.7.3.4 T-Joint Connection

T-joint is one of the most common joint in sandwich structures, which is used wherever one sandwich panel is stiffened transversely by another panel. A T-joint consists of a horizontal base panel, a vertical leg panel, fillers, and triangular fillets. The purpose of the use of the filler and the triangular fillets is to transfer the load and creation of connection between the base panel and the leg. Many types of T-joints have been developed as shown in Figure 7.22. Adhesive bonding is a main method to make a T-joint even though some pins or bolts are used for holding the parts together initially and reinforcing the adhesion force. The mechanical tests and FEA simulation are necessary for confirming the design before making the final decision.

7.7.3.5 Fasten Sandwich Panel to Solid Structural Component

To fasten a sandwich panel to a solid composite, a metal frame, or a component by using screws or bolts is a common practice in the sandwich design. If the sandwich panel is made with regional reinforcing inserts, the bolts or screws should go through the inserts to the solid components. To fasten the uniform sandwich panel to the solid part directly, a set of plug and sleeve, or fastener inserts, should be used for

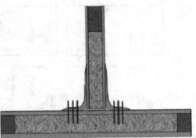

FIGURE 7.22 T-joints for sandwich structures.

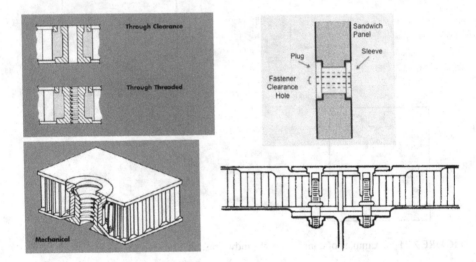

FIGURE 7.23 Fastener inserts and plug/sleeve sets used for fastening sandwich to solid part.[7]

reinforcing the panel as shown in Figure 7.23. Otherwise, the panel will crush when the screw or bolt is tightened down. The fastener inserts could be bonded with the panel by an adhesive or just simply put in the holes that are drilled previously. The large washers should be used for holding the bolt caps and the panel facings together tightly.

REFERENCES

1. Understanding Honeycomb Panels. https://www.plascore.com/honeycomb/honeycomb-panels/.
2. G. Caprino and R. Teti, *Sandwich Structures Handbook*. Padua, Italy: Il Prato Publisher, 1989.
3. X. Li, G. Li, C.H. Wang and M. You, "Optimum design of composite sandwich structures subjected to combined torsion and bending loads," *Applied Composite Materials*, vol. 19, no. 3, pp. 315–331, 2011.

4. L.J. Gibson, "Optimization of stiffness in sandwich beams with rigid foam cores," *Materials Science and Engineering*, vol. 67, pp. 125–135, 1984.
5. T.C. Triantafillou and L.J. Gibson, "Failure mode maps for foam core sandwich beams," *Materials Science and Engineering*, vol. 95, pp. 37–53, 1987.
6. I. Balbin, C. Bisagni, M.R. Schultz and M.W. Hilburger, "Scaling methodology for buckling of sandwich composite cylindrical structures, *AIAA/ASCE/AHS/ASC Structures, Structural Dynamics, and Materials Conference*, 2018. https://doi.org/10.2514/6.2018-1988.
7. Shur-log Advanced Composite Catalog. https://www.shur-lok.com/salesapp/product_dls/Advance_Composites.pdf.
8. E. Greene, "Marine Composites," *American Composites Manufacturing Association*, Tampa Convention Center, October 7, 2004.

8 Sandwich Composite Structure Modeling by Finite Element Method

Guohua Zhou

CONTENTS

8.1 Basics on Finite Element Analysis 296
    8.1.1 Purpose 296
    8.1.2 General Steps and Considerations of Performing Finite Element Analysis When Using Commercial Software 297
8.2 Skin/Face Sheet Modeling: Material Models and Parameter Characterization 302
    8.2.1 Hashin Damage Model and Parameter Characterization 302
8.3 Continuous Surface Core Modeling: Material Models and Parameter Characterization 303
    8.3.1 Crushable Foam Material Model and Parameter Characterization 304
    8.3.2 J2 Elastoplasticity Material Model 306
    8.3.3 End-Grain Balsa Wood Material Model and Parameter Characterization 306
8.4 Discontinuous (Honeycomb, Truss) Core Modeling Approaches 307
8.5 Adhesive and Debonding/Delamination Modeling Approaches 310
    8.5.1 Cohesive Zone Method 310
    8.5.2 Surface Separation Approach 313
    8.5.3 Connector Element Approach 313
8.6 Foam Core Sandwich Composite Modeling Example: Debonding of Foam Core Sandwich Composite Beam 314
8.7 Honeycomb Core Sandwich Composite Modeling Example: Nonlinear Response of Honeycomb Sandwich Composite Beam Subject to Four-Point Bending 320
8.8 Optimization of Sandwich Composite Design Based on Finite Element Analysis 326
8.9 Fatigue Life Modeling of Sandwich Composite via Finite Element Analysis 327
    8.9.1 The S-N Curve Method for Sandwich Composites 327
Appendix: Octave/Matlab code for transferring foam strengths for crushable foam material model in ABAQUS 329
References 331

DOI: 10.1201/9781003035374-8

Sandwich composites are formed by bonding together layers of skin/face sheet and core material in a certain order to fully utilize both strengths. Compared to simulation of commonly used metallic materials such as steel and monolithic composites like concrete, sandwich composite modeling requires special treatments due to the following features: most skin sheet and some core materials such as honeycomb and balsa wood are usually not as simple as isotropic; the mechanical properties of skin sheet and core are dramatically different, resulting in strong heterogeneity in the thickness direction across the bond interface; and failure could occur in one of the multiple complicated modes or mix of them – debonding fracture at the interface, and tensile, shear, and compressive failure in the skin sheet and core. Motivated by these extra complexities, this chapter aims to serve as a basic guideline for beginners on sandwich composite modeling by finite element method. It covers fundamentals and modeling techniques needed for sandwich composite simulation, and a couple of comprehensive numerical examples are provided to demonstrate the modeling process. Specifically, the content is arranged in the following order. The basic steps and considerations for numerical modeling using commercial finite element software are discussed in Section 8.1; how to choose material model for skin sheet, foam, end-grain balsa wood, and honeycomb core and how to calibrate the associated material parameters are covered in Sections 8.2–8.4; Section 8.5 introduces how to model debonding/delaminating between the skin sheet and core and the corresponding parameter characterization; two numerical examples on foam and honeycomb core sandwich composites failure are exercised to show the detailed modeling steps in Sections 8.6 and 8.7, respectively, where numerical results agree well with experimental data, indicating their effectiveness; in the end, some attempts are made to briefly introduce the fatigue and optimization modeling of the sandwich composite in Sections 8.8 and 8.9, respectively. In this chapter, the modeling techniques are chosen to be discussed owing to their simplicity and effectiveness along with the fact that they are also accessible in most commercial finite element software; the content in this chapter does not require the users to write their own codes or subroutines to supplement the modeling and only focus on structural analysis without the thermal effect considered.

Note that the input files of the two numerical examples on the foam and honeycomb core sandwich composite failure in Sections 8.6 and 8.7 are accessible to the reader and can be downloaded at the URL www.routledge.com/9780367441722, which should be helpful for practicing the modeling techniques introduced.

8.1 BASICS ON FINITE ELEMENT ANALYSIS

This section is devoted to introducing some general knowledge on finite element method, and some key steps and considerations on sandwich composite modeling when using commercial software.

8.1.1 Purpose

Finite element method, as a powerful tool, has been widely utilized to assist engineering design and scientific study since its inception in the early 1960s. In general, finite element analysis helps to save both the financial and time cost which would otherwise be spent on many physical tests in order to achieve progressive design iterations. With finite element simulation, these tests can be instead performed virtually on computer, of course provided the numerical model is validated against the experiment or verified first, and numerous design iterations can be conducted until an optimal design is achieved without increasing cost essentially. Specifically, in the field of sandwich composites, finite element method has also been applied intensively for improving the design, such as in References 1–6.

8.1.2 General Steps and Considerations of Performing Finite Element Analysis When Using Commercial Software

"If I had an hour to solve a problem, I'd spend 55 minutes thinking about the problem and 5 minutes thinking about solutions." – Albert Einstein. This quote indicates the importance of understanding a problem thoroughly before starting to figure out the solution. Similarly, when starting a finite element analysis (FEA), the first thing suggested is to understand completely and deeply what the physical problem to be modeled is. Intuitively, it is natural to tell the computer program the following information:

1. What is the geometry of the structure? Different geometries result in different physical problems, e.g., it takes different amounts of force to bend a thin and a thicker plate due to their geometrical difference in thickness.
2. What is the material of the structure? Different materials define different physical problems, e.g., using bare hands, it takes different amounts of force to bend a soft foam plate and a steel plate with identical dimensions.
3. How is the load applied? Different loads define different problems, e.g., the deflection of a plate may be different under self-gravity load and under some concentrated load.
4. What is the displacement boundary condition? Different displacement boundary conditions define different problems, e.g., it takes different amounts of force to bend a plate with one edge fixed to the ground base and a plate with two edges pinned to the ground base.

Finite element analysis is nothing but a process of utilizing a computer program tool to help realize the answers of questions 1 – 4, and therefore a good understanding of the questions is crucial for a successful modeling project. If one experiment is to be modeled, it is good to talk to the test engineer or witness the experiment process to understand the problem better. On the other hand, if the problem does not seem straightforward to you in the first place, it may be worth spending some time ahead to investigate the public domain such as papers, books, and reports to see whether there is any similar simulation work conducted; if there do exist some similar studies, then the knowledge, conclusions, lessons, and suggestions of these studies can be directly borrowed and applied in your coming simulation work such that some modeling time which could be spent on iterations on mistake correction, better efficiency, and/or better accuracy can be saved; actually, this is the exact purpose of this chapter writing – hopefully helping to equip the reader efficiently with some basic knowledge and overall preparation on sandwich composite modeling and reducing their detour effort.

Once the answers of questions 1 – 4 are clear, it is time to work on the specific modeling details using finite element software. The well-developed commercial software, such as ABAQUS,[7] LS-DYNA,[8] and NASTRAN,[9] can help to handle most sandwich composite and other engineering design problems. Which commercial software to choose is a comprehensive question, and the following factors can be considered: the software availability to you, cost difference, the characteristics of the software (i.e., different pieces of software are good at different aspects – linear computation, nonlinear static/implicit computation, or nonlinear dynamic/explicit computation, fatigue, optimization), and specific functionality availability in the software. Under normal conditions, usually different pieces of commercial software should still give very similar simulation results. Some basic steps and considerations when using commercial software for modeling are introduced below.

1. Choose a consistent unit system.

    Using different unit systems does not change the final physical result; it is just a matter of convenience. Two common unit systems used in finite element analysis are provided in Table 8.1 as references; the lengths are both in unit "mm", which makes it easier for meshing and computation for most sandwich composites and other structures with similar length scale.

    Pick the unit system based on your convenience. For example, if the computation is mainly focused on frequency-related analysis such as frequency response analysis, noise, and vibration analysis, it is good to choose unit system #1, since the unit for time is "s" and the frequency is naturally physical as "Hz" (1/s); or if the unite system #2 is picked, then the frequency comes out as 1/ms, i.e., 1,000 times "Hz", which is not so convenient conceptually. On the other hand, if the analysis is devoted for high strain rate events such as high velocity impact, crash, and blast cases, it would be easier if system #2 is chosen; the reason is that these events are occurring within very short time windows, mainly on the level of "ms" or even "µs", and therefore, if "s" is used for time, it takes some effort in our brain to transfer them into "ms" or "µs" by removing some zeros behind the decimal point. Of course, your own different consistent unit system can also be developed for a special

TABLE 8.1
Two Consistent Unit Systems for Finite Element Analysis

Unit System Numbering	Time	Mass	Length	Force	Stress	Energy	Density of Steel	Young's Modulus of Steel	Velocity 35 mph (56.33 km/h)	Gravitational Acceleration
System #1	s	ton	mm	N	MPa	N×mm	7.83E−9	2.07E+5	1.56E+4	9.806E+3
System #2	ms	kg	mm	kN	GPa	kN×mm	7.83E−6	2.07E+2	15.65	9.806E−3

need if necessary. Once one system is decided, stick to it in order to avoid unnecessary errors due to unit inconsistency.
2. Import the CAD (Computer-aided design) file or draw it to define the geometry and mesh it by finite element preprocessor.

As shown in Figure 8.1, a finite element modeling process technically starts with a CAD file, which defines the geometry by points, lines, surfaces, and solids. These geometric components cannot be directly utilized for finite element analysis until they are meshed. If there is a CAD file from the design team such as a CATIA file, it can be directly imported into some finite element preprocessor for meshing (the second step in Figure 8.1); if the model is small and not complicated like a small flat plate component, usually the preprocessors embedded with the solver should be sufficient such as ABAQUS CAE GUI, LS-PREPOST that comes with LS-DYNA. There are some excellent commercial preprocessor tools such as HyperWorks (Hyper Mesh and Hyper View),[10] and ANSA,[11] which also provide corresponding postprocessor products, and they can be used to mesh big models such as a car or an air plane model, which usually involves a few millions of nodes and elements. One thing to keep in mind is that the length unit of the imported CAD file and the mesh should obey the chosen consistent unit system.

Regarding the finite element meshing, attention should be paid to the following items:
- Mesh size

In general, a smaller mesh size leads to higher accuracy; however, if the problem involves damage or material softening, the trend does not necessarily hold – the problem solution becomes mesh size-dependent; in this case, the mesh size represents some sort of physical damage length scale, and it needs to be calibrated with experimental data. On the other hand, the smaller mesh size, the more degree of freedom, and therefore the longer computational time. The mesh size will also be limited by the available computational power, eventually by the affordable waiting time; it is case-dependent, and it may be a good practice to finish a job run within 24 hours, i.e., submit the simulation job today, and the result will be available for evaluation tomorrow. If there is a symmetry in the problem geometry and the boundary condition, then it can be utilized to reduce the model size by half or even more with the same accuracy maintained.

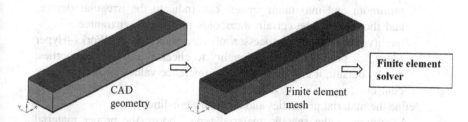

FIGURE 8.1 General steps of a finite element analysis.

- Element type and element order

  Depending on the geometry and the loading condition of the problem, it is able to decide whether a structural element, such as beam, plate, and shell, 2-dimensional plane element, or solid element should be used. Usually, a structural element is preferred if the accuracy is sufficient because it reduces the total degree of freedom compared to the case where 2-dimensional plane element or solid element is otherwise used. If the ratio of span to height is greater than 3, the shell and beam element can be considered: more specifically, if less than 10, the thick shell and Timoshenko beam type of element can be used with shear strain included; if higher than 10, the thin shell and Euler-Bernoulli beam type of element can be used with the assumption of zero shear strain[12]; otherwise, use 3-dimensional solid element or 2-dimensional plane element.

  Regarding element order, usually for large-scale engineering problem, first order (i.e., linear) element is practical and efficient, and if necessary, for some local areas, higher order element can be used to improve local accuracy.

- Element integration scheme

  For a linear 3-dimensional solid element, full integration scheme which needs $8 (= 2 \times 2 \times 2)$ integration points usually improves the accuracy except when the shear locking occurs. Compared to the reduced integration scheme which only requires one integration point, the full integration scheme increases the simulation time significantly due to its seven times of integration point increase. Therefore, for large models, linear 3-dimensional solid element with reduced integration is desired. However, when the reduced integration scheme is used, hourglassing mode may show up and pollute the simulation result, and in commercial software, the hourglassing control should be turned on, and/or use the full integration scheme only within these local area where the hourglassing mode is observed being excited. On the other hand, volumetric locking is another issue of linear 3-dimensional solid element when modeling incompressible materials such as rubber; it is unlikely to be an issue for the focus of this chapter, sandwich composite modeling.

- Mesh quality control

  Regular mesh gives the best accuracy; irregular mesh deteriorates the accuracy: the severer the irregularity or distortion of the mesh, the less the accuracy. The mesh indexes, Jacobian, warpage, aspect ratio, minimum and maximum angles, etc. indicate the irregular degree, and there should be certain thresholds in order to guarantee a high-quality mesh. In the preprocessor software, such as HyperWorks–Hyper Mesh,[10] it provides the functionality to check these mesh properties, and by default, it also provides some reference values for mesh quality control.

3. Define the material properties and boundary conditions.

   According to the specific material type, choose the proper material model in the software; regarding the material models needed for sandwich

composite simulation, the details will be covered by Sections 8.2–8.4. Based on your understanding of the physical problem, define the boundary conditions including loading and displacement constraint types.
4. Select the solver (the third step in Figure 8.1), set up key parameters, and submit analysis.

Some important solver parameters are discussed below.
- Nonlinear geometry switch

    If the problem involves large deformation or large rotation, the nonlinear geometry switch should be turned on; otherwise keep it off (i.e., use the linear option), which speeds up the computation significantly.
- The implicit or explicit solver?

    If the problem is static or quasi-static, where the dynamic inertia effect is not significant, it is good to choose the implicit solver. When using the implicit solver, a bigger time step is allowed, which in the end reduces the simulation time, and the residual balance error is smaller since at each step, the conservation equations are solved to very small error. If the problem is dynamic with high strain rate effect such as impact crash modeling, it is good to use the explicit solver, which usually requires a much smaller time step and the result of every time step moment will be available for detailed response analysis, i.e., a high-resolution result is achieved. It is worth mentioning that the explicit solver can also be used to model quasi-static events, still with a very small time step, and takes a long time to finish the simulation; it is wise to do so when the problem involves high nonlinearity or even element erosion, and the implicit solver encounters trouble to ultimately converge, or when some material models or functionalities are only available in the explicit solver.
- Time step

    This is usually not a problem for commercial software which should have the automatic time step estimation capability after each time step; usually, there should be a scaling factor less than or equal to one, available to the user to further control the time integration accuracy, or the user can also specify a fixed time step which should be within the range of stable time steps. Note that in the explicit solver, the time step represents the physical time moment, whereas in the implicit solver, when modeling static or quasi-static problems with no inertia effect considered, the time step is actually not physically meaningful, and it is a pseudo-time, which is just used to divide the nonlinear problem into a set of smaller consecutive problems to help achieve the final convergence of the problem.
5. Verify the simulation, and validate it if possible

    It is good to check the global internal and kinematic energy output statistic curves to make sure the simulation is stable, i.e., the modeling did not blow up. Refining the mesh in the interested local area where stress concentration is observed should improve the local accuracy. If computationally affordable, higher order element or re-meshing with a smaller mesh size globally can be used to rerun the simulation and check whether there is a convergence trend for some target variables. In mechanics point of view, try to extract the physical meaning out of the simulation result, and the result has to make sense to the analysts before implementing it to design. Of course, it would be great to compare the modeling result with test data if available, for validation purpose.

As an analyst, a deep understanding of finite element theory should be helpful for improving the confidence and efficiency of the simulation. If interested in learning more theory on the finite element method, the following two books are highly recommended: linear finite element theory – *The finite element method: linear static and dynamic finite element analysis* by Hughes, T.J.[13] and nonlinear finite element theory – *Nonlinear finite elements for continua and structures* by Belytschko, T. et al.[14]

8.2 SKIN/FACE SHEET MODELING: MATERIAL MODELS AND PARAMETER CHARACTERIZATION

The skin/face sheet material is usually modeled by shell element due to its geometry feature that the ratio of its width and length to thickness is over 3 – thick shell element and even 10 – thin shell element. In this section, one damage model, Hashin[15] will be introduced. This model has been used frequently, such as in the work,[16,17] and it can be used for the skin sheet shell element.

8.2.1 HASHIN DAMAGE MODEL AND PARAMETER CHARACTERIZATION

- Material model

    In this section, σ_{11}, σ_{22}, and σ_{12} denote stresses in fiber direction, transverse to the fiber direction, and in-plane shear direction, respectively. σ_{11}^+ and σ_{11}^- are tensile and compressive strengths in fiber direction, respectively; σ_{22}^+ and σ_{22}^- are tensile and compressive strengths in fiber transverse direction, respectively; σ_{12}^1 = shear strength in fiber direction; and σ_{12}^2 = shear strength in fiber transverse direction.

    In Hashin damage model,[15] the material failure is described by the following four modes:

    tensile fiber mode,

    $$\left(\frac{\sigma_{11}}{\sigma_{11}^+}\right)^2 + \left(\frac{\sigma_{12}}{\sigma_{12}^1}\right)^2 = 1, \text{ when } \sigma_{11} > 0;$$

    fiber compressive mode,

    $$\sigma_{11} = -\sigma_{11}^-, \text{ when } \sigma_{11} < 0;$$

    tensile matrix mode,

    $$\left(\frac{\sigma_{22}}{\sigma_{22}^+}\right)^2 + \left(\frac{\sigma_{12}}{\sigma_{12}^1}\right)^2 = 1, \text{ when } \sigma_{22} > 0;$$

and compressive matrix mode,

$$\left(\frac{\sigma_{22}}{2\sigma_{12}^2}\right)^2 + \left[\left(\frac{\sigma_{22}^-}{2\sigma_{12}^2}\right)^2 - 1\right]\frac{\sigma_{22}}{\sigma_{22}^-} + \left(\frac{\sigma_{12}}{\sigma_{12}^1}\right)^2 = 1, \text{ when } \sigma_{22} < 0.$$

- **Calibration**
  Obviously, the above four failure modes involve six strength parameters: $\sigma_{11}^+, \sigma_{11}^-, \sigma_{22}^+, \sigma_{22}^-, \sigma_{12}^1,$ and σ_{12}^2, which should be obtained from known references or direct tests. On the other hand, to enable the simulation, the six elastic moduli in *lamina* type also need to be provided: E_1 – Young's modulus in fiber direction; E_2 – Young's modulus in fiber transverse direction; v_{12} – Poisson's ratio in the plane; G_{12} – shear modulus in the plane; G_{13} – shear modulus between fiber direction and out of plane direction; G_{23} – shear modulus between fiber transverse direction and out of plane direction. Usually, all the above elastic moduli and failure parameters should be available from the mechanical data sheet given by the manufacturing company. If proper tests need to be conducted, the following test standards can be referenced: $\sigma_{11}^+, \sigma_{22}^+, E_1, E_2,$ and v_{12} can be obtained by the test, ASTM D3039; σ_{11}^- and σ_{22}^- can be tested by the test, ASTM D6641; and $\sigma_{12}^1, \sigma_{12}^2, G_{12}, G_{13},$ and G_{23} can be provided by the test, ASTM D5379. Note that the test standards listed here and others given in the rest of this chapter are all provided as recommendations based on the testing knowledge and experience of the author of this chapter, and other equivalent test standards may be also applicable to obtain the desired properties; it is always good to confirm the test standards with the professional testing company.
  The Hashin damage material model is implemented in some commercial finite element software, such as in LS-DYNA (*MAT_ENHANCED_COMPOSITE_DAMAGE*) and ABAQUS (*DAMAGE INITIATION, CRITERION=HASHIN*). Note that the symbol "*" used in the last sentence and the rest places of this chapter is to indicate a keyword control card in the input file of some finite element software, e.g., LS-DYNA and ABAQUS. The specific example of involving Hashin damage material model for skin sheet material will be shown in Section 8.7.

8.3 CONTINUOUS SURFACE CORE MODELING: MATERIAL MODELS AND PARAMETER CHARACTERIZATION

Different foam and balsa wood-based products are made from different ingredients and/or by different manufacturing processes. As a result, the core materials show different mechanical behaviors, which correspondingly require suitable material models to capture the correct responses. Foam and balsa wood cores can be modeled by solid elements if the detailed response such as stress, strain, and failure modes are desired. If the analysis is only limited to linear behavior, such as linear stiffness analysis and linear vibration analysis, then isotropic, orthotropic, and anisotropic linear elastic material models are sufficient, and should be accessible in most commercial finite element software; for example, in NASTRAN,[9] isotropic linear elastic material model is *MAT1*; orthotropic one is *MAT8*; and anisotropic one is provided as *MAT9*. Usually linear elastic models are straightforward by referring to the software manual, and they are not covered herein. In this section, one nonlinear foam material model, the crushable foam material model,[18] which is good for capturing the tensile, compressive, and shear strengths, is discussed in Section 8.3.1. This foam material model can be used to model the structural foam core materials such as PET, PVC, SAN, and rigid polyurethane, which have relatively high stiffness and are applied to bear the shear load of the sandwich composite structures. Note that nonstructural foams such as low density foams for cushions, and other open cell and highly compressible and recoverable foams are not covered here; they should be modeled by other suitable material models such as Fu Change foam material model[19] and Hyperfoam material model.[7] Sometimes, if the material properties are not complete yet for the crushable foam material model, J2 elastoplasticity material model can be used to capture the shear strength behavior of the foam core, which is discussed in Section 8.3.2. Finally, in Section 8.3.3, the material model, *MAT_WOOD* in LS-DYNA[20] is introduced to describe end-grain balsa wood. These material models are all or partially provided in commercial software, such as in ABAQUS[7] and LS-DYNA.[20]

8.3.1 CRUSHABLE FOAM MATERIAL MODEL AND PARAMETER CHARACTERIZATION

- Material model

Crushable foam material model was originally proposed by Deshpande et al.[18] It is described as *Crushable foam material with isotropic hardening* in the ABAQUS manual,[7] which assumes the tensile and compressive strengths are the same. Actually, in the manual, other than the original model, another slightly modified model, *Crushable foam material with volumetric hardening* is also provided; this model allows different tensile and compressive strengths, which is typical for most structural foam core materials, and will be discussed in what follows.

The yield surface is defined as

$$F(p, q) = \sqrt{q^2 + \alpha^2 (p - p_0)^2} - B = 0,$$

which is an ellipse in $p = -(\sigma_{11} + \sigma_{22} + \sigma_{33})/3$, which is the pressure, and $q = \sqrt{\frac{3}{2} \mathbf{S} : \mathbf{S}}$, which is the Von Mises stress, with $\mathbf{S} = \sigma + p\mathbf{I}$, the deviatoric stress tensor. $B = \alpha A$ and $A = (p_c + p_t)/2$ are the size of the q-axis and p-axis of the ellipse, respectively; p_c and p_t are the hydrostatic compressive strength and hydrostatic tensile strength, respectively. Also, $p_0 = (p_c - p_t)/2$ is the center of the ellipse on the p-axis.

- Calibration

One set of material parameters are suggested by a paper[18] and the manual of ABAQUS[7] to calibrate the above material model:

1. initial yield strength under uniaxial compression, σ_c^0,
2. initial yield strength under hydrostatic compression, p_c^0, and
3. initial yield strength under hydrostatic tension, p_t^0.

With the parameters in 1, 2, and 3 given, the initial yield function can be fully determined: p_c^0 and p_t^0 are directly provided, and the only unknown α can be obtained by plugging the uniaxial compression stress state, $p = \sigma_c^0 / 3$, $q = \sigma_c^0$ into the yield function,

$$\alpha = \frac{3k}{\sqrt{(3k_t + k)(3-k)}}, \text{ with } k = \frac{\sigma_c^0}{p_c^0} \text{ and } k_t = \frac{p_t^0}{p_c^0}.$$

Usually, the typical material data sheet from the manufacturing company does not necessarily provide exactly parameters (1), (2), and (3); instead, (4) initial yield strength under uniaxial tension, σ_t^0, (5) initial yield strength under shear, τ^0, will be provided together with property (1), σ_c^0.

Of course, (1), (2), and (3) can be additionally requested, which would cost extra time and funding. This difficulty was actually encountered by the author of this chapter, as a computer-aided engineering (CAE) engineer when working on the foam core sandwich composite panel design evaluation. It is desirable to study whether the commonly given parameters (1), (4), and (5) can be equivalently used to calibrate the material model. The given parameters (1), (4), and (5) actually represent the data point F, D, and E, respectively, on the yield surface plot shown in Figure 8.2. These three data points are sufficient for fully determining the yield surface, which has been validated by the work by Carranza et al..[21] To put it simple, the yield surface involves three unknowns, and three given data points are sufficient for solving the solution; in the appendix section, at the end of this chapter, for your convenience, a short code in Octave/Matlab format is attached for converting (1), (4), and (5) into (1), (2), and (3). If these three strengths, (1),

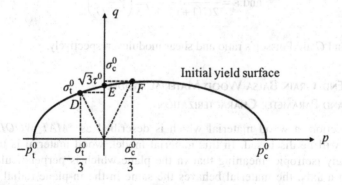

FIGURE 8.2 Yield surface in p–q stress plane of the material model, crushable foam model with volumetric hardening.

(4), and (5) need to be tested, the following test standards can be referenced: shear strength – ASTM C273, compressive strength – ASTM D1621, and tensile strength – ASTM D1623. In Section 8.6, the specific numerical example provided involves this material property-converting approach and the material model, *Crushable foam material with volumetric hardening*. Note that to the knowledge of the author of this chapter, this crushable foam material is equivalently provided in LS-DYNA as *MAT_BILKHU/DUBOIS_FOAM*.

8.3.2 J2 Elastoplasticity Material Model

Note that in *Crushable foam material with volumetric hardening* discussed just now, set the parameter, $\alpha = 0$, and the crushable foam material model reduces to the J2 elastoplasticity material model, which has been frequently used for modeling metallic materials such as steel and aluminum. For more details on J2 elastoplasticity, please refer to an excellent book, *Computational inelasticity* by Simo J.C. et al.,[22] If in reality, only the shear strength is available for the foam core, i.e., the other two strengths are not measured and provided yet, then J2 elastoplasticity material model can be applied to model the foam core material behavior, at least for a preliminary evaluation, which has been done by Rothschild[23] and a reliable result has been obtained. Shear strength is important here since the foam core is desired to bear the shear load of the sandwich composite structural component. In most software, for J2 elastoplasticity material model, it usually requires the stress-strain curve in uniaxial tensile test as a key input to define the plastic behavior, resulting in a need to convert the given shear strength or the shear stress-strain curve into the wanted one by the software. Assuming the shear stress-strain curve is provided by $\tau(\gamma)$, it can be converted to an uniaxial stress-strain curve, $\sigma(\varepsilon)$, via the following relationship,[23,24]

$$\sigma = \sqrt{3}\tau,$$

$$\text{and } \varepsilon = \frac{\sqrt{3}\tau}{2G(1+v)} + \frac{1}{\sqrt{3}}\left(\gamma - \frac{\tau}{G}\right),$$

where v and G are Poisson's ratio and shear modulus, respectively.

8.3.3 End-Grain Balsa Wood Material Model and Parameter Characterization

In this section, a wood material which is described as *MAT_WOOD* in LS-DYNA[20] will be discussed. In this material model, wood material is treated as transversely isotropic, meaning that in the plane which is perpendicular to the wood grain axis, the material behaves the same in the in-plane radial and tangential directions. This transversely isotropic assumption should be reasonable for most wood materials. This model is developed based on the Hashin damage model,[15] which is introduced in Section 8.2.1, and more details on it can be found in LS-DYNA theory manual.[8]

- **Calibration**

    In order to feed this wood material model, the following properties are required.

    Five moduli for elasticity:

    $E_{11}, E_{22}, G_{12}, G_{23},$ and v_{12}

    where E, G, and v denote Young's modulus, shear modulus, and Poisson's ratio, respectively; and the direction is defined as, 1 – the grain direction, 2 – tangential direction in the plane perpendicular to the grain direction, and 3 – radial direction in the plane perpendicular to the grain direction.

    Six material properties for the yield surface:
    1. Tensile strength along the grain direction,
    2. Compressive strength along the grain direction,
    3. Tensile strength perpendicular to the grain direction,
    4. Compressive strength perpendicular to the grain direction,
    5. Shear strength along the grain direction,
    6. Shear strength perpendicular to the grain direction.

    The above properties can be obtained by the test, ASTM D143.

8.4 DISCONTINUOUS (HONEYCOMB, TRUSS) CORE MODELING APPROACHES

The skin/face sheet, as described in Section 8.2, is effective to be modeled by shell element, and the discontinuous core modeling approaches can be chosen depending on the problem scale, loading condition, and interested structure response. Here, without loss of generality, one of the discontinuous core types, honeycomb core is used to introduce the modeling approach, which can be readily applied to other discontinuous cores such as truss core. The potential approaches are discussed in the following.

- Approach 1: model the honeycomb core walls in detail by shell element.[6]

    If the problem size in study is relatively small, and computationally affordable, it is desirable to model the honeycomb core by using shell element to represent honeycomb walls in detail. That way, as an example, assuming the honeycomb core is made of aluminum, J2 elastoplasticity material model is sufficient for describing this aluminum shell wall in the simulation, which is very simple and convenient. With simulation in this scale, the buckling and crushing details of the honeycomb walls can be captured.
- Approach 2: represent the honeycomb core as a homogenized solid, and model it by solid elements.[1,25]

    If big honeycomb core sandwich composite structures such as bus side and roof panels and large airplane panels are to be simulated, it is good to model the honeycomb core component as a homogenized solid since it

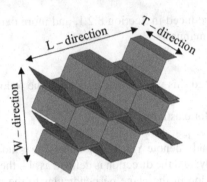

FIGURE 8.3 Honeycomb material direction definition.

is computationally too expensive for approach 1 in this case. The small disadvantage of this approach is that the failure details of the honeycomb walls cannot be captured as accurate as in approach 1; instead, the global response and homogenized local response will be modeled. Of course, if a solid element is used, the mechanical properties of solid material should be representative for the honeycomb walls after homogenization and should be extracted from the honeycomb core coupon tests.[1,25] In LS-DYNA, the material model #126 (*MAT_MODIFIED_HONEYCOMB)[20] is developed for this purpose. The material direction is denoted by L-, W-, and T- direction as shown in Figure 8.3. In this material model, a nonlinear elastoplastic material behavior will be defined independently for each normal and shear stress component, and it requires six uncoupled material test curves[20]:

1. the compression curve in W-direction,
2. the compression curve in L-direction,
3. the compression curve in T-direction,
4. the shear curve in LT-plane,
5. the shear curve in WT-plane,
6. and the shear curve in LW-plane.

More details on this material model can be found in the LS-DYNA material model manual,[20] and in a work by Heimbs et al.,[25] it provides the photo of all the tests for these specific material curves, which are shown in Figure 8.4 for your reference. In Section 8.7, one specific example involves this honeycomb modeling approach.

In reality, if the physical tests in (1)–(6) are not easy to be performed due to time or cost limitation, it could be a good idea to conduct equivalent virtual tests to obtain the curves by simulating tests (1)–(6) using approach 1,[1] i.e., modeling the corresponding coupon tests with shell element representing the honeycomb wall details.

- Approach 3: combine approaches 1 and 2.

  In this approach, within the key area, use approach 1, i.e., put the detailed shell element to represent the honeycomb walls, and in the rest distant

Finite Element Method

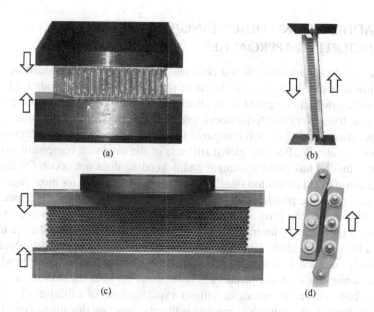

FIGURE 8.4 Honeycomb material curve tests: (a) compression in T-direction, (b) shear in LT- and WT-plane, (c) compression in L- and W-direction, and (d) shear in LW-plane.[25]

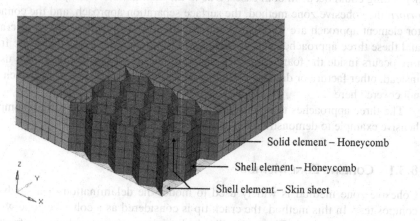

FIGURE 8.5 Honeycomb core modeling: key zone by detailed shell walls; distant area by equivalent solid; and the interface connection between shell and solid by sharing element nodes.

region, apply approach 2, i.e., use homogenized solid element to model the honeycomb walls[3], which is illustrated in Figure 8.5. At the interface of these two areas, define the connection by either sharing nodes or by tie constraints. This interesting and effective approach inherits both advantages of approaches 1 and 2, and the detailed local wall responses can be captured in the crucial region without an unaffordable computational cost.

8.5 ADHESIVE AND DEBONDING/DELAMINATION MODELING APPROACHES

In some sandwich composites, the core and the skin sheet are bonded by a specific, thin adhesive layer. The thickness of the adhesive layer is around 0.1 mm or even smaller, which compared to the thickness of the skin sheet and core is very small. Due to the very small thickness, together with the fact that the adhesive usually has relatively high moduli compared to most core materials, the adhesive layer does not essentially affect the global stiffness of the sandwich composite, provided that the adhesive has enough strength and debonding does not occur. On the other hand, mesh size, 0.1 mm is too small for a regular solid element for most engineering sandwich composite problems and leads to severe mesh irregularity in aspect ratio, which deteriorates the final accuracy. Therefore, if there is no debonding expected to occur at the adhesive layer, it is not necessary to model the thin adhesive layer in sandwich composite simulation, i.e., the core element and the skin sheet element can be directly connected by sharing nodes, or tie constraint. On the other hand, there are also some sandwich composites where the bond between the skin sheet and core is achieved by co-curing processes without a specific layer of adhesive. The debonding/delamination of sandwich composite is the phenomenon that the facture initiates and propagates along the interface between the skin sheet and the core material. No matter whether the interface bond is formed with or without thin adhesive layers, the debonding could occur in both cases. Due to the fact that the crack path is known *a priori*, the cohesive zone method, the surface separation approach, and the connector element approach are well suited and effective for capturing this type of crack, and these three approaches will be introduced in this section. Note that if the fracture occurs inside the foam core, then these are not the proper methods to be used; instead, other facture or damage modeling approaches can be considered, which are not covered here.

The three approaches to be introduced below will be applied to solve a comprehensive example to demonstrate the detailed steps in Section 8.6.

8.5.1 COHESIVE ZONE METHOD

Cohesive zone method is widely used to model the delamination of a sandwich composite.[2,6] In this method, the crack tip is considered as a cohesive zone with a small length as shown in Figure 8.6. Physically, it is consistent with ductile fracture such as fracture in ductile metals; within the small area around crack tip, plastic deformation occurs, and the stress level there is finite, which is different from the linear fracture mechanics where the tip is idealized and introduces singularity in stress field. Most core and adhesive materials are ductile, and therefore, the cohesive zone method is a good fit for modeling the delamination between skin sheet and core material in sandwich composites. Within the cohesive zone, the traction force T between two separating surfaces first grows as the surface separation δ increases; after the separation reaches the critical value δ_0 and T arrives at its peak value T_{ult}, the trend becomes opposite, i.e., the force goes down when separation further increases, representing the zone is cracking and the material is softening;

Finite Element Method   311

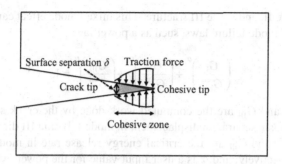

FIGURE 8.6 Schematic of the cohesive zone method.

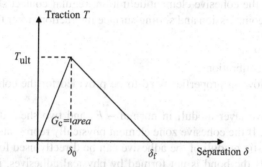

FIGURE 8.7 Linear damage law for cohesive zone.

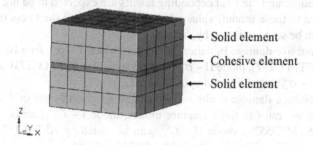

FIGURE 8.8 Three-dimensional cohesive element.

a damage law can be defined to describe this behavior, such as a linear damage law in Figure 8.7. Numerically, the cohesive zone method can be fulfilled by defining the cohesive element with given damage law and other associated material properties at the debonding interface; in Figure 8.8, 3-dimensional cohesive elements are drawn for illustration purpose.

In ABAQUS, *COH3D8* is a 3-dimensional cohesive element type. It can be used to model a three-dimensional cohesive layer, which can be physically treated as a thin layer of adhesive if actually used. In real-world problems, it is likely that composite delamination processes are a mixed mode fracture problem, as a combination of mode I, mode II, and mode III fracture. This mixed mode effect can be taken care by some mixed mode failure laws, such as a power law,

$$\left(\frac{G_\text{I}}{G_\text{I}^\text{cri}}\right)^\alpha + \left(\frac{G_\text{II}}{G_\text{II}^\text{cri}}\right)^\alpha + \left(\frac{G_\text{III}}{G_\text{III}^\text{cri}}\right)^\alpha = 1,$$

where G_I, G_II, and G_III are the computed work done by the crack surface traction along the conjugate separation displacement in mode I, II, and III directions, respectively; G_I^cri, G_II^cri, and G_III^cri are the critical energy release rate in mode I, II, and III directions, respectively; and α is a user input value for the power, which can be set as unity by default. In this chapter, since the cohesive element is defined for realizing the cohesive zone method, it is denoted as the *cohesive element approach*.

In addition to the cohesive element definition, regular contact should also be set up to handle the compression and sliding surface interactions after the cohesive layer is softened.

- Parameter calibration
  The following properties need to be provided for the cohesive element approach.
  1. Cohesive layer moduli in normal – E_n and two shear directions – E_t and E_s. If the cohesive zone element physically represents the adhesive, the elastic moduli of the adhesive can be directly used for the cohesive layer. If the bond is not formed by physical adhesives, instead by the co-curing process without a specific layer of adhesive, the moduli of a general adhesive or that of the sandwich core may serve as a good estimation; and the final debonding results are expected to be not so sensitive to these moduli values due to the very small thickness (thickness can be set to 0.01 mm by default).
  2. Cohesive damage initiation strength in mode I – $\sigma_\text{I}^\text{ini}$ (can be tested by ASTM C297), mode II – $\sigma_\text{II}^\text{ini}$ (can be tested by ASTM C273), and mode III – $\sigma_\text{III}^\text{ini}$ (can be tested by ASTM C273).
  3. Cohesive damage evolution, which can be given by the critical energy release rates in three fracture modes, mode I – G_I^cri (can be tested by ASTM D5528), mode II – G_II^cri (can be tested by ASTM D7905), and mode III – G_III^cri (can be tested by ASTM D7905).

  Note that the debonding does not necessarily occur for all sandwich composites, e.g., if the core is too weak in tensile or shear strengths, then failure in the core may always take place first before any debonding. Therefore, practically, the tensile and shear strength tests via ASTM C297 and ASTM C273, respectively, which are easier and less expensive to be performed, can be used to judge whether the debonding occurs before the foam core failure: if the core fails first, then no need to move to the G_I^cri, G_II^cri, and G_III^cri tests, and it is reasonable to assume that the debonding actually does not take place; if the debonding happens first, then it is confident to proceed to conduct more complicated G_I^cri, G_II^cri, and G_III^cri tests, and the previously tested tensile and shear strengths can be used as the two required input parameters, damage initiation strengths in the above (2).

8.5.2 Surface Separation Approach

The essence of this approach is actually the same as the cohesive element approach discussed in the last section. The only difference appears in how the cohesive behavior is enforced: in the cohesive element approach, it is implemented in the cohesive element, whereas in the *surface separation approach*, the cohesive behavior is embedded in the surface interaction without defining an additional layer of elements. This approach is available in ABAQUS, and other commercial software should have a similar functionality.

- Parameter calibration

  The input parameters of this approach is exactly the same as that of the cohesive element approach in Section 8.5.1, except the contact moduli in normal $E_n^{contact}$ and two shear directions $E_t^{contact}$ and $E_s^{contact}$ are scaled by the thickness of the cohesive layer; *Thickness*:

  $$E_n^{contact} = \frac{E_n}{Thickness}; E_t^{contact} = \frac{E_t}{Thickness}; \text{ and } E_s^{contact} = \frac{E_s}{Thickness}.$$

8.5.3 Connector Element Approach

A connector element can also be used to model the delamination. A connector element is constructed to connect two nodes crossing the interface with a force–displacement curve defined, like a spring when it is in the elastic range; once it is beyond the maximum force capacity, then damage comes into play and scales down the force. The advantage of this method is its simplicity and effectiveness. This approach is called the *connector element approach* in this chapter and is available in ABAQUS; to the knowledge of the author of this chapter, the connector element in ABAQUS is essentially similar to a nonlinear spring element in other commercial software.

- Parameter calibration

  The input parameters of this connector element approach are as follows:
  1. Connector stiffnesses in normal – K_n and two shear directions – K_t and K_s, i.e., the elastic slopes between forces and separations when the force is less than the maximum tensile and shear force capacities.
  2. Connector damage initiation force values – F_n^{max}, F_t^{max} and F_s^{max}, i.e., the maximum tensile and shear force capacity of the connector.
  3. Connector damage evolution curve which defines how the connector force goes down as a function of the separation of the two connected nodes; it can be linear, exponential, or by a table.

     The above parameters (1), (2), and (3) can be equivalently derived from the measured data in the cohesive element approach discussed in Section 8.5.1: $K_n = E_n \times Area/Thickness/Num\_nodes$, $K_t = E_t \times Area/$

$Thickness/Num\_nodes$, and $K_s = E_s \times Area/Thickness/Num\_nodes$ where Num_nodes is the amount of nodes, $Area$ is the finite area value occupied by the Num_nodes nodes, and $Thickness$ is thickness of the cohesive layer; $F_n^{max} = \sigma_I^{ini} \times Area$, $F_t^{max} = \sigma_{II}^{ini} \times Area$, and $F_s^{max} = \sigma_{III}^{ini} \times Area$.

8.6 FOAM CORE SANDWICH COMPOSITE MODELING EXAMPLE: DEBONDING OF FOAM CORE SANDWICH COMPOSITE BEAM

Linear analysis examples are not discussed here. If interested, Chapters 3 and 4 in the book *The Handbook of Sandwich Construction*[24] are a good reference for the linear static and vibrational examples on beam and plate with analytical solutions provided, respectively; and finite element analysis is capable of obtaining a close-enough solution after certain refinement, which can be used to check whether the basic setting and modeling techniques in the finite element model are correct or not. In this section, one nonlinear quasi-static example is discussed. For nonlinear analysis, analytical solution is generally very difficult or mostly impossible to be derived, and hence, the reference solution is usually from experimental data, either your own test data or others published.

One debonding example in a work by Bragagnolo et al.[2] is chosen to be reproduced here using the commercial software, ABAQUS,[7] to demonstrate the modeling techniques discussed in previous sections and show their effectiveness. In this example, we practice the following key techniques:

1. modeling foam core by crushable foam material model – Section 8.3.1;
2. transferring the regularly given parameters from the data sheet by the manufacturing company into required input parameters in ABAQUS – Section 8.3.1;
3. modeling foam core sandwich composite debonding by the cohesive element approach – Section 8.5.1;
4. modeling foam core sandwich composite debonding by the surface separation approach – Section 8.5.2;
5. modeling foam core sandwich composite debonding by the connector element approach – Section 8.5.3.

The setting of the example is shown in Figure 8.9. The consistent unit system #1 in Table 8.1 is used in this simulation. The dimensions are as follows: total length = 130 mm, width = 25 mm, foam core height = 15 mm, the thickness of top and bottom skin/face sheet = 1.5 mm, and the length of pre-existing top interface opening gap = 50 mm. The skin sheet is made of the carbon biaxial NCF material; since its failure is not reported during the debonding test,[2] it is acceptable to model it by an elastic material model with the given properties in Table 8.2. The mesh is shown in Figure 8.10: the skin sheet is modeled by shell element with thickness 1.5 mm and mesh size 1 mm by 1 mm; the foam core is modeled by solid element with a mesh size, 1 mm by 1 mm by 1 mm. The material model *Crushable foam material with volumetric hardening* in ABAQUS with the properties provided in Table 8.3

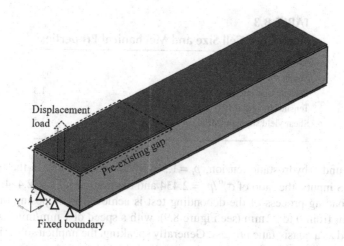

FIGURE 8.9 Schematic of the foam core sandwich composite beam subject to out-of-plane tip displacement loading.

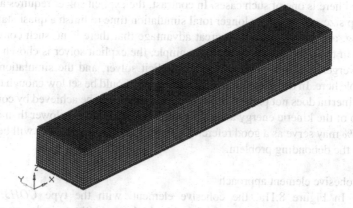

FIGURE 8.10 Finite element mesh of the foam core sandwich composite beam.

TABLE 8.2
Material Properties for the Carbon Biaxial NCF Material[2]

E_1 (MPa)	E_2 (MPa)	Poisson's Ratio v_{12}	Density ρ	$G_{12} = G_{13} = G_{23}$ (MPa)
55.9E+3	56.5 E+3	0.03	1.55E–9	2.4E+3

is chosen to describe the foam core. According to the manual of ABAQUS,[7] these commonly given foam core strengths in Table 8.3 cannot be directly used and need to be transferred into the required ones as discussed in Section 8.3.1. After using the converting code in the appendix section of this chapter, the three needed strengths are obtained: initial yield strength under uniaxial compression, $\sigma_c^0 = 1.3$ MPa, initial yield strength under hydrostatic compression, $p_c = 0.534$ MPa, and initial yield

TABLE 8.3
Foam Core Cell Size and Mechanical Properties[2]

Average cell size (mm)	0.4
Young's modulus (MPa)	92
Compressive strength (MPa)	1.3
Tensile strength (MPa)	2.5
Shear yield strength (MPa)	1.5

strength under hydrostatic tension, $p_t = 1.305$ MPa, which results in the two direct ABAQUS inputs, the ratio of $\sigma_c^0/p_c = 2.434$ and the ratio of $p_t/p_c = 2.444$.

The loading process of the debonding test is achieved by imposing the tip displacement from 0 to 22 mm (see Figure 8.9), with a speed of 1 mm/min,[2] which can be treated as a quasi-static process. Generally speaking, the implicit solver is suitable for static and quasi-static events, where a much bigger time step size can be used to save the total simulation time; however, sometimes it is difficult to converge for highly nonlinear problems, and the composite failure modeling due to debonding discussed here is one of such cases. In contrast, the explicit solver requires a smaller time step size and may take longer total simulation time to finish a quasi-static event modeling, and however, it has a great advantage that there is no such convergence issue as in the implicit solver. In this example, the explicit solver is chosen to avoid the convergence difficulty of using the implicit solver, and the simulation time is affordable here. In the simulation, the loading rate should be set low enough to ensure that the inertia does not play an important role, which can be achieved by controlling the ratio of the kinetic energy and total energy to be equal to or lower than a certain value (5% may serve as a good reference). Next, the three approaches will be applied to solve the debonding problem.

- Cohesive element approach

    In Figure 8.11a, the cohesive element, with the type *COH3D8* in ABAQUS and a size 1 mm by 1 mm by 0.01 mm (0.01 mm is the size in the

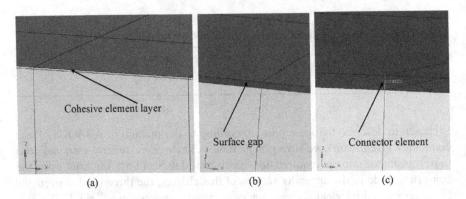

FIGURE 8.11 Bond modeling approaches: (a) by cohesive element, (b) by surface separation, and (c) by connector element.

Finite Element Method

TABLE 8.4

Fracture Properties of the Interface between Foam Core and Skin Sheet

Initial shear strength (MPa)	1.5
Initial tensile strength (MPa)	2.5
Normal direction modulus (MPa)	92
Shear direction modulus (MPa)	92
Fracture energy release rate in mode I, G_I^{cri} (N/mm)	0.31
Fracture energy release rate in mode II, G_{II}^{cri} (N/mm)	0.77

thickness direction) is defined between the foam core solid element and the skin sheet shell element. The input parameters involved are listed in Table 8.4, which are derived from the work by Bragagnolo et al.[2] In the table, the initial shear and tensile strengths of the cohesive element are set as the values of foam core with the assumption that the interface debonding initiates when the strengths of foam core are just reached. The moduli in normal and shear directions are also set as Young's modulus of foam core. The power law (see Section 8.5.1) is adopted to govern the mixed mode behavior with the power, $\alpha = 1.0$.

- Surface separation approach

   For the surface separation approach, only a surface interaction needs to be defined between the surface of the foam core and the skin sheet, no special treatment on the finite element mesh – as can be seen in Figure 8.11b, there is a gap between the foam core solid element and the skin sheet shell element. The input parameters are essentially the same as in the cohesive element approach, i.e., the data in Table 8.4 are copied here, and the required stiffnesses are derived by scaling the distance, 0.01 mm: normal direction stiffness = 92 MPa/0.01 mm = 9,200 MPa/mm, and similarly, shear direction stiffness = 9,200 MPa/mm.

- Connector element approach

   Figure 8.11c shows that connector elements are constructed to connect the foam core solid element layer and the skin sheet shell element layer. Force-based damage initiation and motion-based damage evaluation mechanisms are specified. The corresponding input parameters involved are all derived from those given in Table 8.4 by considering the connector element representative area and the element length: initial tensile force for damage = initial tensile strength × element area = 2.5 MPa × 1 mm × 1 mm = 2.5 N; normal direction stiffness = normal direction modulus × element area/length = 92 MPa × 1 mm² / 0.01 mm = 9,200 N/mm, and similarly, shear direction stiffness = 9,200 N/mm; the relative maximum axial displacement for the damage evaluation is set as 0.0005 mm, which is determined numerically by trial and error to fit the tested force–displacement curve.

The numerical force-displacement curves solved by the three approaches, which are processed with the Butterworth filter with frequency, 600 Hz to filter out very high frequency noise influence, are compared with experimental data[2] in Figure 8.12. As can be seen, the results match well with the test data in terms of the damage initiation moment, peak force value, and the debonding trend, which are important for engineering evaluation. The stress contours by the three approaches when the tip loading displacement reaches 22 mm are shown in Figures 8.13–8.15, respectively, where it

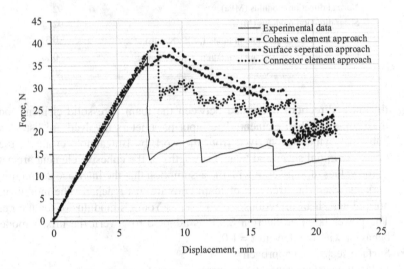

FIGURE 8.12 Force–displacement curve from numerical simulations and experimental data.

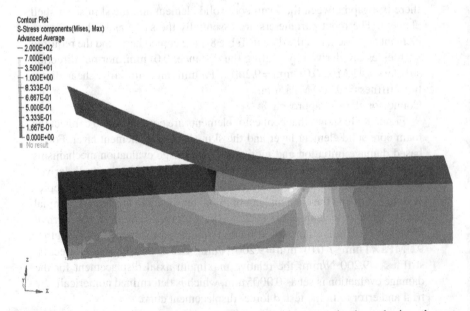

FIGURE 8.13 Stress contour of foam core sandwich composite by cohesive element approach during the debonding when the tip loading displacement is 22 mm.

Finite Element Method 319

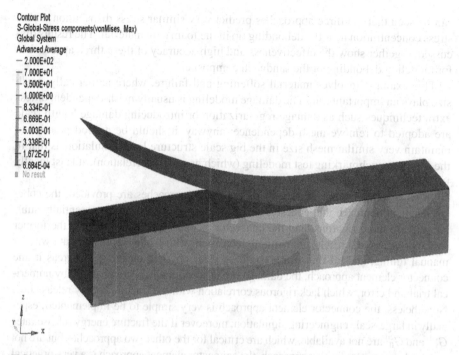

FIGURE 8.14 Stress contour of foam core sandwich composite by surface separation approach during the debonding when the tip loading displacement is 22 mm.

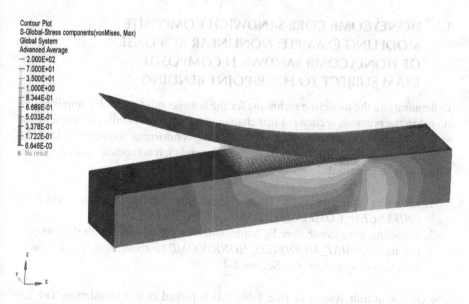

FIGURE 8.15 Stress contour of foam core sandwich composite by connector element approach during the debonding when the tip loading displacement is 22 mm.

can be seen that the three approaches predict very similar stress distribution with a stress concentration near the debonding tip in the foam core material. The results discussed, together show the effectiveness and high accuracy of these three approaches for modeling debonding of the sandwich composite.

This example involves material softening and failure, where numerically, mesh size plays an important role. The damage modeling is usually mesh-dependent unless extra techniques such as damage regularization or introducing damage length scale are adopted to remove mesh dependence; anyway, it should be a good practice to maintain very similar mesh size in the big scale structure level simulation as that in the smaller benchmarking test modeling (which has good correlation), at least locally for the region where damage is likely to occur.

In what follows, some comments on the three approaches are provided: the cohesive element approach and cohesive surface separation approach are essentially similar, and are more rigorous than the connector element approach in that in the former two approaches, the results are naturally consistent with the experimental data without manual tuning, and all the parameters have clear physical meanings, whereas in the connector element approach, the damage evolution parameters are obtained by numerical trial and error, which lack rigorous correlation with the fracture energy release rate. Nonetheless, the connector element approach is very simple to be implemented, especially in large-scale engineering simulation; moreover if the fracture energy release rates G_I^{cri} and G_{II}^{cri} are not available, which are critical for the other two approaches but are not required parameters for this approach, the connector element approach is a very practical choice in order to get a preliminary numerical evaluation on the debonding performance.

Note that the input files of this numerical example using ABAQUS are accessible to the reader and can be downloaded at the URL www.routledge.com/9780367441722.

8.7 HONEYCOMB CORE SANDWICH COMPOSITE MODELING EXAMPLE: NONLINEAR RESPONSE OF HONEYCOMB SANDWICH COMPOSITE BEAM SUBJECT TO FOUR-POINT BENDING

To demonstrate the modeling technique for the honeycomb sandwich composite discussed in the previous sections of this chapter, the four-point bending example in the work by Menna et al.[4] will be exercised here using commercial software LS-DYNA.[8] In this example, the following key honeycomb sandwich composite modeling techniques will be practiced:

1. Modeling skin sheet by shell element and material model *MAT_ENHANCED_COMPOSITE_DAMAGE* – Section 8.2.1;
2. Modeling honeycomb core by homogenized solid elements with the material model, *MAT_MODIFIED_HONEYCOMB* (without modeling all the honeycomb wall details) – Section 8.4.

The consistent unit system #1 (see Table 8.1) is picked in this simulation. The setting of the problem is shown in Figure 8.16, and the dimensions are as follows: total

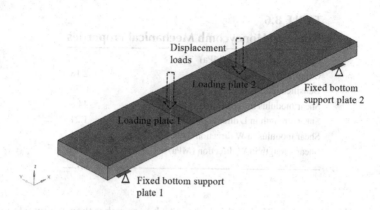

FIGURE 8.16 The four-point bending problem setup.

length = 490 mm, support span = 420 mm, loading span = 140 mm, width = 100 mm, honeycomb core height = 21.5 mm, top and bottom skin sheet thickness = 1 mm, and the width of the two top loading plates and two bottom support plates = 25 mm. The composite consists of E-glass phenolic fabric skin and hexagonal Nomex® honeycomb core.[4] E-glass phenolic fabric skin is 1 mm thick, modeled by shell element with a mesh size 4 mm by 4 mm and the material model, *MAT_ENHANCED_COMPOSITE_DAMAGE (*MAT_054–055).[20] This material model, as mentioned in Section 8.2, can reduce to the discussed Hashin damage model. The mechanical properties for the material model are listed in Table 8.5.

The Nomex® honeycomb is modeled by solid element with a mesh size 4 mm by 4 mm by 3.6 mm (3.6 mm is the mesh size in the thickness direction) and the material model *MAT_MODIFIED_HONEYCOMB (*MAT_126)[20] discussed in detail in Section 8.4. With this material model, the honeycomb core is modeled as a homogenized solid, and following the work by Menna et al.,[4] the required stress–strain curves for all the stress components (see Figure 8.18) are mainly extracted from Reference 25 with slight modification on the initial slopes by applying the moduli data (which is provided in Reference 26) in Table 8.6. The two top loading plates are loaded by prescribing vertical displacement, and the two bottom plates are fixed; all the four plates are modeled by shell elements with a rigid material model, and there are contacts defined between the plates and the skin shells. On the other hand, since

TABLE 8.5
Mechanical Properties of Composite Skin, E-Glass Phenolic Fabric[4] (Refer to Section 0 for the Meaning of the Variables in the First Row)

E_1 (MPa)	E_2 (MPa)	G_{12} (MPa)	Poisson's Ratio v_{12}	σ_{11}^+ (MPa)	σ_{11}^- (MPa)	σ_{22}^+ (MPa)	σ_{22}^- (MPa)	σ_{12}^1 (MPa)	σ_{12}^2 (MPa)
25.54E+3	22.97E+3	3.41E+3	0.15	325.77	293.19	288.21	259.39	41	41

TABLE 8.6
Nomex® Honeycomb Mechanical Properties

Compressive modulus (MPa)	136
Compressive strength (MPa)	2.18
Densification strain	0.76
Shear modulus in L direction (MPa)	44.8
Shear strength in L direction (MPa)	1.21
Shear modulus in W direction (MPa)	24.1
Shear strength in W-direction (MPa)	0.69

it is reported in Reference 25 that this honeycomb core rather than the interface bond always fails first during the composite tensile and shear strength tests, the interface debonding is not expected to occur; therefore, the skin shell element and the honeycomb core solid elements are connected by sharing nodes, i.e., no need to model the debonding in this case. Actually, among available solid element types in the software, the corotational solid element, element type 0, is recommended to work with *MAT_MODIFIED_HONEYCOMB* by the LS-DYNA material model manual,[20] and it is used in this simulation. The meshed model is shown in Figure 8.17. This experiment is a quasi-static process, where the loading speed is very slow – 6 mm/min,[4] and the simulation is conducted as a slow-motion dynamic event using the explicit solver in LS-DYNA, and the attention is paid to ensure that the kinetic energy is small enough compared to the total energy as in the example in Section 8.6.

The force–displacement curves measured at the loading point by simulation and experiment are compared in Figure 8.19. The initial slopes of the testing data and the simulation match well, meaning that the modeling is able to reproduce the rigidity of the sandwich composite structure when all the components are still in the elastic stage. The first initial yield point force is captured well, and however there is some noticeable difference observed in the peak force. This peak force difference is

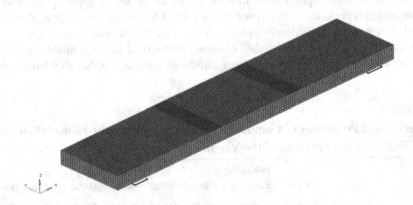

FIGURE 8.17 Finite element mesh of the four-point bending problem.

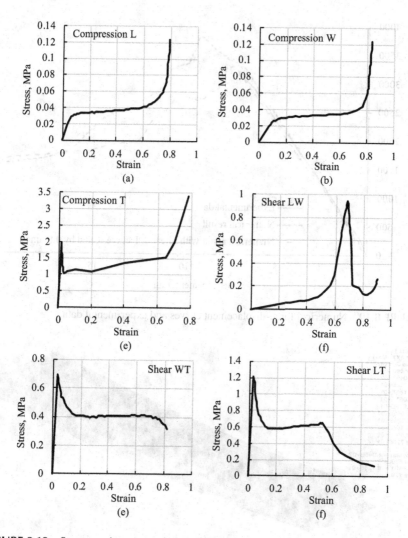

FIGURE 8.18 Stress-strain curves of Nomex® honeycomb core.

acceptable in terms of magnitude and could be due to the test variation error and certain small inaccuracy of the material curve input, among other possible reasons. One attempt is tried to figure out the factors which the force–displacement curve result is highly sensitive to, and it turns out that the shear LT material curve contributes significantly; as can be seen in Figure 8.19, if the shear LT material curve is scaled in stress magnitude, i.e., times vertical axis value by a factor of 0.8, the result is getting very close to the tested curve. On the other hand, conceptually, among components of the sandwich composite beam, the skin sheet is strong in in-plane tension and should bear the stress σ_{xx}, and the shear stress σ_{zx} should be mainly taken by the honeycomb core; the contours of the two stresses are shown in Figures 8.20 and 8.21,

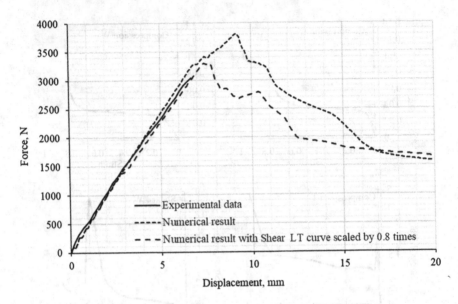

FIGURE 8.19 Numerical force-displacement curves and experimental data.

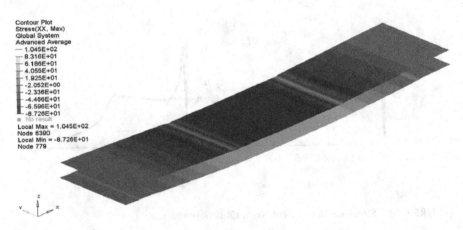

FIGURE 8.20 Stress contour of component σ_{xx} in the skin sheet when the loading displacement reaches 8 mm.

respectively, and are consistent with the expectations. The test photo in Figure 8.22[4] shows that the Nomex® honeycomb fails in large shear deformation; and the simulation captures this failure mechanism well; as shown in Figure 8.23, the final deformation mode is very similar to that in the photo. To sum up, in general, the results prove the effectiveness and accuracy of the techniques for honeycomb core sandwich composite modeling.

Again, please note that the input files of this numerical example for LS-DYNA are accessible to the reader and can be downloaded at the URL www.routledge.com/9780367441722.

Finite Element Method 325

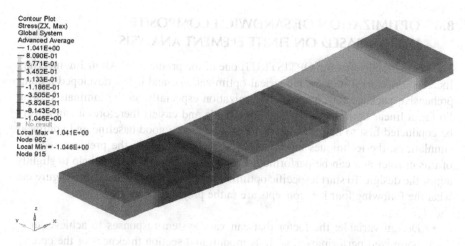

FIGURE 8.21 Stress contour of component σ_{zx} in the honeycomb core when the loading displacement reaches 8 mm.

FIGURE 8.22 The photo of Nomex® honeycomb sandwich composite subject to shear failure during four-point bending test.[4]

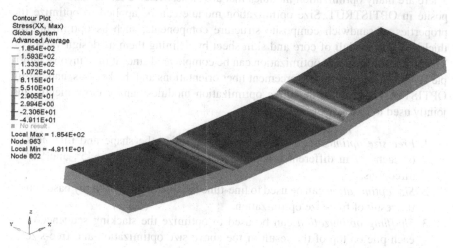

FIGURE 8.23 The stress contour of component σ_{xx} in the honeycomb core composite beam when the loading displacement reaches 20 mm.

8.8 OPTIMIZATION OF SANDWICH COMPOSITE DESIGN BASED ON FINITE ELEMENT ANALYSIS

The commercial software, OPTISTRUT, one of the products by Altair Engineering, Inc.[27] is a powerful tool for numerical optimizations, and it has developed a comprehensive package for composite optimization especially on the laminate design. So far, a linear optimization is usually reliable and easier; therefore, it can always be conducted first to help the design to land on a very good baseline and nonlinear simulations, the techniques of which have been discussed in the previous sections of this chapter and can be performed subsequently to confirm or to help to slightly adjust the design. To start a specific optimization project, the first task is to figure out what the following four key concepts are in the project:

- Design variable: the factor that can vary system responses to achieve an optimized performance, such as moduli and section thickness of the core in a sandwich composite optimization problem. Of course, the sensitivity of the variables differs.
- Response: any value or function that depends on the design variables during the problem-solving process, such as deflection and strain of the composite structure under certain loadings.
- Objective: the targeted responses to be minimized or maximized; one example can be the total weight or cost of the sandwich composite structure.
- Constraints: the condition on design variables or response that needs to be satisfied during the optimization process; usually it is the operating range defined by lower and upper bounds, such as the maximum and minimum limit on the skin sheet and core thickness, which could come from the designing packaging point of view or manufacturing technology limitations.

There are many optimization modules that are closely related to the sandwich composite in OPTISTRUT. Size optimization module can be applied to optimize the properties of sandwich composite structure components, such as optimizing the thickness and moduli of core and skin sheet by defining them as design variables. The composite laminate optimization can be complicated since it may involve multiple layers with different reinforcement fiber orientations and thicknesses, and within OPTISTRUT, the following three optimization modules can be conveniently and jointly used to tackle the problem:

1. *Free-size optimization* can be used to optimize the shape and thickness of each ply in different orientations such as in 0°, −45°, +45°, and 90° directions.
2. *Size optimization* can be used to fine-tune the thickness of each ply based on the result of free-size optimization.
3. *Shuffling optimization* can be used to optimize the stacking sequence of each pile on top of the result of the above size optimization and free-size optimization.

Finite Element Method

Here, only a general capability of OPTISTRUT on the sandwich composite is briefly touched, and please refer to OPTISTRUT manual[28] and tutorials for more technical details.

8.9 FATIGUE LIFE MODELING OF SANDWICH COMPOSITE VIA FINITE ELEMENT ANALYSIS

Fatigue failure evaluation is important for sandwich composites; it is a challenging task, however, and is still an active open research topic. The fatigue failure of sandwich composites could occur due to the skin sheet breakage likely by tensile stress,[28] the core failure mostly in shear modes,[29] or interface debonding[30,31]; three corresponding examples on the fatigue failure modes are shown in Figures 8.24–8.26, respectively.

8.9.1 THE S-N CURVE METHOD FOR SANDWICH COMPOSITES

The S-N curve method has shown great success in industry applications to predict fatigue life, especially for metallic structures; however, to the best knowledge of the author of this chapter, it is not yet so mature for the fatigue evaluation of a relatively

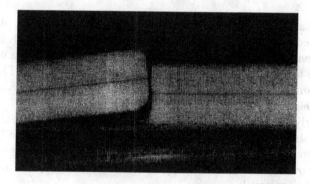

FIGURE 8.24 Sandwich composite fatigue failure mode: tensile failure in the skin sheet.[27]

FIGURE 8.25 Sandwich composite fatigue failure mode: shear failure in the core material.[28]

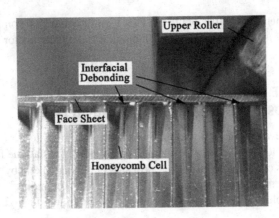

FIGURE 8.26 Sandwich composite fatigue failure mode: interfacial debonding failure.[29]

complex sandwich composite, especially in the current commercial software, mainly due to the fact that in general, the skin sheet material is not isotropic, and the interface debonding caused by fatigue is usually under mixed mode, which adds extra complexities. Nonetheless, it can still be adopted for the following simple load cases:

1. If the sandwich composite structure is expected to fail in the skin sheet subjected to the simple loading mode, only in-plane tensile loads along one direction, then the needed S-N curve can be obtained from a consistent tension-tension fatigue coupon testing of the skin sheet – ASTM D3479, *Standard test method for tension-tension fatigue of polymer matrix composite materials*.
2. Similarly, if it is expected to have a fatigue failure mode in core shear, then a core shear fatigue test can be conducted to predict the fatigue life of the structure via ASTM C394, *Standard test method for shear fatigue of sandwich core materials*, which is applicable to foam, balsa wood, and honeycomb core materials.

For the above two simple cases, the steps for the fatigue evaluation by finite element analysis (FEA) are similar to that for metallic structures and are given as in the following flow chart in Figure 8.27.

Note that N-CODE software (https://www.ncode.com) in Figure 8.27 is known as one of the leading fatigue simulation software (based on the knowledge of the author of this chapter); its fatigue solver has also been implemented and is available in other commercial software such as in NASTRAN. If the loading mode is more complex than the abovementioned two cases, e.g., the skin sheet is not only subject to in-plane tensile loading but also significant shear loading, or the fatigue failure is expected to occur at the interface, it is challenging to assess the fatigue life by finite element analysis for two main reasons, respectively: first, the fatigue failure mechanism of the skin sheet, which is not an isotropic material as metallic materials, under mixed loading modes is complicated; and second, it is very difficult to obtain

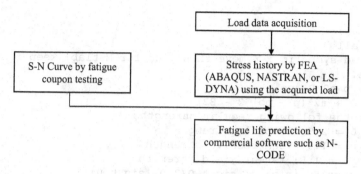

FIGURE 8.27 Flow chart of fatigue life evaluation by FEA.

a representative S-N curve for the interface between core and skin sheet. Since the fatigue evaluation by FEA is not yet so reliable in these situations, it is desirable to conduct component-level or even sub-system level fatigue test to ensure the fatigue life, and of course, the ultimate criterion will be the full-scale product-level durability fatigue test. Even though so far FEA is not able to accurately predict the fatigue life for those complex cases, it can still be utilized to assist to reach a design that likely can pass the component-level fatigue test; for example, one approach could be that by controlling the strain level in the skin sheet (tensile and compressive strain) and foam (shear strain) under certain portion of the strain limit at breakage or failure, maybe 30% as a starting point, which can be adjusted later during progressive iterations, it is more likely to obtain a good baseline design that could be close enough to satisfy the component level fatigue test without significant overdesigning.

APPENDIX: OCTAVE/MATLAB CODE FOR TRANSFERRING FOAM STRENGTHS FOR CRUSHABLE FOAM MATERIAL MODEL IN ABAQUS

```
%OCTAVE CODE for converting input strengths for crushable foam
material model in ABAQUS
%by Guohua Zhou, ghzhou2016@gmail.com, 2/27/2021; thanks for
using the code.
%download free OCTAVE software code: https://github.com/
NexMirror/Octave
%it can be used for MATLAB readily.
%crushable foam strengths inputs conversion: regular strengths
--> required strengths
%https://github.com/cbm755/octsympy/releases
%download this package: symbolic-win-py-bundle-2.9.0.tar.gz
%enter the following in the octave command window
%  >> pwd
%ans = D:\xxx\xxx\crushable_foam
%copy the downloaded file - symbolic-win-py-bundle-2.9.0.tar.
gz to the above folder
%>> pkg install symbolic-win-py-bundle-2.9.0.tar.gz
%>> pkg load symbolic
%Code starts --------------------------------------------------
```

```
clc
clear all
close all
pkg load symbolic % remove this line for Matlab; keep it for
Octave.
syms q p a2 B2 po
F = q^2 + a2*(p - po)^2 - B2;
%input the following regular strengths:
sig_t_0 = 2.5; %tensile strength
sig_c_0 = 1.3; %compressive strength
tao_xy_0 = 1.5; %shear yield strength
eq1 = subs(F, {p,q}, {-sig_t_0/3.0, sig_t_0});
eq2 = subs(F, {p,q}, {sig_c_0/3.0, sig_c_0});
eq3 = subs(F, {p,q}, {0.0, sqrt(3)*tao_xy_0});
eq21 = simplify(eq2 - eq1);
eq31 = simplify(eq3 - eq1);
soln = solve(eq21 == 0, eq31 == 0, a2, po);
a22 = eval(soln.a2);
po2 = eval(soln.po);
if(a22<0.0)
disp('ERROR: strength data is not consistent, please check and
adjust the strength data')
stop
endif
eqc1 = eval(subs(eq21, {a2, po}, {a22,po2}));
eqc2 = eval(subs(eq31, {a2, po}, {a22,po2}));
B2_form = solve(eq1,B2);
B = sqrt(eval(subs(B2_form, {a2, po}, {a22,po2})));
a = sqrt(a22);
pc = B/a + po2;
pt = B/a - po2;
sig_c_0;
% three pts data confirmation
leftpt = eval(subs(F, {p,q,a2,B2,po}, {pc, 0.0,a22,B^2,po2}));
rightpt = eval(subs(F, {p,q,a2,B2,po}, {-pt, 0.0,a22,B^2,po2}));
sigma_c_pt = eval(subs(F, {p,q,a2,B2,po}, {sig_c_0/3.0,
sig_c_0,a22,B^2,po2}));
disp("below key parameters:")
error_tol = sqrt(leftpt^2 + rightpt^2 + sigma_c_pt^2)
if(error_tol<0.001*sqrt(sig_t_0^2 + sig_c_0^2 + tao_xy_0^2))
disp("error is small enough: SUCCESSFUL!")
% visulize and plot the initial yield surface in q(p)
px = [-pt: (pc+pt)/100: pc];
qy2 = B^2 - a22*(px - po2).^2;
qy = sqrt(abs(qy2));
plot(px, qy)
xlabel('p')
ylabel('q')
title('Initial yield surface plot q(p)')
else
disp("error is not small enough: NOT SUCCESSFUL!")
```

```
endif
pc % initial yield strength under hydrostatic compression
pt % initial yield strength under hydrostatic tension
sig_c_0 % initial yield strength under uniaxial compression
disp("ABAQUS INPUT: Two ratios:")
disp("sig_c_0/pc,")
sig_c_0/pc
disp("pt/pc,")
pt/pc
%Code ends ---------------------------------------------------
```

REFERENCES

1. S. Boria and G. Forasassi, "Honeycomb sandwich material modelling for dynamic simulations of a crash-box for a racing car," *The 10th International Conference on Structures Under Shock and Impact*, 2008.
2. G. Bragagnolo, A.D. Crocombe, S.L. Ogin, I. Mohagheghian, A. Sordon, G. Meeks and C. Santoni, "Investigation of skin-core debonding in sandwich structures with foam cores," *Materials & Design*, vol. 186, p. 108312, 2020.
3. B.L. Buitrago, C. Santiuste, S. Sánchez-Sáez, E. Barbero and C. Navarro, "Modelling of composite sandwich structures with honeycomb core subjected to high-velocity impact," *Composite Structures*, vol. 92, no. 9, pp. 2090–2096, 2010.
4. C. Menna, A. Zinno, D. Asprone and A. Prota, "Numerical assessment of the impact behavior of honeycomb sandwich structures," *Composite Structures*, vol. 106, pp. 326–339, 2013.
5. M. Rinker, M. John, P.C. Zahlen and R. Schäuble, "Face sheet debonding in CFRP/PMI sandwich structures under quasi-static and fatigue loading considering residual thermal stress," *Engineering Fracture Mechanics*, vol. 78, no.17, pp. 2835–2847, 2011.
6. W. Wang, "Cohesive zone model for facesheet-core interface delamination in honeycomb FRP sandwich panels," Ph.D Dissertation, West Virginia University, Morgantown, WV, 2012.
7. Abaqus v6. 12 Documentation-ABAQUS analysis user's manual, Dassault Systèmes Simulia Corp, 2012.
8. J.O. Hallquist, LS-DYNA theory manual, Livermore Software Technology Corporation, 2006.
9. Nastran, M.S.C. User's Manual, MSC. Software, USA, 2010.
10. HyperWorks User's Manual V14.0, Altair, 2016.
11. ANSA v12.1.5 User's Guide - BETA CAE Systems S.A., 2008.
12. E. Ventsel, T. Krauthammer and E.J.A.M.R. Carrera, *Thin Plates and Shells: Theory, Analysis, and Applications*. NewYork: Marcel Dekker, Inc., 2001.
13. T.J. Hughes, *The Finite Element Method: Linear Static and Dynamic Finite Element Analysis*. Chelmsford, MA: Courier Corporation, 2012.
14. T. Belytschko, W.K. Liu, B. Moran and K. Elkhodary, *Nonlinear Finite Elements for Continua and Structures*. Hoboken, NJ: John Wiley & Sons, 2013.
15. Z. Hashin, "Failure criteria for unidirectional fiber composites," *Journal of Applied Mechanics*, vol. 47, pp. 329–334, 1980.
16. S. Long, X. Yao, H. Wang and X. Zhang, "Failure analysis and modeling of foam sandwich laminates under impact loading," *Composite Structures*, vol. 197, pp. 10–20, 2018.
17. R. Sriram, U.K. Vaidya and J.E. Kim, "Blast impact response of aluminum foam sandwich composites," *Journal of Materials Science*, vol. 41, no. 13, pp. 4023–4039, 2006.

18. V.S. Deshpande and N.A. Fleck, "Isotropic constitutive models for metallic foams," *Journal of the Mechanics and Physics of Solids*, vol. 48, no. 6–7, pp. 1253–1283, 2000.
19. F. S. Chang, "Constitutive Equation Development of Foam Materials," Ph.D. dissertation, Wayne State University, Detroit, MI, 1995.
20. "LS-DYNA® Keyword User's Manual Volume II Material Models," Livermore, CA, USA, 2018.
21. I. Carranza, A.D. Crocombe, I. Mohagheghian, P.A. Smith, A. Sordon, G. Meeks and C. Santoni, "Characterising and modelling the mechanical behaviour of polymeric foams under complex loading," *Journal of Materials Science*, vol. 54, no. 16, pp. 11328–11344, 2019.
22. J.C. Simo and T.J. Hughes, *Computational Inelasticity*. Berlin: Springer Science & Business Media, 2006.
23. Y. Rothschild, *Nonlinear Analyses of Sandwich Panels*, Stockholm: Deptartment Aeronautics, Royal Institute of Technology, Report 92-14, 1992.
24. D. Zenkert, *The Handbook of Sandwich Construction*. Worcester: Engineering Materials Advisory Services, 1997.
25. S. Heimbs, P. Middendorf and M. Maier, "Honeycomb sandwich material modeling for dynamic simulations of aircraft interior components," *9th International LS-DYNA Users Conference*, Dearborn, MI, USA, 2006.
26. A. Zinno, A. Prota, E. Di Maio, and C.E. Bakis, "Experimental characterization of phenolic-impregnated honeycomb sandwich structures for transportation vehicles," *Composite structures*, vol. 93, no. 11, pp. 2910–2924, 2010.
27. OptiStruct 11.0 user manual, Troy, MI: Altair Engineering Inc., 2011.
28. R.A. Shenoi, S.D. Clark and H.G. Allen, "Fatigue behaviour of polymer composite sandwich beams," *Journal of Composite Materials*, vol. 29, no. 18, pp. 2423–2445, 1995.
29. E. Ayorinde, R.A. Ibrahim, V. Berdichevsky, M. Jansons and I. Grace, "Development of damage in some polymeric foam-core sandwich beams under bending loading," *Journal of Sandwich Structures & Materials*, vol. 14, no. 2, pp. 131–156, 2012.
30. Y.M. Jen, and L.Y. Chang, "Effect of thickness of face sheet on the bending fatigue strength of aluminum honeycomb sandwich beams," *Engineering Failure Analysis*, vol. 16, no. 4, pp. 1282–1293, 2009.
31. M. Manca, A. Quispitupa, C. Berggreen and L.A. Carlsson, "Face/core debond fatigue crack growth characterization using the sandwich mixed mode bending specimen," *Composites Part A: Applied Science and Manufacturing*, vol. 43, no. 11, pp. 2120–2127, 2012.

9 Application of Sandwich Structural Composites

Wenguang Ma

Russell Elkin

CONTENTS

9.1 Marine Industry 334
    9.1.1 Recreational Boat Building 335
    9.1.2 Military Ship Constructions 338
    9.1.3 Commercial Marine Industry 339
9.2 Wind Energy Industry 341
9.3 Airplane and Aerospace 347
    9.3.1 Applications in Airplanes 348
    9.3.2 Applications in Helicopters 351
    9.3.3 Applications in Space 353
    9.3.4 Future of Aeronautic Sandwich Structures 354
9.4 Transportation 357
    9.4.1 Rail Car Application 358
    9.4.2 Bus Body Application 364
    9.4.3 Truck and Semitrailer Body Application 367
    9.4.4 Automotive Industry 368
9.5 Building and Civil Industries 373
    9.5.1 Housing and Building Construction 373
    9.5.2 Bridge Building 379
    9.5.3 Bridge and Dock Protective Systems 385
    9.5.4 Challenges and Issues 387
9.6 Miscellaneous 388
    9.6.1 Sandwich Structure for Radome Construction 388
    9.6.2 Medical Equipment 395
    9.6.3 Acoustic Barriers 399
        9.6.3.1 Sound Transmission through Sandwich Structures 399
        9.6.3.2 Sound Transmission Reduction by Adding a Layer of Membrane-Type Acoustic Meta-Materials 401
        9.6.3.3 Reduce the Sound Transmission by Acoustic Separation of the Layers of Sandwich 405

DOI: 10.1201/9781003035374-9

9.6.3.4 Introduce Air or Sound Insolation Gap to Core or between Panels 407
9.6.3.5 Honeycomb Sandwich Panels with Micro-perforated Facings 410
9.6.3.6 Use Damping Core and Perforated Facing 412
9.6.4 Sports and Leisure 413
9.6.4.1 Sporting Boards on the Water 413
9.6.4.2 3D Printing Sport Boards 415
9.6.4.3 Skis and Snowboard 416
9.6.4.4 Sandwich Construction for Making Canoes, Kayaks and Paddleboards 419
References 420

Sandwich structures have been used for building aircraft fuselage and wings in World War II, which was the first composite application for a balsa core sandwich. Today, sandwich is still one of the major materials systems for the aerospace industry. After World War II, sandwich has been incorporated into mostly every type of moving vehicle including power and sail boats, commercial and military marine vessels, rail vehicles, trucks, trailers, buses, and automobiles. Since the 1990s, the wind energy industry has employed the sandwich structures to fabricate large wind turbine blades and now, the largest market application of sandwich constructions by volume. Sandwich composites have also been used in a wide variety of structures and products such as sporting goods, buildings and temporary shelters, antenna radomes, medical equipment process storage tanks, and infrastructure.

9.1 MARINE INDUSTRY

Sandwich structures, with specific reference to fiber-reinforced composite sandwiches, have been extensively incorporated in the marine industry following their early applications during and after World War II when composites were designed to overcome corrosion issues experienced with steel, aluminum, and wood. Weight reduction was, and remains, a key aspect, particularly for topside weight (above the water line). For over 75 years, the sandwich structures have been utilized in marine applications to produce boats and yachts and provide light, stiff, and strong constructions. A variety of marine vessels have been produced using foams, end-grain balsa wood, honeycombs, and other core materials, along with metal and fiber reinforcement composite facings. Marine products made by sandwich composites include hulls, stringers, bulkheads, flat deck panels, containers and furniture for racing boats, lifeboats, sailing boats, and leisure yachts. Some military ships and submarines also use sandwich construction for structural parts and components such as walls, topside superstructures, tanks, doors, electrical enclosures, table and worktops, shower units, and other installations.

The largest structural part of the marine vessel made by a sandwich structure is the hull and deck. The nautical industry has experienced an evident technological evolution, and today, the manufacturing of hulls is based on latest-generation production systems. The introduction of advanced sandwich composite materials represented a milestone innovation in the manufacturing of boats and vessels. Sandwich composite technology has allowed manufacturers to improve the quality of products obtaining stiff and light structures, with benefits in terms of sailing performances and working life. Stiffer hulls and decks are the main applications where the shipbuilding industry has adopted composite sandwich structures. The weight reduction results in larger cargo capacity, fuel-saving, lower inertia, and increased ship stability and buoyancy. In addition, the composite facings show satisfactory corrosion resistance in the marine environment and require less maintenance.

9.1.1 Recreational Boat Building

Recreational boat building is the biggest area to use the sandwich composites in marine industries, which has been stimulated by evolution of recreational boat construction techniques. From the 1950s, advances in materials and fabrication techniques used in the pleasure craft industry have helped to reduce production costs and improve product quality. Early fiberglass boat building produced single-skin structures with stiffeners to maintain reasonable panel sizes. Smaller structures used isotropic (equal strength in x and y directions) chopped strand mat laid up manually or with a chopper gun. As strength requirements increased, fiberglass cloth and woven roving were integrated into the laminate. An ortho-polyester resin, applied with rollers, was almost universally accepted as the matrix material of choice.[1]

Later, boat builders of custom and higher-end craft have used a variety of other resins that exhibit better performance characteristics. Epoxy resins have long been known to have better strength properties than polyesters. Their higher cost has limited use to higher performance vessels, but this is changing particularly due to adoption of vacuum infusion. Iso-polyester resin has been shown to resist blistering better than ortho-polyester resin, and some manufacturers have switched to this entirely or for use as a barrier coat. Vinyl ester resin has performance properties somewhere between polyester and epoxy and has recently been examined for its excellent blister resistance. Cost is greater than polyester but less than epoxy.

The marine industry has evolved from predominantly woven fabrics to widespread adoption of knitted materials. The use of carbon fiber has also increased, and aramid fibers are also used when it demonstrates a clear advantage. Some low-cost reinforcement materials that have emerged lately include polyester and polypropylene fibers. These materials combine moderate strength properties with high strain-to-failure characteristics. Lower cost reinforcements are limited to small vessels but are beginning to see use in interiors and other lightly loaded sandwich parts.

In marine sandwich structures, E-glass composite materials are the most commonly used with carbon being the choice of high-performance vessels. Polymeric foams and light-density balsa wood are mainly used as core materials. Some vessels utilize plastic honeycomb, and Nomex® is only found in a few high-performance boats. The basic sandwich structure of the hull is shown in Figure 9.1, which has the core at the center, and inner and outer fiber-reinforcing plastic (FRP) facings, and a layer of outside protecting gel coat that provides UV and weather resistance, and gloss and color decorating surface.

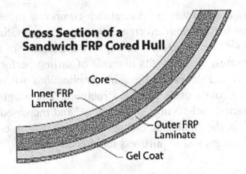

FIGURE 9.1 Boat hull structure of sandwich composite.[2]

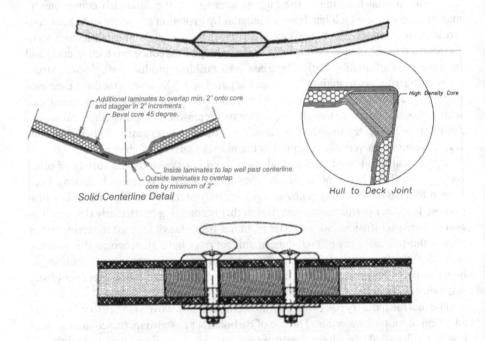

FIGURE 9.2 Examples of transition from solid to sandwich, from hull to deck, and reinforcement.[3]

The transitions from solid to sandwich, from hull to deck, and reinforcements are important for the sandwich structural boat hulls. Some examples are shown in Figure 9.2, in which one way to stiffen the keel is to use a double sandwich made from high density foam or use a solid composite section. A high-density foam also is used for the transitions from the hull to the deck or used for screw bolt fastening.

Figure 9.3 shows a couple of boat hull constructions made with the sandwich structures. Today, resin infusion has greatly increased the use of sandwich in hull bottoms. The sandwich structural deck and sidewalls can be made with the hull or made separately and then assembled together.

Sandwich Structural Composite Applications

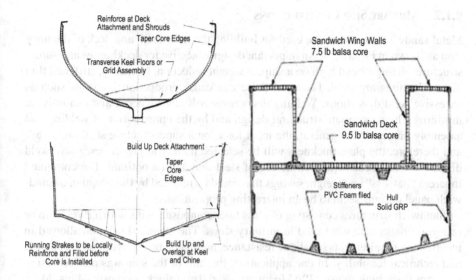

FIGURE 9.3 Boat hull sandwich structural constructions.[1]

FIGURE 9.4 Boat hull vacuum infusion process.[4]

The boat hulls are usually built by using female molds and either open molding or the vacumm infusion process as shown in Figure 9.4, presented in detail in the Chapter 4. Today, smaller crafts are still mainly laminated by hand/spray; infusion becomes favorable for lengths above 10 m depending on the level of performance. Some builders are infusing their entire product line, even as small as 6 m.

9.1.2 MILITARY SHIP CONSTRUCTIONS

Metal sandwich structures are used for building the hull, shell, and deck of the navy ship as shown in Figure 9.5. An important design objective for deckhouses and superstructures, being placed high on a ship, is weight reduction. Normally, stiffened thin plate structures are used. However, these can cause production problems such as excessive weld distortions. Welding distortions will always occur and can only be minimized by appropriate structural design and by the optimization of welding and assembly processes. In general, the local loads on a superstructure shell are small, and therefore, the plate thickness will be selected mainly to prevent excessive weld distortions. As a consequence, the use of steel may not be optimal. The combined inherent "flatness" and weight savings that could be realized by the adoption of sandwich panel designs seem to be an interesting proposition.

Sandwich structures consisting of glass fiber composite skins with a PVC foam or end-grain balsa core were used in military ships. The technical solution allowed in meeting the requirements of fatigue resistance, nonmagnetic behavior, and economic and technical feasibility. In one application, the hull of the ship was realized using a 60 mm thick high-density PVC foam core and 8 mm thick composite skins. More specifically, the skin panels were manufactured combining chopped strand mat and woven roving reinforcements, in order to maximize the bond strength, fire resistance, and interlaminar shear strength.

In another example, the navy ship designed for low visibility, radar cross-section, and infrared signature, was built entirely in composite materials using sandwich panels with carbon and glass reinforcement and vinyl resin matrix. Indeed, composite materials provided robustness, resistance to underwater shock loads, and low magnetic capacities combined with lightweight nature. Although the cost of carbon fibers was at least five times higher than that of glass limiting their use in large naval structures, it was found that by using some carbon fibers in the composite, besides the adequate electromagnetic shielding, the weight of the hull was reduced by about 30%, without significantly increasing the manufacturing cost. The achieved weight-saving translated into a reduction in fuel consumption.

Nonstructural ship components are being considered as candidates for replacement with composite parts. Two types of advanced non-structural bulkheads are in

FIGURE 9.5 Superstructure module of offshore patrol vessel made with metal sandwich structures.[5]

Sandwich Structural Composite Applications

FIGURE 9.6 Deckhouse (top) and helicopter hangar made with sandwich composite structure for military ships.[4,6]

service in navy ships. One of these consists of aluminum honeycomb with aluminum face sheets and the other consists of E-glass composite skins over an aramid core material. Deckhouses and superstructures like helicopter hangar are the big components made by sandwich composite structures as shown in Figure 9.6. Other topside enclosures, decks, bulkheads, doors, ballast, and storage tanks also can be made by the sandwich structures.

9.1.3 COMMERCIAL MARINE INDUSTRY

The use of sandwich structural construction in the commercial marine industry has flourished over time for a number of different reasons. Initially, long-term durability and favorable fabrication economics were the impetus for using sandwiched FRP. More recently, improved vessel performance through weight reduction has encouraged its use. Since the early 1960s, a key factor that makes the sandwiched FRP construction attractive is the reduction of labor costs when multiple vessels are fabricated from the same mold.

In small to meddle boat sectors, the sandwich structural composites are worldwide used to produce vessels of fishing boats, hovercraft, and catamarans. The applications for other boats include barges, fishing boats, and lifeboats. The sandwich structure is the predominantly used material representing approximately 80% of the hulls, for ships long up to 45 m.[7]

Boats built for utility vessels are usually modifications of existing recreational hulls. Laminate schedules may be increased or additional equipment added, depending upon the type of service. Local and national law enforcement agencies, including natural resource management organizations, compromise the largest sector of utility boat users. Other mission profiles, including pilotage, firefighting, and launch service, have proven to be suitable applications of sandwich FRP construction. To make production of a given hull form economically attractive, manufacturers will typically offer a number of different topside configurations for each hull.[1]

Hovercraft ships represented the main application of fiber-reinforced plastics in the marine sector for a long time. For instance, the 15.5 m hull of the hovercraft boat was manufactured using sandwich structures made of fiber-reinforced composite laminates with PVC foam cores. A European company produced a boat by using FRP sandwich reaching a length of 32.2 m, a gross tonnage of 290 tons, and a maximum speed of 42 knots. The typical cargo configuration of the air cushion vehicle with a minimal payload of 550 kg in size from 7 to 16 m is fabricated from shaped solid foam block, which is covered with FRP skins. The volume of foam gives the added value of vessel unsinkability.[1]

Conventional ferries are being replaced by fast ferries, due to improved economic conditions, increased leisure time, demands for faster travel, and more comfort and safety, air congestion, reduced pollution, and higher incomes. To date, sandwich composite construction has been utilized more extensively by overseas builders of commercial vessels. An air cushioned ferry vehicle was structured entirely of FRP sandwich. The air cushion carries about 85% of the total weight of the ship with the remaining 15% supported by the hulls. The design consists of a low-density PVC cellular plastic core material with closed, non-water-absorbing cells, covered with a face material of glass fiber-reinforced polyester plastic. The complete hull, superstructure and foundation for the main engines and gears are also built of FRP sandwich. Tanks for fuel and water are made of hull-integrated sandwich panels.[1]

Catamarans are mainly used for fast ferries, off-shore and oceanographic sailing ships, and small pleasure boats due to their good passenger capacity, fuel efficiency, and high reliability. An example is a catamaran ferry of 37 m length with a load capacity of 184 passengers, which was made by using aramid-reinforced sandwiches. The combination of aramid fiber laminates (skin) and PVC foam (core) provided an increase in hull tensile strength of the five times if compared to steel and in a reduction in noise, crucial for seismic investigation operations. Carbon reinforcement is also used in the production of catamarans and trimarans.[1]

Foam cored sandwich structures are routinely being used as buoyancy materials in commercial submersibles. An unmanned submersible has an operating depth of 500 m, which uses high crush point closed cell PVC foam material for buoyancy. A manned submersible with operating depths of 6,000 m was built with high crush point foam for buoyancy and FRP materials for nonpressure skins and fairings. The oil industry is making use of a submersible that not only utilizes foam for buoyancy but uses the foam in a sandwich configuration to act as the pressure vessel. The use of composites in the hull allowed the engineers to design specialized geometries that are needed to make effective repairs in the offshore environment.[1]

Advanced sandwich composites on large ships have the potential to reduce fabrication and maintenance costs, enhance styling, reduce outfit weight, and increase reliability. The sandwich composites have been used for cruise liner stacks, such as the $3 \times 5 \times 12$ m funnels for a cruise ship that represented a 50% weight and 20%

Sandwich Structural Composite Applications

cost savings over aluminum and stainless steel structures they replaced. Sandwich composite materials in ship applications also can be topside superstructure, stacks, doors, bulkheads, tanks, motor housings, condenser shells, electrical enclosures, tables, insulation, showers, etc.

9.2 WIND ENERGY INDUSTRY

The history of wind turbines for electric power generation started in Cleveland Ohio, USA in 1888 and in Askov, Denmark in 1889. In 1941, electricity production from wind was made using turbines with steel blades. One of the steel blades failed after only a few hundred hours of intermittent operation. Thus, the importance of the proper choice of materials and inherent limitations of metals as a wind blade material was demonstrated early in the history of wind energy development. The next, a quite successful example of the use of the wind turbine for energy generation is the so-called Gedser wind turbine with three composite blades built from steel spars, with aluminum shells supported by wooden ribs, installed at Gedser coast in Denmark in 1956–1957. The turbine (24 m rotor, 200 kW) had run for 11 years without maintenance. After the 1970s, wind turbines were mainly produced with composite blades.

The main requirements to wind turbine blades can be summarized as follows:

- High strength to withstand even extreme winds, as well as gravity load,
- High fatigue resistance and reliability to ensure the stable functioning for more than 20 years and 10^8 cycles,
- Low weight to reduce the load on the tower, and the effect of gravitational forces,
- High stiffness to ensure the stability of the aerodynamically optimal shape and orientation of the blade during the work time, as well as clearance between blade and the tower.

Wind turbines are manufactured in a wide range of vertical and horizontal axis. The smallest turbines are used for applications such as battery charging for auxiliary power for boats or caravans or to power traffic warning signs. Larger turbines can be used for making contributions to a domestic power supply while selling unused power back to the utility supplier via the electrical grid. Arrays of large turbines, known as wind farms, are becoming an increasingly important source of intermittent renewable energy and are used by many countries as part of a strategy to reduce their reliance on fossil fuels.

A wind turbine consists of a rotor that has wing-shaped blades attached to a hub; a nacelle made by sandwich structural construction also, which houses a drivetrain consisting of a gearbox, connecting shafts, support bearings, the generator, plus other machinery: a tower, and ground-mounted electrical equipment. The turbine blades play a very important role in the wind turbines. Blades are required to preserve an optimum cross-section for aerodynamic efficiency to generate the maximum torque to drive the generators. The efficiency of the wind turbine depends on the material of the blade, shape of the blade, and angle of the blade.[8]

Generally, a wind turbine should work for 20–25 years without repair and with minimum maintenance. The problem of ensuring high reliability of wind turbines becomes especially important for large and extra-large turbines, because of high wind and gravitational loads, on the one side, and the difficulties of repair of large turbines, on the other side. In view of these requirements, only materials with the very high strength, fatigue resistance, and stiffness, i.e., composites, can be used in wind turbine blades. No other materials, neither metals, nor alloys, nor wood, can satisfy this list of requirements fully.

A wind turbine blade consists of two halves (on the suction side and the pressure side), joined together and stiffened either by one or two integral (shear) webs linking the upper and lower parts of the blade shell or by a box beam as shown in Figure 9.7. The aeroshells, which are made of sandwich structures, are primarily designed against elastic buckling. The box beam or shear webs inside the blade are adhesively joined to the shell. Certain manufacturers do mold the entire blade structure together, without bonding. The different cyclic loading histories that exist at the various locations at the blades suggest that it could be advantageous to use different materials for different parts of the blade. Figure 9.8 shows the schema of the section of the blade.

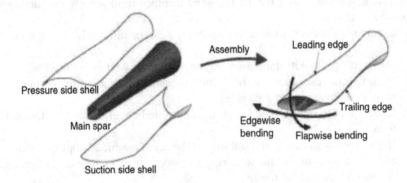

FIGURE 9.7 Wind turbine blade components.[9]

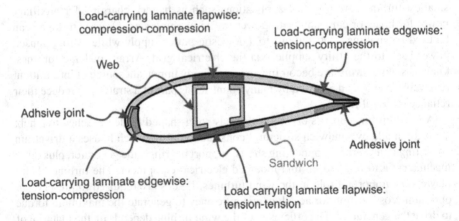

FIGURE 9.8 Schema of the section of the wind turbine blade.[10]

Sandwich Structural Composite Applications

In unsupported parts of the wind shell, sandwich composites are used, which ensures the blade shape. The sandwich structures ensure much higher stiffness than the monolithic composites. The sandwich core materials, placed between two composite plies, are typically PET foams, balsa wood, or in some cases, PVC and SAN foams.

Present-day designs are mainly based on glass fiber-reinforced plastics (GFRP), but for very large blades, carbon fiber-reinforced plastics (CFRP) are being used increasingly to reduce the weight. The manufacturing techniques used may differ very much with different manufacturers, but generally, the production of wind turbine blades is based on either composite prepreg technology or vacuum-assisted vacuum infusion (VARTM or variations of this process).

Today, the blades of an average onshore wind turbine are about 50–70 m long; the latest offshore blades even exceed 80–100 m. This makes great demands on the materials used in the composite structures to have the high mechanical stiffness and strength, and high rigidity without too much extra weight. Beyond material performance, the blade manufacturers' choice of core materials is driven by processability, costs, and availability. The dominant core materials are end-grain balsa core and PET core materials. Composite sandwich structures in rotor blades are divided into two main groups: shear webs and shell panels. The main purposes of these sandwich sections are to keep the aerodynamic shape of the blade, which determines how well it can extract energy from the wind, and to prevent blade buckling. The crucial core material properties concerned are shear and compression strength. The gravity loads induced by the rotation of the rotor blade make high demands on the material's fatigue properties. To resist the fatigue load, the core not only needs good compression and shear strength but also excellent fatigue properties.

An airfoil section of a wind turbine blade with current typical design details of the main spar and the outer shell is shown in Figure 9.9. The main spar and the wing shells appear as constituent parts that are manufactured separately and then joined in a separate bonding process. Some wind turbine blades are manufactured this way, but it is not always the case. Alternative designs may involve that the two wing shells are joined with two or more internal webs (stiffeners). In this conceptual design, the wing shells are manufactured with relatively thick so-called spar-caps,

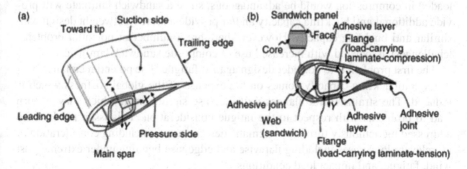

FIGURE 9.9 Wind turbine blade airfoil section with main spar and outer shells (a) and design details (b).[9]

which are usually monolithic composite laminates. Regardless of the one or the other of the design concepts, the main structural principles described earlier apply, i.e., the flapwise bending load is carried by a main spar or a "main spar-like" structure (constituted by the spar caps and internal webs/stiffeners), and the edgewise load is carried by the shells.[9]

As mentioned early, it is clear that sandwich structures or structural elements already play a very important role in modern wind turbine blade design. Thus, sandwich structures/elements are presently used for the wing shells and the webs of the main spar, see Figures 9.9. In all cases, the principal rationale behind the use of sandwich elements is to enhance the buckling resistance and minimize the weight at the same time.

An interesting question is whether it would be advantageous to use composite sandwich materials/structures for more structural parts than is practiced in current day wind turbine blades. Since most blade parts are already based on sandwich materials, this reduces to the question whether it will be advantageous if the main spar flanges are also manufactured as composite sandwich materials/structures.[9]

A study using a parametric FEA model was conducted to analyze two basic designs with monolithic composite and sandwich flanges, respectively. The analyses show that buckling of the spar flange loaded in compression is by far the governing criterion for the monolithic design, which is in good agreement with the findings reported with the results of other studies.

It is further shown that introducing sandwich laminates in the spar flanges results in a globally more flexible structure making tower clearance the critical criterion. The reason for this is that the airfoil dimensions are fixed, which implies that a substantial amount of the stiff composite laminates (e.g., the sandwich lower face) are moved closer to the airfoil center (axis of flapwise bending), thereby reducing the bending rigidity. However, significant weight reduction and increased buckling capacity is predicted for the proposed "sandwich" design. Moreover, the study shows that a proper choice of sandwich core material is important to prevent face wrinkling of the compressive loaded.[10]

The conclusions are local buckling of the main spar flange (or spar cap) on the suction side of a wind turbine blade is the dominating failure mode, and the use of sandwich rather than monolithic composite laminates for the flange (or spar cap) loaded in compression would be advantageous, since a sandwich laminate will provide additional anti buckling capacity and/or provide a more lightweight design with similar anti buckling capacity. However, there are a number of potential problems (challenges) associated with increased use of composite sandwich laminates.

The first problem is to consider design against fatigue. The potential enhancement of the antibuckling capacity comes on the expense of the global stiffness which is reduced. The strains in the flanges will increase significantly, which may in turn lead to problems with respect to the fatigue consideration. This issue needs to be addressed meticulously through systematic design studies including consideration of all relevant load cases including flapwise and edgewise bending under extreme gust wind, fatigue, and impact load conditions.[10]

Damage tolerance is another very important issue to be addressed. A prerequisite for replacing the most important primary load-carrying structural parts of a wind turbine blade, e.g., the flanges of the main spar, with composite sandwich laminates is that the overall structural reliability is not compromised in doing so. Wind turbines are generally much less safety critical than aircraft or ships. However, wind turbine blades present a special challenge in that they are produced in large numbers, similar to aircraft (or larger), but the possibilities for regular in-service inspection are much more limited or even nonexistent, because of accessibility problems.

Thus, the ideal approach would be safe-life design accounting for the worst combination of production defects that is likely to go undetected during production, and the worst in-service damage that is likely to occur without being noticed. This presents a major challenge for the industry since competition between manufacturers limits the amount of information they are able to share while differences in production techniques make the production defects more manufacturer-dependent than in many other industries. With high production volumes, improvement of all aspects of production control, including nondestructive inspection (NID) capabilities, should be a cost-effective means of reducing the incidence of such defects and of the uncertainties associated with them.

Then with the information of maximum allowable defects/damages at hand the issue of detectability becomes important. A major disadvantage of sandwich structures is that manufacturing defects and damages (especially in deep in the core and the interfaces) cannot always be detected by common methods. One side access only (in some cases) and large area inspection poses further challenges. A number of NDI technologies are available and under development that better fit the needs to inspect sandwich panels, where the most promising nondestructive inspection techniques for composite sandwich structures are based ultrasound, shearography, and X-ray principles.

Major challenges in regard to the detection of defects and damages in relatively thick sandwich structures include:

- The ability to detect deep defects/damage in thick sandwich structures remains limited,
- Sandwich structures with cores of end-grain balsa are especially difficult to inspect because the defects are masked by the many joints between core blocks and by natural features in the core material leading to large local density variations,
- It is generally not possible to detect far-side defects with one-sided inspection methods,
- At present there is insufficient knowledge about the sensitivity and reliability of many NDI systems when applied to composites to enable detectability limits and probabilities of detection to be quantified,
- Nondestructive testing standards are not yet well developed for application on sandwich composites. For some of the newer NDI techniques they do not exist at all.

A development that is likely to improve the situation for wind turbines is the widespread use of structural health monitoring of blades using, for example, fiber-optic sensors. These may be used both to detect abnormal events and to detect changes in dynamic response associated with the incidence of damage or major growth of a defect.

Naturally, any approach to introduce composite sandwich materials as a primary structure depends heavily on the design, architecture, and material selection of the sandwich structure itself. A major issue (disadvantage) of sandwich materials/structures compared with monolithic composites is that they are more prone to delamination and failure due to the presence of large weak interfaces between adjacent materials with very different stiffness and strength properties.

This means that sandwich materials/structures with homogeneous (e.g., foam core, balsa core) or nonhomogeneous (e.g., honeycombs, corrugated cores) support cores are notoriously sensitive to failure by interlaminar shear or through the application of concentrated loads, at joints and points or lines of support, and due to localized effects induced in the vicinity of geometric and material discontinuities. The reason for this is that although sandwich structures are well suited for the transfer of overall bending and shearing loads, localized shearing, and bending effects, as mentioned above, induce severe through-thickness shear and normal stresses. These through-thickness stress components can be of significant magnitude and may approach or exceed the allowable stresses in the sandwich constituents as well as in the material interfaces.

Wind turbine blades include numerous joints (at leading and trailing edges of the wing shells, between wing shells and main spar, between spar cap and internal stiffeners/shear webs), and in the vicinity of these, localized effects as described above cause the inducement of stress concentrations that may significantly affect the static and fatigue strengths of the sandwich. Moreover, buckling phenomena, as discussed in the previous section, also induce severe interlaminar and through-thickness normal stresses, which in many cases determine the ultimate load-carrying capability of wind turbine blade structures. Thus, composite sandwich material systems with improved/enhanced damage tolerance as well as innovative crack stopper and load introduction concepts will be key issues.

A way of improving the damage tolerance as well as the skin/core interface properties is to develop composite sandwich materials systems with structural elements in the form of fibers, pins, stitches, or even structural plate elements extending in the through-thickness direction of the sandwich laminate. These "z-direction" elements should provide stiff and strong connections between the face sheets/skins that also allow for load redistribution if local damage occurs, and at the same time, the in-plane stiffness and strength properties of the sandwich should not be compromised.

Various sandwich material systems with such performance characteristics are being used or are under development at present time. An example of this can be seen in Figure 9.10, which shows the so-called X-Cor™ sandwich material system. X-Cor™ is produced by reinforcing lightweight polymer foam with a truss network of pultruded carbon fiber rods and then laminated between composite face sheets. During processing the tips of the rods penetrate both face sheets, and the result is a sandwich element with improved damage tolerance and a superior skin/core bond.

The truss network carries both shear and compressive loads, and the foam core provides support against local buckling (wrinkling). The X-Cor™ sandwich material system was originally developed for helicopter fuselages and rotor blade systems, and it is based on a patented manufacturing process. It is probably not of direct interest for application for wind turbine blades, but sandwich material systems aimed

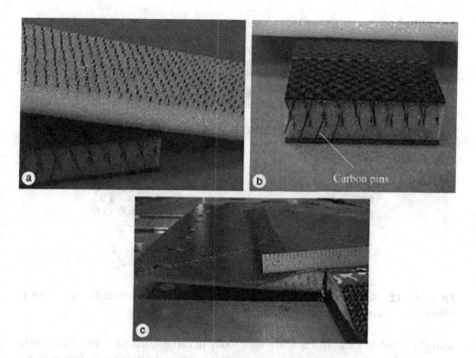

FIGURE 9.10 Core and sandwich panel of X-Cor™ truss sandwich.[11]

specifically for wind turbine blade applications, adapted for the manufacturing processes used for wind turbine blades, and with similar enhanced performance features, could be developed.

9.3 AIRPLANE AND AEROSPACE

One of the key features of sandwich construction is a high ratio of stiffness to weight. Aircraft and aerospace vehicles all have the requirements for light weight, and stiff and strong construction. Civil engineering has used sandwich construction since 1849 and several sources claim that a patent may have been taken out in 1915 a sandwich structure with honeycomb core. In 1924, a patent for a glider fuselage was filed and is cited in the papers. A four-engine transatlantic mail plane able to carry 22 passengers, which made its first flight in 1937 (see Figure 9.11a). The sandwich was designed with plywood skins and a balsa core. For a patent, dating from 1934 detailed in Figure 9.12, a sandwich is made up of two plywood skins and a cork core drilled with holes to optimize the mass. This process is believed to have been applied to an aircraft in 1938.

The plane that is most famous and most cited for its plywood skin and balsa core sandwich structures is the de Havilland "Mosquito" DH 98 (see Figure 9.11b). It turned out to be one of the best planes of the Second World War, both for its pure performance and for the extraordinary missions it achieved. Manufacturing was a one-shot process, which is now sought by manufacturers to reduce costs (see Figure 9.11c). There are

FIGURE 9.11 Pictures of some aircraft with sandwich structures.[12,13]

FIGURE 9.12 Sandwich wing made by plywood skins and a cork core drilled with holes to optimize the mass.[12]

also glued/bolted joints that are still used today in certain structures of military helicopters and are the subject of active research to reduce the number of fasteners and bring down costs. For these reasons, beyond just sandwich structures, the Mosquito is one of the most important precursors of modern, composite-structure planes.[12]

9.3.1 Applications in Airplanes

The use of honeycomb constructed panels in application of airplanes saved weight while not compromising strength. Initially, aluminum core with aluminum or fiberglass skin sandwich panels were used on wing panels, flight control surfaces, cabin floorboards, and other applications. A steady increase in the use of honeycomb and foam core sandwich components and a wide variety of composite materials characterizes the state of aviation structures from the 1970s to the present. Today, a large variety of sandwich cores are being applied in aircraft structural engineering (see Figure 9.13).

On aircraft with stressed-skin wing design, honeycomb-structured wing panels are often used as wing skin panel.[13] Aluminum core honeycomb with an outer skin of aluminum is common. But honeycomb in which the core is an Aramid® fiber and the outer sheets are coated phenolic resin is common as well. In fact, a myriad of other material combinations such as those using fiberglass, plastic, Nomex®, Kevlar®, and carbon fiber all exist. Each honeycomb structure possesses unique characteristics depending upon the materials, dimensions, and manufacturing techniques employed. Figure 9.14 shows the locations of honeycomb construction wing panels on a jet transport aircraft. Figure 9.15 shows an entire wing leading edge formed from a honeycomb structure.

There is a broad range of composite sandwich structures application in Airbus aircraft. Typical external structures are aerodynamic fairings, covers and doors.

Sandwich Structural Composite Applications

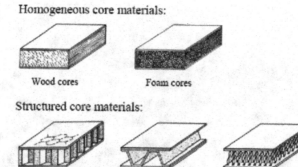

FIGURE 9.13 Different sandwich core types used in aircraft engineering.[14]

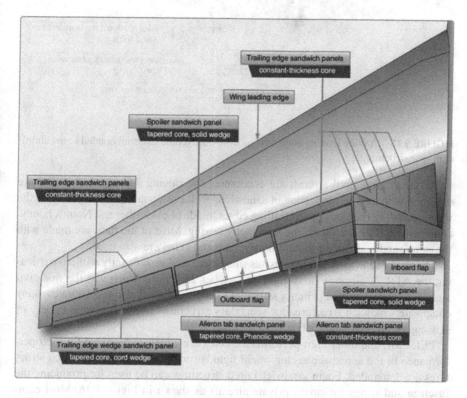

FIGURE 9.14 Honeycomb core panel wing construction on a large jet transport aircraft.[13]

Examples are radomes, belly fairings, leading and trailing edge fairings, engine cowlings and landing gear doors. Moreover, there is a variety of composite sandwich control surfaces throughout the Airbus fleet (e.g., rudder, aileron, and spoiler). Examples for the application of composite sandwiches inside the aircraft are fairings and floor panels in the passenger compartment.

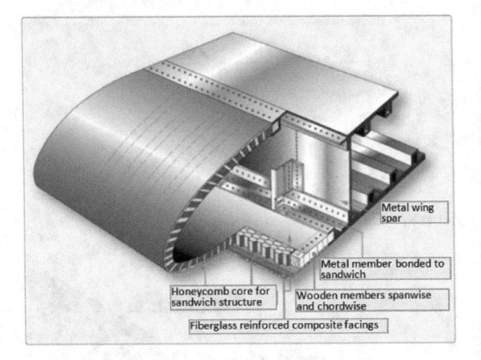

FIGURE 9.15 A wing leading edge formed from honeycomb material bonded to the aluminum spar structure.[13]

As far as large civil aircraft are concerned, the Boeing 747 and Airbus 380 are designed with a large proportion of sandwich.[12,14] It has about half the surface of the wing, including the leading and trailing edges, made of glass fiber and Nomex honeycomb, which is also used for the large belly fairing. Most of the flaps are made with the same sandwich but aluminum honeycomb and skins are also used.

The use of sandwich composites has since increased significantly with, in particular, the ATR 72 airplane (first flight on October 27, 1988), which was the first civil aircraft to have a carbon primary structure (the wing box). It also incorporates many composite sandwich structures for secondary structures but with a wide variety of skins: glass, Kevlar, and carbon.[12]

PEI, PMI, and other high-performance cores are widely recognized for their performance in the aerospace sector where light, strong, and resilient sandwich structures are demanded. Foam sandwich cored structures can be used for producing the fuselage and wings for small, private aircraft as shown in Figure 9.16. Most components of large aircraft, such as the floor, interior ceiling, bulkheads and side wall paneling, window sets, luggage stow bins, seat sets, serving carts, and landing gear doors, also are fabricated of sandwich structures.

The application of sandwich structures in large commercial airplanes is currently restricted to secondary structures. For a more widespread application and in order to introduce sandwich in primary structures several challenges must be met. For primary structures the possibility of a failure that leads to a catastrophic failure of

Sandwich Structural Composite Applications 351

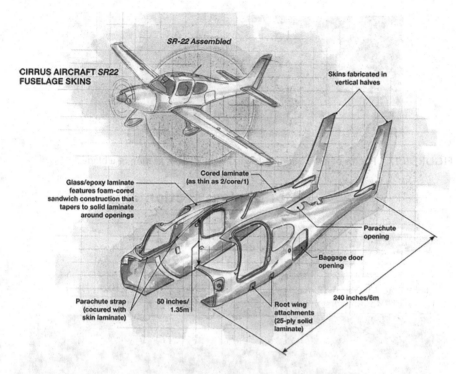

FIGURE 9.16 Airplane fuselage and wing made by using foam core sandwich.[15]

the aircraft must be prevented. Therefore the structure needs to be evaluated in order to prove that damage occurring during the service life of the aircraft will not lead to failure or excessive structural deformation until the damage is detected. Full compliance to this requirement needs to be shown by simulation and test. The following works need to be done for meeting the full compliance:

1. Verified simulation methods and tools for simulating the damage behavior of complex sandwich structures;
2. Cost-effective, in service nondestructive testing methods for complex sandwich structures;
3. Introduction of structural health monitoring;
4. Advanced in-service repair methods for existing and new sandwich concepts;
5. Novel load introduction and reinforcement concepts
6. New material systems with improved toughness and impact behavior.

9.3.2 APPLICATIONS IN HELICOPTERS

The first application was rotor blades made of honeycomb or foam cores with fiberglass skins for helicopter as seen in Figures 9.17 and 9.18. The first composite sandwich blades were tested in 1959 then, following research programs, all the 4,130 steel

FIGURE 9.17 Photos of helicopter blades and typical cross section of a blade.[12]

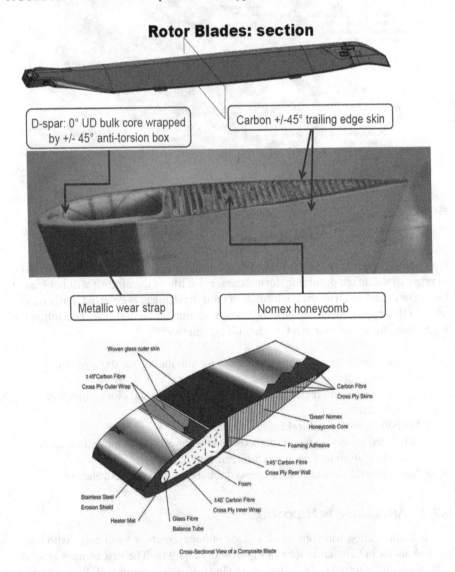

FIGURE 9.18 Helicopter rotor blade structures.[16]

blades of one type of helicopter had been replaced by composite blades by the mid-1970s. The lifespan of a composite helicopter blade is longer than the lifespan of the helicopter. In addition, the possibility of producing optimized aerodynamic shapes (cambered and twisted sections) by molding makes it possible to increase the take-off weight and reduce fuel consumption. Another advantage of these composite blades was their tolerance to damage, which had been emphasized since their introduction in the 1970s. The new designs make it possible to absorb hard projectiles launched at 150 m/s, whether in frontal or razing impact. They are also resistant to the detachment of ice blocks from the fuselage in the event of flights in icing conditions.

The main part of the structure was in Nomex honeycomb/metallic skin sandwich because this solution is economical and has better vibratory qualities, especially for the tail boom. We can also note that the floor was made of honeycomb with aluminum skins because it is also a more economical solution. The weight saving with a carbon/ Nomex honeycomb floor would be 20%, but the cost would be increased by 70%. In general, the introduction of sandwich and composite parts into helicopter structures has resulted in weight reductions of 15%–55% and cost reductions of 30%–80%.

9.3.3 Applications in Space

The use of sandwich structure in the Apollo project that successfully landed on the moon in 1969 showed the high potential of sandwich structure in the field of aerospace. With the help of this unique technology, it was possible to construct the Apollo capsule and its heat shield, which was light and yet strong enough to sustain the stresses of acceleration during the start and re-entry phase. Honeycomb sandwich structures are used in a wide variety of critical structures in aerospace systems. These include components of space shuttle, payload fairings (shrouds) for launch vehicles, and adapters for mounting of satellite payloads, solar array substrates, antennas, and equipment platforms.

However, since 1964, there have been several known or suspected failures of honeycomb structures. These failures have been attributed to the lack of venting in the panel design/manufacture. On the other hand, based on available information, vented honeycomb sandwich panels never have experienced failure during flight. In the cases documented herein, the consequences of the failures have been significant and costly.

Honeycomb sandwich panels that are not vented will contain air (and possibly volatiles, including moisture) which causes a pressure differential during launch into orbit. When the sandwich structure is heated and no air pressure out of space, the internal pressure will rise further. In any case, each individual unvented honeycomb cell acts as a tiny pressure vessel imposing stresses on the skin-to-core bonds. If these stresses are high enough, panel failure (i.e., skin-to-core debonding) will occur. Certain defects introduced during panel manufacture would make failure more likely.

Vented honeycomb sandwich structures use perforated, slotted, or porous honeycomb cores through which air can flow readily from cell to cell at a rate corresponding to the pressure drop during the ascent phase of a launch vehicle. Venting to the exterior is provided either through the skin or panel edge members. The changes in pressure within the panel ideally should occur at a rate corresponding to the external atmospheric pressure change during launch vehicle ascent. A few of vented cores are shown in Figure 9.19.

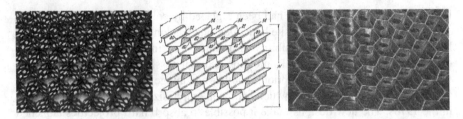

FIGURE 9.19 Examples of vented honeycomb cores: leno fabric honeycomb core (a),[17] composite honeycomb core (b),[18] and aluminum honeycomb core (c).[19]

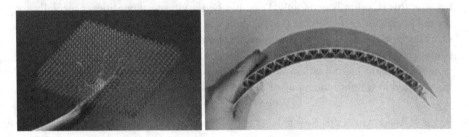

FIGURE 9.20 3D printed truss lattice core and its curved sandwich panel.[20]

Other ventable cores are truss lattice structures that make for stronger materials, especially in shear. This means they are better than honeycomb core in resisting sliding forces along the surface of a material and in bending. One example is that a 3D printed lattice core is made from nanocrystalline nickel faced with carbon fiber-reinforced plastic as shown in Figure 9.20. The 3D-printing technique can grow compound shapes and curves because you can grow the structure into the shape desired, no machining is needed. Also, the density of the lattice can be adapted to match local stress – less dense where less strength is needed, higher density where it is needed. One more example of the truss lattice core and sandwich is shown in Figure 9.21. The concept is designed to provide a maximum of weight saving while still offering attractive protection against impact and noise. Again with the outer skin providing the aerodynamic surface. In between the two skins, the core material is ventable in order to avoid moisture accumulation.

Sandwich structures are used in satellite systems. The satellite dish antennas are made from Kevlar or aluminum honeycomb core with carbon fiber prepreg skins.[21] Because the parts will ultimately operate in the vacuum of space, honeycombs are vented so every hole is connected to the next and no air is trapped in the part when it is launched into space. Otherwise, the vacuum would just pop the skins off.

9.3.4 Future of Aeronautic Sandwich Structures

New development is mainly focused on structural improvement, the integration of functions and the multifunctionality of sandwich structures. Regarding structural improvement, many innovative cores have been developed or rediscovered in recent years. A few of these possibilities could be interesting to replace Nomex or aluminum

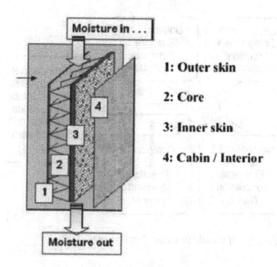

FIGURE 9.21 Ventable truss lattice core and sandwich structure.[14]

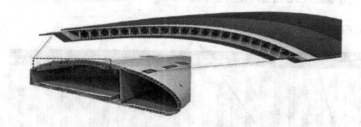

FIGURE 9.22 Kagome core and sandwich structure for an aileron.[12]

honeycombs, which are very efficient. One study unveiled a titanium Kagome core that outperformed traditional honeycombs in shear and compression. This solution is proposed for ailerons that are a hinged flight control surface usually forming part of the trailing edge of each wing of a fixed-wing aircraft (see Figure 9.22). It also has the advantage of being ventilated, which eliminates the potential problems of moisture ingression.

A multifunctional material system will be next target for new sandwich structure, which should integrate in itself the functions of more than two different components increasing the total system's efficiency. Many of the sandwich structures presented in the previous sections are at least two multifunctional in that, generally, they naturally integrate two physical functions passively: mechanics + thermal insulation; mechanics + stealth; mechanical + moisture ingression, mechanical + acoustic absorption, and mechanical + vibration damping. Figure 9.23 shows a concept of multifunction structure that integrates the structural, drag reduction, electrical conduction, and sensing functions.

There are a few real multifunctional applications where the sandwich is designed a priori to fulfil a wide variety of functions. Figure 9.24 shows a possibility of using

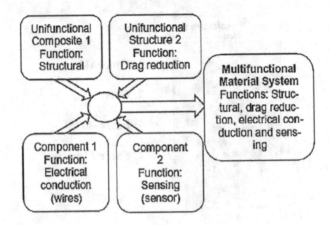

FIGURE 9.23 Concept of a multifunctional structure.[12]

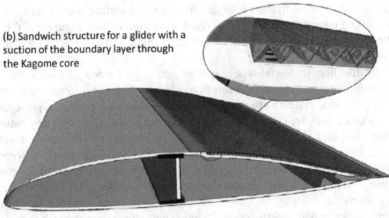

FIGURE 9.24 Two examples of multifunctional sandwich structures.[12]

Sandwich Structural Composite Applications

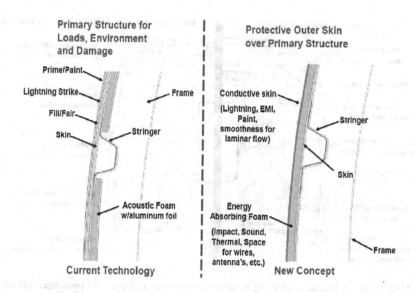

FIGURE 9.25 Protective skin concept of sandwich structure.[12]

solar cells as working skins for the Solar Impulse project. A sandwich structure is constructed with a core that is a radio frequency antenna in addition to playing a structural role in a drone application. A sandwich allowing suction of the boundary layer for a glider (see Figure 9.24b). The suction is provided by a pump and the folded sandwich is perforated.

It should be noted that, in many other areas, sandwich structures serve as mechanical supports for other functions: energy harvesting, heat exchange, microwave absorption, integrated electronic device, battery integration, damping with resonator integration, and fire protection. A concept of "protective skin" is a solution of asymmetric sandwich from the functional and mechanical point of view as shown in Figure 9.25. The inner skin plays the structural role, and the core and the outer skin integrate a large number of functions, including aerodynamic optimization, acoustic and thermal insulation, protection against lightning strikes, moisture insulation, damage protection, and installation of ice protection systems, wires, antennas, or other sensors. The sandwich structure with the protective skin will be the concepts of the next-generation fuselage of airplanes and other aerospace vehicles.

9.4 TRANSPORTATION

In ground transportation, sandwich structures can be found in trucks, cars, busses, and trains. Since the 1980s, front cabs of locomotives have been built with sandwich technology because of its high strength and good impact and energy absorption properties. There are also some examples of sandwich paneled rail vehicles. Some advantages of using sandwich composites over traditional materials in applications of buses and passenger rail vehicles are shown in Figure 9.26.

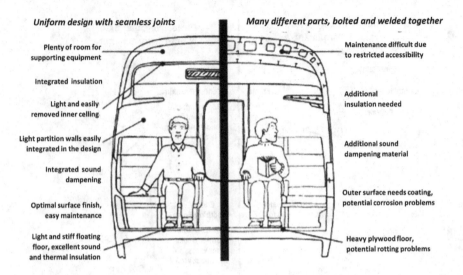

FIGURE 9.26 Advantages of using sandwich composites in buses and passenger rail vehicles (left: multifunctional sandwich; right: traditional metal design).[22]

To use the sandwich structures in the transportation industries, the following aspects should be paid attention to:

1. Requirements with respect to stiffness, strength, stability, insulation behavior, fire rating, assembly and manufacturing, etc. have to be taken into account right from the beginning.
2. Special focus must be put on the edge stiffeners, the joining techniques, and the local load introduction, since they strongly determine the mass of the structure and the costs for the manufacturing and for the assembly. A few of panel edge reinforcement, roof and wall joining, and assembling technologies are shown in Figure 9.27.
3. Sandwich structures usually trigger larger efforts in the design phase, in the structural analysis as well as in the manufacturing than conventional engineered structures (like welded sheet design with local stiffeners).
4. The increased functionality of the sandwich structures as well as the overall cost reduction justifies the application of hybrid design.

Figure 9.28 shows that strength, stiffness, thermal and acoustic insulation, and surface finish can be combined into one sandwich structural panel. This would drastically reduce the number of parts, complexity and also assembly time of a bus and rail vehicle bodies. These are some of the goals with the so-called multifunctional body panel design.

9.4.1 Rail Car Application

Modern rail vehicles are designed with the aim to minimize mass and thus operational energy demand. The use of composite materials, especially sandwich structures, in

Sandwich Structural Composite Applications

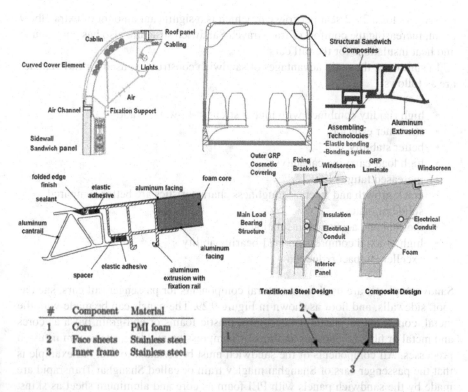

FIGURE 9.27 Sandwich composites roof and wall assembly joining and edge stiffing technologies.[23]

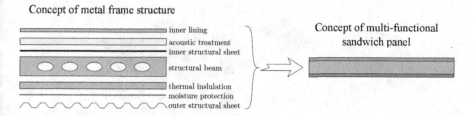

FIGURE 9.28 Conceptual cross-section of traditional rail car and bus body and the multi-functional sandwich panel design.[24]

vehicle structure could reduce the weight and thereby the fuel consumption without compromising the stiffness of the structure.

Using sandwich structures for the vehicle also reduces manufacturing complexity, by reducing the number of parts needed and integrate several functionalities in to one single panel. Another benefit of a sandwich paneled car body is reduced wall thickness. Reducing wall thickness gives extra interior space. This may seem insignificant, but for any car, if the wall thickness was reduced by 120 mm by using a sandwich panel replacing traditional structure, it will give an extra 6 cm between passengers for a 2×2 seat per row car, which is a significant amount of extra elbow room, increasing the comfort of train travel. Sandwich structure also has good sound and heat insulations for the rail cars.

In summary, the main advantages of sandwich construction used for rail vehicles are as follows:

- high rigidity combined with higher strength-to-weight ratio
- smoother exterior
- better stability
- high load-carrying capacity
- increased fatigue life
- crack growth and fracture toughness characteristics are better compared to solid laminates
- thermal and acoustical insulation
- high bi-axial compression load-bearing ability
- excellent impact reduction

Sandwich panels are used for structural components for passenger rail cars, such as roof, sidewalls, and floor as shown in Figure 9.29. The panels can be made with the metal, composite or plastic honeycombs, plastic foams or end-grain balsa as cores and metal or fiber composites as skins by compression molding, or vacuum infusion processes. All components of the sandwich must be resistant to fire. An example is that the passenger cars of Shanghai maglev train or called Shanghai Transrapid are made by the sandwich panels with PEI foam as core and aluminum sheet as skins. Figure 9.30 shows that the sandwich panels are used as floor, sidewalls, and roof.

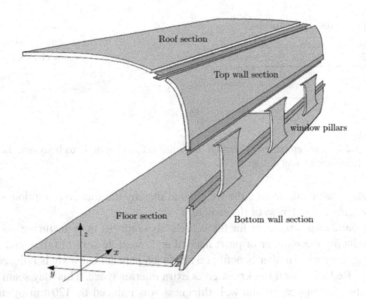

FIGURE 9.29 Module sandwich rail car body.[24]

Sandwich Structural Composite Applications

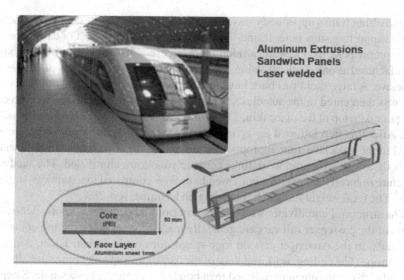

FIGURE 9.30 Structure of the car body of Shanghai maglev train.[23]

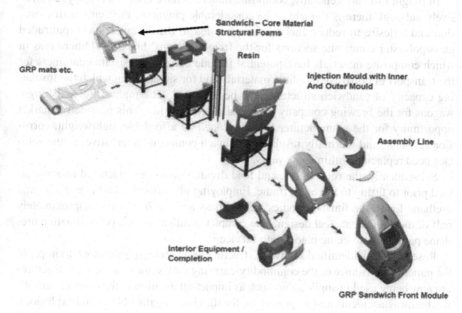

FIGURE 9.31 Vacuum infusion process for making a front cap of fast train.[23]

Figure 9.31 shows a cap made by a vacuum infusion process. The foam cores are cut and assembled into a cap shape before being laminated into the mold with the precut fiber fabrics. The rigid inner and outer mold can be repeatedly used for producing multiple parts. Auxiliary parts are assembled with the cap after being removed from the mold.[24,25]

One tilting train's upper body was constructed of a lightweight sandwich structure with a supporting steel inner frame. The sandwich elements consist of carbon fabric/epoxy prepregs for the faces and an aluminum honeycomb core. The entire car body is manufactured as one single structure. This was accomplished by means of large-scale autoclave. A large mold was built in which the outer face was first laid out. The outer face was then cured in the autoclave. Secondly, the inner frame and honeycomb core was placed on top of the outer skin. The core and skin were bonded by use of an adhesive. After this step followed lay-up of the inner face. Lastly, the entire structure was cured in the autoclave after appropriate vacuum bagging. By constructing the entire car body as one structure, weak links between panels are eliminated. The sandwich structure reduced the upper car body weight by 39% compared to a stainless-steel car body. The total weight reduction, including underframe, was 28%.[26, 27]

The structural and divider wall panels, exterior and decorating roofs, doors, and floors of the passenger rail cars are generally made of sandwich structure composite materials, in the passenger cars on high-speed railways in China, Japan, America, and Europe. The ceiling, wall, and floor of the bathroom and toilet boxes are all made of sandwich composite materials and then bonded or connected as a whole. Sleeping berths, tea tables, and luggage racks are also made using sandwich.[28]

In freight rail cars generally, composite materials have not been employed extensively although there is thought to be considerable potential. Problems with corrosion and a desire to reduce fuel consumption led to the specification of pultruded glass/polyester composite sections for the freight wagons' housings. Other areas in which composite materials have potential include corrosion resistant containers for the transport of corrosive or edible materials and for situations in which the insulating capacity of sandwich structures can be exploited. Thermally insulated freight wagons for the brewing company exploit all these benefits. This provided a market opportunity for the manufacturer that developed an affordable, lightweight, corrosion resistant, and thermally insulated sandwich composite alternative, as the vehicles need replacing within some years.

Subsequently, the roof, doors, and load dividers were manufactured and assembled prior to fitting to the underframe. Employing glass fibers, vinyl ester resin, and urethane foam, the finished bodies have a mass around 7t which is approximately half of the equivalent steel design. The completed railcar is then painted with a urethane paint prior to being placed into service.[29]

Research work identified sandwich structure as a potential technology to improve the puncture resistance of the commodity-carrying tank vehicle. Sandwich structures are now being used in applications such as impact-attenuation in the event of run-off-the-highway accidents and as protection for ship hulls against blasts and explosions. The potential merits of using sandwich structures investigation for railroad applications have been conducted.

One of the concepts under development encases the pressurized commodity-carrying tank in a separate car body. Moreover, this improved tank car concept treats the pressurized commodity-carrying tank as a protected entity. Welded steel sandwich structures are examined as a means to offer protection of the commodity tank against penetrations from impacting objects in the event of a collision. Sandwich can provide greater strength than solid plates of equal weight. Protection of the tank is realized through blunting of the impacting object and absorption of the collision energy. Blunting distributes impact loads over a larger area of the tank. Energy absorption reduces the demands on the commodity tank in the event of an impact. In addition, the exterior car body structure made from sandwich panels is designed to take all of the in-service loads.

An example shows that a sandwich panel with a square egg-crate core is shown in Figure 9.32. The egg-crate geometry is used here as an exemplar. A variety of core geometries are available to be used within sandwich structures, such as tubular cores, square diamond core, double corrugated core and solid surface core materials as shown in Figure 9.33. The face sheets and the core of the sandwich structures described in this example are steel, which are constructed into the curved shell by welding.

The design example treats the conventional tank as a protected entity which is encased by a separate car body. The conventional tank is reinforced with stiffeners around the head. The reinforced tank with insulation sits within a structural car body, which is made of sandwich panels. Between the reinforced and insulated tank

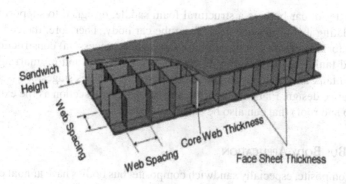

FIGURE 9.32 Annotated egg-crate sandwich panel for tank carbody.[30]

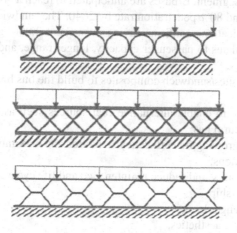

FIGURE 9.33 Tubular cores, square diamond core, and double corrugated core.[31]

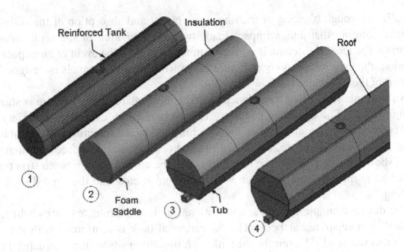

FIGURE 9.34 Progression of design assembly (tub and roof are sandwich structures).[32]

and the exterior car body is a structural foam saddle, designed to support the tank while isolating it from service loads from the car body. Therefore, the construction sequence for this design is a three-phase fabrication process: (a) construction of the reinforced tank, (b) construction of the exterior car body, and (c) marriage of the reinforced tank and car body. Figure 9.34 shows the various stages of assembly for an alternative design. Flat panels form an octagonal cross-section as the exterior car body (tub and roof) that can also be curved structures.

9.4.2 Bus Body Application

The all-composite, especially sandwich composite, bus bodies have annual expanded production volumes. The continued growth of the electric bus market needs even more light materials for the industry. The buses transfer to electric faster than any other major vehicle segment. E-buses are anticipated to reach a 40% penetration rate in the US by 2030 and 80% penetration rate by 2040. The sandwich composite body solution in bus applications confirms saving over about 2,000 kg in mass per bus. This reduction translates to passenger capacity, longer range, and increased overall operational efficiency.

Using the composites/sandwich composites to build the bus body can

- Reduce weight to enable alternative fuel sources such as electric, hybrid electric, or natural gas;
- Comply with growing fuel efficiency standards and maximum gross vehicle weight regulations;
- Increase durability and reduce maintenance and life cycle costs;
- Optimize ridership;
- Increase passenger load;
- Improve vehicle aesthetics;
- Reduced up-front production investment.

Sandwich Structural Composite Applications 365

The first applications are developed using a hybrid design concept combined with sandwich and solid laminated composite, and metal structures. The sandwich construction is considered for application to primary structures such as the body shell, roof, and floor, while solid laminated composites are applied only for components with a relatively high curvature and complex geometry, which are more troublesome to manufacture using the sandwich panels. Figure 9.35 shows the manufacturing concept of a low-floor bus made of sandwich composites and other materials.

The sandwich panels used in the body shell, floor, and roof structures of the low-floor bus are composed of woven glass fabric/epoxy laminate face sheets and aluminum 5052 honeycomb core. The outer face sheet of the sandwich panel has a thickness that is twice the thickness of the inner face sheet to save on additional cost and weight. The face sheets of the sandwich panel are laminated with woven glass fabric/epoxy prepreg. The core used in the sandwich panels is aluminum honeycomb with a thickness of 25.4 mm.

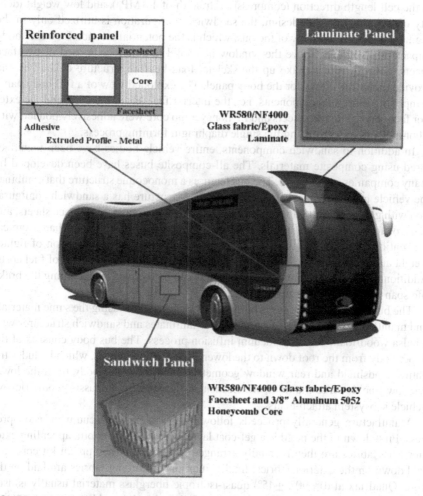

FIGURE 9.35 Design concept of a low-floor bus with sandwich composites.[33]

Another example of using sandwich structures is in bus structures. A weight reduction of up to 160 kg (−20%) can be achieved by using sandwich roofs. Lightweight bus structure with sandwich structural roof and sidewalls offers quiet rattle-free riding comfort and superior corrosion resistance for long life.[34,35]

All thermoplastic sandwich composites also have been used for bus components. An example shows that a design featured a sandwich composite panel with E-glass fiber/polypropylene (PP) face sheets and PP honeycomb core as bus body panels, which provide high strength and energy absorption benefits. The thermoplastic composite body panel exhibited excellent weight saving of more than 55% compared to a conventional bus with aluminum skin and supporting steel bars. A sandwich panel feature was chosen as a viable option because of its potential to resist dynamic impact in addition to carrying structural loads.

In the concept of the thermoplastic composite body panel,[36] PP honeycomb was chosen as the core material due to its excellent ability to absorb energy, high strength in the cell length direction (compressive strength of 1.3 MPa), and low weight (density of 80 kg/m^3). In this design, the sandwich configuration is utilized only in the segment below the window of the bus, which is the potential location for a side body impact. The segment above the window is a solid laminate composed of two face sheets. Several frames make up the skeletal load-bearing structure of the bus and provide mounting points for the body panel. The exploded view of a full body panel comprising three subcomponents, i.e., the interior face sheet, the core, and the exterior face sheet. The face sheets and PP honeycomb core were adhesively bonded with a hot-melt adhesive by using the single diaphragm forming process.

In addition to sandwich components, entire vehicle bodies have been manufactured using composite materials. The all-composite buses have been developed by many companies in the world. The bus features a monocoque structure that combines the vehicle body with its chassis elements. The structure has a sandwich configuration with glass or carbon fiber-reinforced, vinyl-ester or epoxy resins face sheets, and balsa or foam cores. The non-metallic bus is more than 30% lighter than a typical conventional bus and requires 60% less power to run. The combination of lighter weight and reduced power consumption leads to a significant reduction of fuel costs. Additionally, the maintenance cost of the bus can be reduced by extending the brake life span and diminishing tire wear.

The bus in an example case is built in four basic sections using the same materials and manufacturing process, in a mix of solid laminates and sandwich structures with a balsa wood/foam cores by vacuum infusion process. The bus body consists of the upper body from the roof down to the lower edge of the windows, which includes the entire windshield and rear window geometries; and the lower body from the lower window line down to the floor, which includes all structural chassis geometries for vehicle subsystem attachments.

Manufacture generally proceeds following the standard vacuum infusion process. First down in the mold is a gel-coat layer to produce a smooth, appealing exterior. Dry fabrics are then manually arranged in the mold; then precut kit cores are laid down on the exterior fabrics; finally, interior reinforcing fabrics are laid on the core. Quadriaxial (0°, 90°, ±45°) quasi-isotropic fiberglass material usually is used where high strength and stiffness are especially critical, notably where suspension

FIGURE 9.36 All-composite bus structure after assembling.[38]

components interface, where mirrors and lights and other exterior elements are connected, and in areas subject to high compression – for example, where airbags interface with the body and at the major drive-axle connection. The layup is vacuum bagged, infused with a catalyzed vinyl ester resin system, and cured at room temperature. An upper body is ready for vacuum infusion. An all-composite bus structure is shown in Figure 9.36.[37]

9.4.3 Truck and Semitrailer Body Application

The sandwich panels are characterized by a particularly low weight with the highest technical characteristics. Sandwich composite panels are thus particularly suitable for the construction of load-optimized vehicles. Tailor-made solutions for emergency vehicles, event vehicles, and truck superstructures for special applications are key strengths for the composite panels. The sandwich panels are especially favored for courier, parcel, and express services trucks as well as for similar vehicles (roof / wall / floor).[39]

The technological innovation of design composite lies in the unique and flexible manufacturing process which makes it possible to offer extremely light panels. Highly durable elements with high stiffness use foam, polypropylene, or aluminum honeycomb cores and glass fiber-reinforced plastic or aluminum facings for a wide range of applications. The sandwich panels with thermoset resin composite skins can be made by compression molding, vacuum infusion, and pultrusion. The all-thermoplastic honeycomb sandwich panels also are widely used for making truck body boxes, which have 20% less weight than that of the thermoset panel with similar structural properties. A truck body made of the all-thermoplastic sandwich panels and a fixture for assembling are shown in Figure 9.37.

The sandwich composite panels are being used for building the highly "customized" 53 feet lightweight semitrailer van for regional and long-distance delivery, and

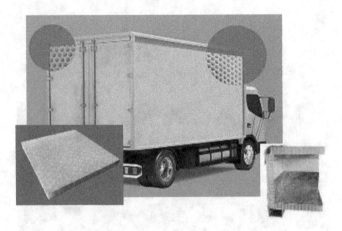

FIGURE 9.37 A delivery truck body made of all-thermoplastic sandwich panels.[40]

a 40' beverage van featuring an overhead door with lift gate, custom E-track system, insulation package, and thermal heater unit for local delivery. One type of sandwich panels that are made with thermoplastic polyethylene terephthalate (PET) foam core by the pultrusion process is used for building the trailers. The panels are seamless, snag-free, and recyclable for the full-length sidewalls and roof and floor of the trailer body, which provide insulation, along with the advanced fatigue and mechanical properties, resulting in increased strength, performance, durability, and in most cases, weight reduction. The panel will not warp, rot, swell, or deform at nearly any moisture level. Whole piece seamless structures simply remove all alloy components and eliminate the concern of any future rust or corrosion.[41,42]

Other trailer OEMs are introducing sandwich composite technology in their refrigerated trailer and containers. A reefer trailer body series with advanced molded sandwich structural composite technology, which makes up the trailer's walls, floor, and roof, consists of a high-efficiency foam core encapsulated in a fiber-reinforced polymer shell and protective gel coat. It is reported to boost the thermal performance over conventional wall material by up to 28%, double puncture resistance and reduce weight by up to 20%.[43]

The sandwich structural technology is bringing new levels of thermal performance to refrigerated semi-trailers and refrigerated truck bodies. Improving thermal efficiency and reducing weight more than traditional designs, it is revolutionizing operations for refrigerated carriers and will be used in new applications beyond commercial transportation.[44]

9.4.4 Automotive Industry

While a great deal of sandwich structures for the potential for their use within automobile and development of tools to ease their implementation have been ongoing since at least the mid-1980s, the use of pressed and spot/laser welded steel structures is still the praxis within the automotive industry. Not so many works exist describing full-scale mass production of automobiles which use sandwich structures

Sandwich Structural Composite Applications

as load-bearing or structural components. The primary reason for this is cost; if a new material is to be used to replace an old, its entire cost including raw material, manufacture, design, and development must be less than that of the material it replaces. Even the simplest sandwich construction using a metallic face sheet and a polymer foam core becomes a very complex problem when attempting to predict face-core de-bonding or buckling phenomena, particularly when dynamic loading is involved. Experimental methods are often necessary to assess the crashworthiness of sandwich structures.

New methods of production are being developed which show promise, and it is likely only a question of time and resources before satisfactory production rates and levels of automation sufficient for the automotive industry are achievable. Sandwich constructions also have the potential to eliminate problems of corrosion and increase the lifetime of components, provided they are designed and manufactured correctly.

Even though the sandwich composites are not used for making structural components of automobiles in mass production yet, many non-structural parts have been made by using sandwich composites now. A sandwich laminate of natural fiber-reinforced thermoplastic and thin PET foam produced by compression molding is currently in production by Mercedes-Benz. Figure 9.38 shows that the components made of polypropylene honeycomb sandwich composites are used or will be used for a passenger car, including under body panels, door module, battery housing, inner bonnet, front separation wall, seats and seat back, and backend module, etc.

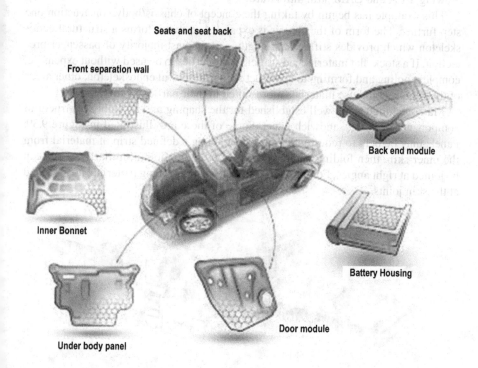

FIGURE 9.38 Passenger car components made of honeycomb sandwich composites.[44]

A sandwich panel made of select-density, balsa wood core, and E-glass fiber composite skins was used as the floors of Chevrolet Corvette sport car from 1996, which works better than metals for reducing weight and part count, improving sound and vibration, while adding structural stiffness. The compression molded floors were used until 2020; the balsa wood core was replaced by 130kg/m^3 density PET foam core in 2008. Some special models during this time used carbon fiber for the skins.[45]

A multifunctional sandwich concept for a roof system was proposed in which all components from the outer sheet metal to the interior trim were replaced by a single multifunctional sandwich panel. The concept car roof panel consists of four components; external face sheets of isotropic material, a structural foam layer of typical polymeric sandwich core foam, a single layer of lightweight, open–celled acoustic foam, and an interior face sheet which provides both structural and aesthetic functionality In the panel configuration, the interior face sheet was perforated to allow fluid interaction.[46]

Research works have been ongoing on how to use sandwich structure composite materials to manufacture automotive structural components, such as vehicle chassis. One example presents an alternative concept for the production of a lightweight vehicle chassis, using preformed flat panels assembled into a primary structure by the simple processes of CNC routing, folding and adhesive bonding. The assembly of structures from flat sheet as described here has a number of advantages: no tooling is required, existing technology CNC equipment may be used, material quality is assured by the supplier, not the manufacturer, and direct integration between CAD drawing, FEA and CAM straightforward.

This example has begun by taking the concept of chassis/body construction one step further. The form of the chassis is expanded so that it forms a structural endoskeleton which provides stiffness, strength, and the vast majority of passenger protection. If a stock, flat material had sufficient properties to be used without expensive, complex tooling and forming to construct such a structural endoskeleton, other areas of the vehicle could use materials more suited to their particular function.

Many techniques are well established for the shaping and assembly of structural components from flat sandwich panel. Some of these are illustrated in Figure 9.39. Panels may be bent to required angles by removing a defined strip of material from the inner skin, then folding and adhesively bonding the joint. Similarly, panels can be joined at right angles. For additional strength, a reinforcing material can be added at the skin joints.

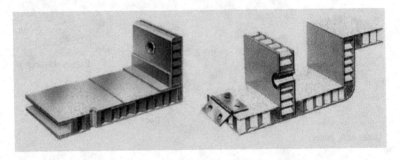

FIGURE 9.39 Some jointing methods for sandwich structural chassis.[47]

FIGURE 9.40 Design model of the car chassis made of sandwich flat panels.[47]

The chassis was designed to be assembled using the cut-and-fold methods, the sandwich panel components were drawn using AutoCAD, which was based on a beam design which provided simplicity with efficient use of the material and a structure which was continuous along the vehicle length, offering stiffness and a degree of front and rear impact resistance as shown in Figure 9.40. A CNC router was used to machine the panels. The twin longitudinal beams were bonded first and then suspension components fitted to the panels. The panels were then bonded, and the remaining components fitted. Bonding was achieved throughout the structure with a structural adhesive applied by a special application gun.

One more example is that a light vehicle concept is developed, with a body structure of only 90 kg and a high level of damage tolerance, in case of accidents. The structural concept is a consequent implementation of hybrid materials, resulting in a lightweight structure made of few parts with a relatively simple shape. This is achieved by adapting materials and using a sandwich architecture for structural components. Especially structural polymer foams and honeycomb for cores in combination with metallic sheets as facings are qualified. The goal of the project with the sandwich structures is the development of a highly efficient, affordable urban vehicle, with excellent passenger safety. The concept is a two-seated urban vehicle with an overall weight of 450–500 kg. The maximum speed will be 100 km/h, with a range of 100 km.

For achieving this structural concept, a lightweight car body structure is developed that is almost completely made of sandwich elements. The use of high-cost materials such as CFRP is avoided. Aluminum-sandwich parts are used for the main structural components. Some of the sandwich-elements are three-dimensionally shaped and form a metal monocoque structure, incorporating highly innovative deformation mechanisms to achieve maximum passive safety for a given weight. The combination of extremely light weight with excellent crash behavior is achieved by making use of the specific properties of metal structures reinforced with light core materials.[48]

The use of sandwich panel construction techniques can be widely seen in modern-day Formula 1 chassis. The most common face material in this category by far is sheet metal, which offers good properties at a reasonable cost. Several types of core shape and core material have been applied to the construction of sandwich structures. One type of core shape is honeycomb, which consists of very thin foils in the form of hexagonal cells perpendicular to the facings.

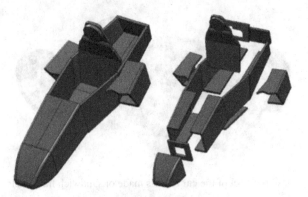

FIGURE 9.41 Formula 1 car chassis made of sandwich structures.[49]

For one F1 car, aluminum was the material choice for facings and core. Using aluminum sandwich panel was based on the requirements to account for costs in the design and manufacture and the requirement for a high strength-to-weight ratio (essential for a competition car). A further advantage is the ease of recycling the chassis at the end of life.

The chassis design for one F1 car was deconstructed into component parts as shown in Figure 9.41. The component parts were constructed from "shaped" panels that have been cut from larger panels. The cutting of the shaped panels was performed with a computer numerical control (CNC) router–cutter using a file generated from a three-dimensional (3D) computer model of the chassis. To facilitate folding, the shaped panels were routed along the fold lines. Routing consists of removing part of the facing panel and removing a section of the core material beneath. The width of the face plate removed defines the angle to which the panel can be folded. Therefore, the panel is self-jigging (i.e., no supplementary jig is required to position the panel). The reason for removal of the core material is to prevent "stacking" of the core material when the panel is folded. Assembly of the component parts was by folding the shaped panels along the fold lines to the specified angle. To restore load path continuity, a reinforcement plate was bonded to the inner facing plates with a two-part epoxy.

Another Formula 1 racing car, Honda racing F1, consists of two parts: a top section and a bottom section. Sandwich construction and low-density materials such as carbon composites provide an ideal solution for these parts. The floor presents a greater design challenge in that is required to fulfil a range of different tasks. The layer of aluminum or aramid (Nomex) honeycomb material sandwiched between the two skins varies in thickness depending on the structural design. A film adhesive is applied between the skins and the core creating a strong bond when cured.

Points where bolts and other fasteners pass through the honeycomb core are locally reinforced using inserts. Inserts are solid composite or metallic pieces which are supplied to the laminators precut and are positioned in predetermined cut outs in the honeycomb core. The inserts serve to spread point loading over a larger surface of the composite and thus reduce stress concentrations. They also prevent the bolts, etc. from moving under load which crush the honeycomb core and ultimately fracture the composite skin.

The majority of structures are assembled by adhesively bonding a number of parts together. Adhesive bonding is a particularly effective method of assembling complex structures, especially those made from different materials. Before the main chassis is closed by bonding the top and bottom, the two internal bulkheads are bonded in with an epoxy paste adhesive. One is the seat back, a sculpted sandwich panel located directly behind the driver's seat, forming a partition between the fuel cell and the cockpit, and the other is the "dash" bulkhead positioned slightly forward of the cockpit aperture, through which the driver's legs pass.[50]

9.5 BUILDING AND CIVIL INDUSTRIES

Fiber composite sandwich has become the new generation of materials used in civil infrastructure in the last decades. Because of its special feature, the sectional area is increased with consequently an increase in its flexural rigidity. The evolution of sandwich structures with enhanced material systems provided an opportunity to expand the application of this material in civil infrastructure. At present, there is a strong interest in the development and applications of fiber composite sandwich structures for civil engineering and construction. The light weight of sandwich composites facilitates handling during assembly and reduces installation and transportation costs. They also offer corrosion-resistant structures requiring less maintenance. Worldwide, fiber composite sandwich structures are now being used as structural panels in residential and industrial buildings, boardwalks, bridge decks, and timber replacement girder in a number of infrastructure projects. Its application for a railway sleeper (crossties) is also now trialed.

The advantages composite sandwiches have to offer seem obvious. Compared to steel, wood, and concrete they have a higher stiffness and strength-to-weight ratio and stand up well against the elements, resulting in a longer life span and lower maintenance costs. They're also shapeable practically without limitations. However, the sandwich composites are still held back in civil engineering because their possibilities aren't widely known yet. After their advantages are widely understood, the sandwich composites are bound to get more applications in building and civil industries.

9.5.1 Housing and Building Construction

Composite sandwich structure has numerous advantages in housing and construction. The light weight of sandwich composites facilitates handling during assembly and reduces installation and transportation costs. This can significantly speed up construction, especially in the rebuilding of structures in calamity-affected areas. Consequently, an increasing range of fiber composite sandwich structure is now available like roof, wall, floor, and subfloor system for housing and construction. Thermal insulation, waterproofing, and acoustics were integrated into the system during the prefabrication of the sandwich roof structure enabling easy transportation and rapidly installation.

Today's architects are faced with the urgent task of achieving energy-efficient and high-performance building enclosures. Structural insulated panels (SIPs) are an option for part of the enclosure assembly that can help achieve these goals. SIPs do an impressive job of slowing down the transfer of heat, air, and vapor through the assembly. They also dramatically reduce the drying potential of the enclosure, lessening its ability to recover from inadvertent water intrusion. Such an airtight assembly with great thermal resistance can result in a high-performance and durable enclosure if detailed and built correctly.

In 1952, Alden B. Dow created the first foam core SIPs which were being mass-produced by the 1960s. Today, SIPs are prefabricated building components for use as walls, floors, roofs, and foundations. SIPs provide a continuous air and vapor barrier as well as increased R-value when compared to traditional construction. Construction costs associated with SIPs are comparable to more conventional building methods when savings associated with labor costs, material waste, and energy efficiency are considered.

SIPs are composed of an insulated foam core between two rigid board sheathing materials. The foam core is generally one of the following: expanded polystyrene (EPS), extruded polystyrene (XPS), and polyurethane foam (PUR). With EPS and XPS foam, the assembly is pressure laminated together. With PUR and PIR, the liquid mixture is injected into the facing assembly and then foamed and cured under high pressure. The most common sheathing boards are oriented strand boards (OSB). Other sheathing materials include sheet metal, plywood, fiber-cement siding, magnesium-oxide board, fiberglass mat gypsum sheathing, and composite structural siding panels. Figure 9.42 shows three different SIPs panels.

SIPs are designed to resist not only axial loads, but also shear loads and out of plane flexural loads. The panels' ability to resist bi-axial bending and lateral shear allow them to be used as roofs and floors. SIPs panels are acceptable to use as shear walls in all seismic design categories. A structural engineer should determine if a secondary structural system is required based on the design loads. Figure 9.43 shows all elements of a SIP building.

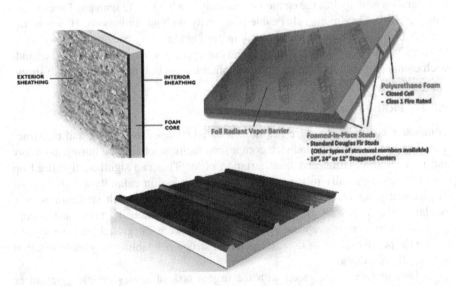

FIGURE 9.42 SIPs panels – EPS foam core panel (a), PUR foam core panel (b), and PUR core/metal skins roof panel.[51,52]

Sandwich Structural Composite Applications 375

FIGURE 9.43 Exploded axonometric of a SIP building.[53]

To date, the tallest structure constructed exclusively of SIPs is four stories. Taller structures are possible; however, design limitations are due to the fact that SIPs are bearing walls and therefore open spaces at lower floors are more difficult to achieve. Often large SIPs structures rely on a secondary framing system of steel or timber to satisfy requirements for unobstructed spaces. Unique screw connections are available to attach SIPs to wood, light gage steel, and structural steel up to 1/4 inch thick. It is imperative for foundations for SIP panels to be leveled.

Joint design is imperative for structural and long-term durable performance. Proper joint design should be given special attention and if properly executed in the field, will eliminate the air infiltration problems. The primary joint design generally includes seals within the thickness of the panel, typically spray foam or gaskets as shown in Figure 9.44.

Two of the most widely used panel joint connections are the surface spline and the block spline. The surface spline joint connection consists of strips of OSB or plywood inserted in slots in the foam just inside each skin of the SIP. The block spline is a thin and narrow SIP assembly that is inserted into recesses in the foam along the

FIGURE 9.44 Three different joints for connecting SIPs panels – block spline (c), surface spline (b), and cam lock (a).[53]

panel edges. Another joint connection, mechanical Cam locks, create a tighter joint between panels, but make up only a small percentage of the market.

Typical SIP wall panel thicknesses are 115 and 165 mm. The largest panel size to date is 2.74 m × 7.32 m. Curved panels are possible although not common, and it is often more practical to use stud framing for non-orthogonal geometries. Roof panels are typically 260 and 310 mm thick. Roof panel thickness depends upon the required R-value and span. EPS and XPS panels can be made up to 310 mm thickness. PUR and PIR panels can be made up to 210 mm thick. End wall panels for various roof profiles can be achieved with SIPs.[53]

A new structural composite wall and structural composite sheathing system is targeting not only mainstream residential housing but also light commercial applications. A composite structural insulated panel (C-SIP) wall panel system is targeted to go head-to-head with conventional above-foundation structural materials, including wood studs, plywood, OSB panel and house wrap in the construction market.

These highly advanced materials comprising the C-SIP wall panel are chemically bonded together by a chemical reaction during manufacturing – no adhesives are used – eliminating the risk of delamination. The strength, energy performance and efficiency of the C-SIP wall panels are the result of bringing together fiberglass reinforced thermoset polymer skins and the moisture resistant, rigid closed cell polyurethane foam insulation that has excellent compressive strength and long term, stable R-values. The panel is made in 115 mm thick with 6 mm thick skin each side. Figure 9.45 shows the panel sample and one with foam cut-out for connection.

The C-SIP wall panel's interior and exterior FRP skins create far superior strength and energy efficiency, while eliminating thermal transfer and dramatically reducing any air, water or vapor infiltration. In a single product, the C-SIP replaces six different materials or envelope treatments: exterior sheathing; foam insulation, air barrier – no house wrap required; vapor retarder – stops hot air at the outside of the wall, especially valuable for humid climates; moisture barrier – continuous, sealed exterior building envelope; and structural walls that eliminate wood or metal framing.

Pultruded solid composite H-Stud is used for connecting the C-SIP wall panels. The connector is made from this high-performance solid composite and designed to firmly attach the adjacent C-SIP wall panels together. The H-Stud is mounted in the pregrooved C-SIP wall panel, and the adjacent panel grooves fit snugly over the

FIGURE 9.45 Sandwich panel and sample with foam cut-out for connecting.

H-Stud to create a continuous stud running the length of the panel joints from the bottom to the top plate.

Like the H-Stud, continuous pultruded composite C-channels are used for covering the top and bottom edges. The composite channels are engineered to eliminate thermal bridging while creating a structurally strong and stable connection that blocks air and moisture infiltration and won't degrade over the life of the building.[54]

Precast concrete sandwich panel (PCSP) is another SIP being used in building construction. The PCSPs consist of two or more high strength layers referred to as wythes separated by a low-strength material known as insulation. There are three major component parts of precast concrete sandwich panels, namely, the wythes, shear connectors, and the insulation or void. The wythes are referred to as the concrete external rigid/solid body of considerable strength sufficient to resist an imposed load or self-weight of the structure. Shear connectors play a very importance role in determining the structural strength of the composite system. The material, shape, thickness, embedment length and spacing of shear connectors influence the behavior of panels significantly. The most commonly used shear connector material is steel. More recently, other materials such as fiber-reinforced composites are studied in many investigations. Concrete studs are also used as a shear connection in reinforced concrete sandwich systems.

Insulations are low strength materials with high thermal resistance used as a means of separating the concrete wythes in precast concrete sandwich panels. The thickness of the insulator depends on the thermal efficiency required by the manufacturers. Sometimes, the space between wythes is designed as vacuum-insulated, hollow core or filled with insulator material, such as plastic, rubber, rigid foam, expanded polystyrene, extruded polystyrene, polyurethane, phenolic foam, etc. Figure 9.46 shows a structure diagram of PCSP. The PCSPs are similar to other SIPs that are used as walls for residential and commercial constructions.

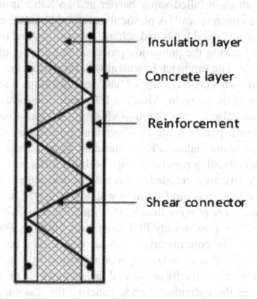

FIGURE 9.46 Section sketch of precast concrete sandwich panel.[55]

FIGURE 9.47 Foam cored sandwich composite basement foundation walls.[54]

A foam-cored fiberglass reinforced sandwich composite panel, including an interior system of reinforcing glass webbing and integral studs, spaced on along the wall's length is developed for house basement foundation walls as seen in Figure 9.47. A key to its success, the sandwich composite basement wall is engineered to present features very familiar to construction personnel accustomed to conventional foundation materials and methods of construction with a size of 86 mm thick by 7.32 m long by 2.74 m wide. The foundation wall design that not only incorporated the advantages of composites, but also fit within the conventional, residential house design and building process, making it easy for contractors to understand and install.

Compared to a concrete wall, the composite wall will not crack and provides home buyers with an as-installed vapor barrier and an R16.5 insulation value, vs. an R1.5 rating for a concrete wall. A plastic foam cored laminate could carry those allowable transverse and axial loads and achieve fire, smoke, and toxicity (FST) ratings, yet ensure that during the pultrusion process, total wet-out of the entire glass schedule was feasible. The sandwich basement wall eliminates the dampness, mold, and chill associated with a concrete-walled basement and turns it into totally livable space from day one of the move in. Also, the 180 mm thick sandwich wall provides up to 25 m^2 of additional space compared to a 300 mm thick concrete wall.

Structural sandwich panels made of fiberglass/ polyester resin skins and low-density polyethylene terephthalate (PET) foam core with fire-retardant gel coat are developed for facade cladding panels. To cope with the different challenges such an eye-catching façade structure entailed, the developer was searching for a sandwich solution, combining high structural strength with low weight. The panels were built for covering a surface area of more than 40,000 square meters building surface area as shown in Figure 9.48. Low density PET foam core, made of 100% postconsumer PET, turned out to be the core material of choice due to its reduced weight, while providing the required stiffness and strength properties.

The lightweight nature of the foam allowed facade panels of up to 15 m in length. All in all, the longer the individual facade panel is, the less supporting structure is required, the easier and faster the handling during assembly and the lower the

FIGURE 9.48 Building with sandwich structural composite facade.[56]

transport and installation costs. In addition, fewer joints and seams mean enhanced aesthetic appeal for the facade. Stiffness and, at the same time, lightweight and dimensional tolerances were not the only challenges the designers and architects had to cope with the middle eastern climate, with its extremely large variations in temperature, from subzero at night to +55°C during the summer, in combination with wind and sand, required the use of very robust materials.

A lightweight sandwich composite roof structure was designed and built in Switzerland.

The sandwich construction allowed for an integration of static, building physical and architectural functions that enabled the prefabrication of the entire roof in only few lightweight elements that were easily transported to the site and rapidly installed. The roof must be lightweight due to the limited load-carrying capacity of the supporting glass walls, and at the same time, it must provide thermal insulation and waterproofing for the building. Consideration of the complex double-curved geometry, furthermore, led to the use of a sandwich composite structure of variable depth. The face sheets have thicknesses from 6 to 10.5 mm and consist of several layers of glass fiber fabrics and mats. E-glass fibers and a polyester resin were used. The polyester is a filled, low-viscosity, and self-extinguishing resin that shows low flammability and medium smoke formation.

The core consisted of a polyurethane (PUR) foam of three different densities and strengths. Since the shear load-carrying capacity of even the densest core type was not sufficient, the foam core had to be reinforced by an internal system of orthogonal fiberglass composite webs. CNC cutting of foam blocks and adhesive bonding proved to be an advantageous procedure for the fabrication of the complex roof shape, without the use of expensive molds. The installed roof of the campus main gate and one element of the roof are shown in Figure 9.49.

9.5.2 BRIDGE BUILDING

A sandwich panel made with a rigid polyurethane core and steel plate facings is called the sandwich panel system (SPS) that is an innovative bridge deck system and can be used for both new bridge construction and bridge rehabilitation applications.

FIGURE 9.49 Sandwich composite roof for a main gate building of a campus.[57]

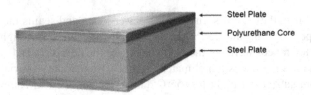

FIGURE 9.50 Structure of the sandwich plate system.[58]

The panel structure is shown in Figure 9.50. The steel plates are designed to resist the loads resulting from flexure while the core resists the transverse shear. A typical bridge application utilizes SPS primarily as a bridge deck acting compositely with conventional support girders, but other applications have also been given consideration, such as in ship, stadium and building constructions. It has been used over 450 projects and 300,000 m² in service in 30 countries. An SPS bridge deck is typically constructed from a series of panel segments, matching the width of the bridge, connected together along the span of the bridge.

SPS is similar to a conventional orthotropic plate solution, but without the required intermediate stiffeners, the polyurethane core serves the same purpose as intermediate stiffeners by providing continuous support to the steel plates. Since the core is continuous, the local buckling effect resulting from discretely spaced stiffeners is eliminated. Additionally, the lack of intermediate stiffeners eliminates the "hard spots" inherent to orthotropic decks resulting in a more continuous system. The versatility of SPS allows for implementation in a wide variety of bridge applications including new construction, deck replacement, and deck rehabilitation. Figure 9.51 shows a comparison of the SPS bridge deck structure and the conventional stiffened steel structure.

When compared to conventional bridge solutions such as reinforced concrete and orthotropic systems, SPS offers a number of advantages including weight savings due to the light weight of the panels, speed of construction, impact resistance, and acoustic damping without sacrificing strength. The design of SPS panels can be tailored to specific applications by varying the plate and core thicknesses. For new construction applications, the panels can be designed with support girders significantly smaller than a conventional solution. In rehabilitation applications, the SPS panels can be

Sandwich Structural Composite Applications

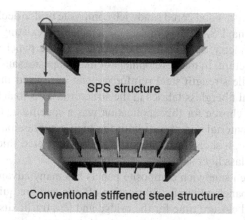

FIGURE 9.51 SPS and conventional bridge deck structures.[58]

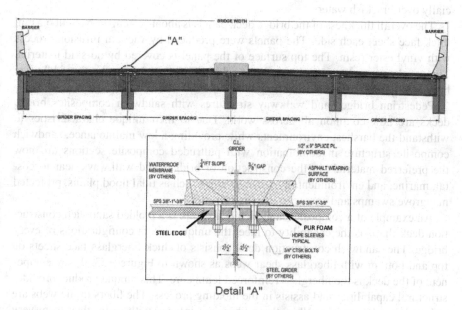

FIGURE 9.52 SPS bridge deck example and connection detail.[58]

designed to work with the existing superstructure; this would typically result in additional capacity due to the reduction in dead load of the deck system.

By using SPS decks, a bridge construction can have the following advantages: simpler, faster, reduced risk fabrication delivering, improved designs and in-service performance, enhanced protection and safety, and transforming construction to onsite assembly. A SPS deck is made with steel girders through continuous longitudinal bolting. The top surface is covered by asphalt or lightweight wearing materials. A bridge deck built by SPS plates and a detail of connection of plates are shown in Figure 9.52. Steel edge bars are used at the edge of sandwich panels for reinforcing panel edges that are fastened to steel girders by bolts.

The end-grain balsa wood cored sandwich composite bridge decks were developed and used in Louisiana, USA. Being replacement panels for existing steel construction means that the geometry, specifically the thickness of the panel, is dictated by the existing bridge structure. In the development of the composite sandwich bridge deck, layers of high-tensile strength steel reinforcements were used in conjunction with conventional biaxial fiberglass fabrics in the structural skins to achieve the required stiffness. The core chosen for this application was a specific density range of end-grain balsa. This material can absorb both the high compression loads required of a bridge, and the high shear loads imposed by the restricted thickness and highly loaded metal and glass fiber skins.

Using balsa-cored sandwich composite panels has many advantages as the lightweight composite panels allow for fabrication off site and are quick to install. This results in a reduced closure time for the bridge and less traffic disruption, on a road which is an important evacuation route during hurricane season. In addition, these panels are immune to corrosion which shortens the life span of steel structures, especially over brackish waters.

The overall thickness of the bridge deck panels is about 130 cm with about 6.5 mm thick face sheet each side. The panels were produced by vacuum infusion process with vinyl ester resin. The top surface of the panel is covered by no-skid material. The panels were adhesively bonded to steel girders of the bridge. For a middle size bridge, only need 3 days closing time to install all decks.[59,60]

Pedestrian bridges and walkway structures with sandwich composites bridge decks are now common all across world. Due to their unique characteristics to withstand the harshest environments while providing a low maintenance, sandwich composite structure in combination with pultruded composite sections are now the preferred materials in the construction of bridges and walkways near to costal, marine, and environmentally sensitive areas such as tidal flood plains, protected mangrove swamps and corrosive mining facilities.

An example of a composite sandwich boardwalk is a molded sandwich construction deck that has the capability to meet the unique design configurations of every bridge. The sandwich construction deck consists of thick fiberglass face sheets on top and bottom with fiberglass shear webs as shown in Figure 9.53. A key component of the deck is the fiberglass reinforced foam core. This unique product provides structural capabilities and assists in the molding process. The fibers in the webs are oriented at ±45° angles. When these fibers are infused with resin, they form very strong, stiff shear webs for the sandwich cross-section.

The closely spaced webs provide good crushing resistance to concentrated loads. There is no local skin deflection since the skins are so well supported by the webs. The redundancy of the multiple webs provides improved dam-age tolerance over thick, wider-spaced webs. The fiberglass webs can be used directionally or as a bi-directional grid for design flexibility. Closed cell foam is used inside the sandwich as a processing aid and prevents water from building up in internal cavities. Deck panel edges are encapsulated in fiberglass plies and could be made with curbs together. Common depths used for pedestrian bridges include 3, 4, and 5 inches, but can also be built to custom depths. Depth selection depends on load requirements and support structure geometries. Fabricating deck panels in a factory allows for superior quality

Sandwich Structural Composite Applications

FIGURE 9.53 Cross section of sandwich structural deck for pedestrian bridges.[61]

control during the fabrication process, and substantial reduction of bridge installation time. The lightweight panels allow the use of light equipment which significantly reduces installation time and labor costs.[61]

The development of structural beams from sandwich structures is now gaining interest in bridge construction as girders. The concept of gluing a number of sandwich composite panels to form a structural beam is highly practical. The structural composite sandwich panel is made up of glass fiber composite skins and phenolic foam core with an 18 cm thickness as shown in Figure 9.54. A number of these composite sandwich panels were assembled and glued together to produce the glue-laminated sandwich beams. Two section configurations for full-size composite beams can be considered: a beam section with composite sandwich panels glued together in the flatwise position and a section with sandwich panels glued together in the edgewise position. The sandwich composite beams were fabricated by gluing a number of 18 cm thick sandwich panels using a structural epoxy resin to form a 150 cm deep and 230 cm wide section. The beams in the flatwise position was produced by laminating 8 sandwich panels while in the edgewise position with 13 sandwich panels. Figure 9.55 shows the full-size composite beams.

FIGURE 9.54 The sandwich composite panel for making glulam beam.[62]

FIGURE 9.55 Full-size composite beams made with sandwich panels.[62]

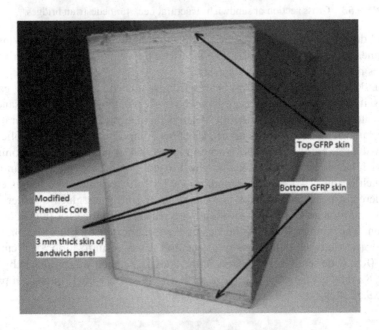

FIGURE 9.56 Hybrid glulam sandwich composite beam.[63]

A hybrid sandwich beam composed of glue-laminated sandwich panels oriented in the edgewise position at the middle with top and bottom glass fiber-reinforced composite skin plates as shown in Figure 9.56 is now being developed. The beam with this configuration has an improved shear strength due to the vertically oriented sandwich panels and an improved bending and compressive strength due to composite skin plates. The potential application of this hybrid beam is for bridge girder, railway transom, or sleepers used in a railway construction.[63]

The composite railway sleeper made by gluing layers of sandwich structure together in flatwise (horizontal) and in edgewise (vertical) orientations have the

Sandwich Structural Composite Applications

strength and stiffness properties as well as the resistance to hold screw spikes. The composite railway sleeper also has better mechanical properties than most of the commercially available sleepers and showed comparable properties with the existing timber rail sleepers. The composite sleepers are performing to expectations, and it is estimated that their serviceable life should be well in excess of 50 years.

9.5.3 Bridge and Dock Protective Systems

The low weight, high-strength, and corrosion resistance of sandwich composite systems make them suitable for the development of floating and impact-resistant structures. In China, floating and energy absorptive elements made of sandwich composite systems were developed for a collision avoidance structure. An innovative large-scale composite bumper system (LCBS) for bridge piers against ship collision was recently proposed. The modular segment of LCBS is made of fiberglass composite skins and lattice webs, polyurethane foam cores in which vacuum infusion process is adopted in the manufacturing process.

This novel bumper system offers several remarkable advantages, such as self-buoyancy in water, modular fabrication of segments, efficiency for on-site installation, excellent corrosion resistance, as well as ease in replacing damaged segments. An in-depth analysis of performance evaluation results indicated that LCBS can effectively increase the impact time of ship-bridge collision and reduce the peak collision forces to a nondestructive level, leading to a good effect in energy dissipation. The results suggest that LCBS is an effective bumper system for protecting bridges and ships in ship-bridge collisions. There were more than 50 completed design projects, which used this structure in China. Figure 9.57 shows that structure, manufacturing and assembling processes, and application of LCBS.

FIGURE 9.57 Structure, manufacturing and assembling, and application of large-scale composite bumper system.[64]

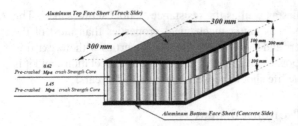

FIGURE 9.58 Design of double layer aluminum honeycomb sandwich I-Lam system (a) and product sample (b).[65]

A similar concept of the floating collision-avoidance structure is the collision-protection/scarifying impact laminate (I-Lam) system for protecting concrete bridge girders. This system uses sandwich structure with a crushable core. The system has smart sensors and actuators for remote sensing, triggering, and monitoring. Numerical simulations and full-scale impact tests on reinforced concrete beams showed that about 60%–70% of the kinetic energy was absorbed by crushing of the core in the I-Lam system.

Based on the design requirements and design protocol, a double-layer aluminum honeycomb I-Lam sandwich system is designed for the over height collision protection. The double-layer I-Lam system (Figure 9.58a) consists of two 10 mm thick aluminum face sheets (top and bottom) and two 90 mm thick hexagonal honeycomb layers with crushing strength of 0.62 Mpa (6.35 mm cell size low density) and 1.45 Mpa (5.08 mm cell size high density), respectively. The lower crushing strength honeycomb layer is designed for the impact/collision of lighter over height materials and low speed; while the higher crushing strength honeycomb layer is intended for the case of heavier over height sheared-off materials and high speed. The purpose of using the two-layer honeycomb configuration is to improve structural efficiency. The final product of I-Lam system is shown in Figure 9.58b.

One of the common damages in existing highway bridges is the damage at the bottom corners or edges of the reinforced concrete beams or box girders induced by an impact of trucks exceeding the allowable height clearance of the bridges. The I-Lam system is bolted and/or bonded to the bottom portions or edges of concrete girders. Design criteria and guideline for I-Lam are developed, and the analysis, optimal design, and quality control tests of the system are conducted. Smart piezoelectric sensors are integrated with the I-Lam panels for monitoring the performance of the collision protection system. The developed smart bilayer honeycomb I-Lam sandwich is capable of reducing the transferred contact force dramatically, absorbing/mitigating impact energy, protecting the underneath concrete structures by system scarifying and core crushing, and monitoring the impact incident with smart piezoelectric sensors, and it is applicable to protecting other structures (e.g., steel girders, columns) from accidental vehicle impact in the highways. Figure 9.58 shows installation details and a final installed sandwich structural I-Lam system on the bridge.

9.5.4 CHALLENGES AND ISSUES

A number of issues contributing to the slow uptake of composite sandwich structures in civil infrastructure have been well documented. These important aspects have to be addressed in order to advance their use of sandwich structures in civil engineering applications.

New product development is needed to encourage the application of sandwich composites in civil engineering. Unlike in aerospace and transportation applications, the sandwich cores for building and construction need to be very strong to withstand high concentrated and impact loads. Moreover, a thicker composite sandwich panel is usually used in structural than in industrial applications where the shear strength of the core is a critical parameter to efficiently transfer the shear between the top and bottom skins. The shear cracking of the core is the dominant failure mode for a sandwich structure with a thick core. Thus, it is anticipated that the evolution of composite sandwich structure with lightweight, high-strength core and with good capacity for mechanical connections could provide wider opportunity to increase the acceptance and utilization of this type of construction in civil infrastructure. However, the method of enhancement of the core structure should not involve a complex manufacturing process and increase the cost of production.[63]

Broader acceptance of sandwich composites in civil infrastructure will require new or revised design guidelines. The poor understanding of the overall behavior of sandwich structures is commonly claimed to place them at a disadvantage against traditional construction and building materials. This problem, combined with the lack appropriate design codes and standards, is recognized as a significant barrier to broad utilization. Without an established design method and data, it is unlikely that structures utilizing sandwich composites will be used beyond the scope of research and demonstration projects. Similar design method should therefore, be developed so that sandwich structures could gain wider acceptance in civil infrastructure. New and simple analysis techniques should be able to analyze the overall behavior of sandwich structures within acceptable levels of confidence.

That sandwich structure is an acceptable alternative material, now used for floors, roofs, walls, bridge decks, and other innovative structural applications, shows the potential of this type of construction has not been fully explored yet despite civil engineers having access to a wide range of fiber composite sandwich panels.

The growth of sandwich composite construction in civil infrastructure can be further realized by developing innovative structures which exploit its many advantages. Fiber composite sandwich structures will generally be feasible in infrastructure when the need for corrosion resistance, high strength, reduced weight, or fast installation is a driver for the system. As seen by the recent development of innovative sandwich composite bridge decks, this effort is driven by the need to replace the heavy weight and corrosion prone reinforced concrete decks and the opportunity to upgrade the load-carrying capacity of the existing bridge. Such composite structures are needed in order to address the need in the construction industry for more durable and cost-effective infrastructure. Moreover, typical infrastructure prototypes need to be developed to demonstrate its practical application, increase its acceptance and to build a market volume.

9.6 MISCELLANEOUS

9.6.1 SANDWICH STRUCTURE FOR RADOME CONSTRUCTION

Radomes are housings or enclosures that protect a radiating antenna against the elements of environment. The word radome is derived from the words RADar and dOME. Originally, radome referred to radar transparent, dome-shaped structures used to protect radar antennas on aircraft. Over time, radome has come to mean almost any structure that protects a device that sends or receives electromagnetic radiation, such as that generated by radar, and which is substantially transparent to the electromagnetic radiation. Radomes are required to have the necessary structural strength and be designed so as not to exceed some specified maximum deterioration in the electromagnetic performance of the antenna under operational conditions.

Radomes can be classified into ground-based, shipborne, and airborne radomes as shown in Figure 9.59. Shipborne radomes are normally required due to the fact that the antenna has to operate in a wet environment. Airborne radomes are crucial and are designed so that the aerodynamics of an aircraft or a missile will not be degraded. For ground-based applications, however, it is not always necessary to use a radome. They are required in locations where strong winds, and heavy rainfall or snow are frequent, as well as providing physical security for the antenna.

A radome is an integral part of a radar system because the thickness of the radome and its properties affect the effectiveness of the radar and must be compatible with the specific properties of the radar set. Major design criteria of a radome include electromagnetic radiation transparency, structural integrity, environmental protection (e.g., protection from rain erosion and lightning strikes) and, especially for aircraft, an aerodynamic shape, and light weight. Economics also require that the cost should be as low as possible and the service life as long as possible.

Radomes with a diameter of 4–5 m for X-band radar (frequency between 8.0 and 12 GHz) applications are considered large. Radomes for L-band surveillance radar systems (frequency of 1–2 GHz) have diameters of the order of 20 m. Hence anything from 4 m up to 20 m constitutes a large radome. However the size of the radome is determined by the size of the antenna, which in turn is governed by the frequencies under interest. A radome 0.5 m in diameter cannot be considered large even if it is intended for an antenna at 20 GHz, which is the type of radomes that are found in aerospace industry.

Currently, a common type of radome is one having a fiberglass reinforced sandwich construction. For a small-size radome, it can be made into one whole sandwich

FIGURE 9.59 Radome illustration (a), and ground-base (b), shipborne (c), and airborne (d) radomes.

sphere structure without connecting flanges. However, for a large diameter radome, it consists of multiple pieces of sandwich panels that are physically connected with the flanges, which means the individual sandwich panels are joined together by means of the flanges. Planar sandwich panels or doubly curved ones can be used. The resulting radome geometry will be that of a polyhedron or a perfect sphere.

The geodesic radomes can be classified into three categories although this classification is purely arbitrary and is based on common practice more than anything else. There exist geodesic radomes, which have three different types of panels for constructing a radome and sometimes are called the 3-panel radomes. There are also the radomes, which have five different types of panels, and they are called the 5-panel radomes. Finally there exist radomes, with a number of different types of panels, which are mainly irregular in shape and radomes made of these are known as quasi-random radomes. A sphere provides maximum rigidity and structural integrity out of all the possible shapes. Hence any attempt to design a radome as close as possible to a perfect sphere would be mechanically ideal. Figure 9.60 shows four different geodesic radomes.

What determines the choice of geometry is the size of the radome and the requirements. Normally as the radome diameter is increasing, the radome geometry changes from the 3-panel one to the 5-panel one. The reason for this transition is related to the panel size. This size has to be kept smaller than a certain value. For, long flanges are associated with very large panels, which are difficult to laminate by hand, are awkward in their transportation, and most importantly, if the weather conditions are bad, very difficult to install. Consequently if the radome diameter is sufficiently large, then one needs to cover the sphere with a larger number of panels, which inevitably leads to more different types of panels so as to keep the flange length within acceptable limits. Radome wall panels can be a plain thin dielectric composite laminate or one of five types of sandwich construction. Figure 9.61 shows five types of sandwich panels.

A-Sandwich form core: Three layers consisting of high dielectric constant skins and low dielectric constant foam core, Figure 9.61a. Normally the skins are made of solid glass fiber composite. However, in some applications, one can meet Kevlar or Quartz fiber instead of glass fiber. A thin outer layer of hydrophobic coating is most of the times required. This is true not only for the A-sandwich but for all radome walls.

A-Sandwich honeycomb core: It is the same as the A-sandwich, but the core is made of low dielectric constant honeycomb material, Figure 9.61b. It is very preferential

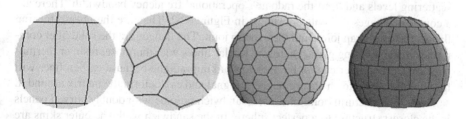

FIGURE 9.60 Radome framework with orange peel shadow (left), quasi-random shadow (central two) and symmetric orange shadow (right).

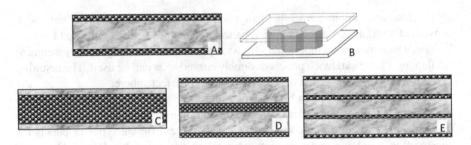

FIGURE 9.61 Different types of sandwich constructions, (a) A-sandwich foam core, (b) A-sandwich honeycomb core, (c) B-sandwich, (d) C-sandwich, and (e) more than five layers.[67]

for aerospace applications where weight is of paramount importance. The honeycomb material has a low cut-off frequency, and its modelling requires more complicated procedures.

B-Sandwich: It is the opposite of an A-sandwich. The skins are normally thin, and of low dielectric constant, and the core is of higher electric permittivity, Figure 9.61c.

C-Sandwich: It consists of two A-type sandwiches back to back forming five layers, Figure 9.61d. The thicknesses of the five layers may be chosen randomly and it is not necessary for the middle layer to be twice as thick as the outer skins. C-sandwich walls are very broadband and where cost is not the driving factor it is much preferred in wide band applications.

More than five layers: A multilayer construction with very thin skins and suitable cores, Figure 9.61e. In large radomes only the A-type sandwich is used, and the reason is the cost. The labor for laying more than three layers is increased significantly especially when the thicknesses of the layers must be kept within very tight tolerances. These cases occur in wide-band applications where manufacturing errors can destroy the tuning for some frequencies. More than three layers can sometimes be used in radomes as large as 4 m.

The frequency bandwidth of the radome, in terms of transmission losses and sidelobe perturbation performance, is determined by the combined effect of the panels and the seams. The panels, usually sandwich A type exhibit a wide frequency bandwidth, which ordinarily does not limit the overall radome performance. The physical dimensions of the seams are determined by the stresses they have to withstand due to all environmental and physical loads including extremely high wind loading. Those seams may degrade the total system performance by introducing high scattering levels and limit the radome's operational frequency bandwidth. There are a couple of flanges or joints are shown in Figure 9.62. They are the inward turning flange, the over strap joint, and the overlap joint. The flanges are the solid fiber composites made by using glass, Quartz or Kevlar fibers with thermoset resin or thermoplastics. Carbon fiber cannot be used for making radome because carbon fiber will heat up in reaction to an electromagnetic signal and can deform the matrix around it.

From the structural point of view, a sandwich radome with doubly curved panels is the closest structure to a perfect sphere. In the sandwich walls the outer skins are in tension and the inner skins are in compression. Hence, shear forces are exerted, and the stresses generated must be efficiently transferred into the ground. During the

Sandwich Structural Composite Applications

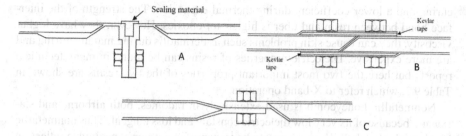

FIGURE 9.62 Different types of radome joints, (a) inward turning flange, (b) over strap joint, and (c) overlap joint.[67]

transfer of the shear loads, the latter are confronted with discontinuities, which are the joints and the flanges. When the flanges are integrally molded with the sandwich construction then the discontinuity is much smoother. In the case of the space frame radomes all the loads are transferred through the hubs or the nodes where the flanges meet.

Sandwich composite materials provide a great advantage for radomes. By combining foam or honeycomb as core, fibers, and resin for skins and flange, a bulk material is produced with a strength and stiffness close to that of the fibers and with the chemical resistance of the plastics. Another advantage of composite materials is the very high values of specific modulus and specific strength. Hence the higher specific value implies a great strength for a relatively light material. Since sandwich composite materials are made of more than one component their electrical properties should be the weighted average of the electrical properties of the components.

The skins of a sandwich for the radomes would normally be made of GFRP. For radomes with a diameter of the order of 4 m or less, E-glass can be substituted by Kevlar or even Quartz for some applications. However, cost considerations prohibit the use of any other reinforced plastic other than E-glass for larger radomes. Table 9.1 provides a very good indication of the values of the relative permittivity and loss tangent of fibers. The values correspond to X-band operation.

There are several types of resins, and two mostly used for making skins of the sandwich panels and flanges, which are epoxy and polyester resins. They are usually isotropic in their properties. Epoxy resins are superior to polyester resins in this respect. They have better strength and elastic properties, with lower shrinkage on

TABLE 9.1
Dielectric Properties of Basic Fibers for Sandwich Composites of Radome[67]

Fiber	Relative Permittivity	Loss Tangent
E-glass	6.06	0.004
Kevlar	4.1	0.02
Quartz	3.8	0.0001
Polyethylene	2.25	0.0004

curing and a lower coefficient during thermal expansion. The strength of the interface bond between resin and fiber is higher for epoxies. However, they have higher viscosity, they can cause skin problems such as dermatitis during manufacturing and are more expensive. Dielectric properties of resins can be found in many technical reports but here the two most important properties of the two resins are shown in Table 9.2, which refer to X-band operation.

Nonmetallic honeycomb is used extensively in radomes, both airborne and stationary, because of its very low dielectric constant and loss tangent. Thus nonmetallic honeycomb allows the wave energy to be transmitted with only negligible reflection and absorption. Nomax honeycomb is the most common core material. Honeycomb core cells for aerospace applications are usually hexagonal. Apart from hexagonal cell shape honeycombs are available in over expanded core and flex core that can be used with ease to perform curve sphere radome. Figure 9.63 shows the dielectric constant as a function of core density for several Nomex honeycomb types. The values

TABLE 9.2
Electrical Properties of Resin Systems[67]

Resin	Relative Permittivity	Loss Tangent
Epoxy	3.6	0.04
Polyester	2.95	0.007

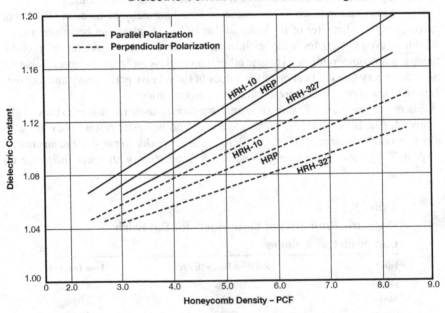

FIGURE 9.63 Dielectric constant as a function of core density for several honeycomb types.[68]

were obtained for both polarizations and with the electric field vector E perpendicular and parallel to the ribbon direction. Testing was conducted at 9375 MHz. In addition to the electric field polarization, the dielectric constant is a function of the incidence angle and the thickness of the honeycomb.

HRH-10® is manufactured from NOMEX® aramid fiber sheets. A thermosetting adhesive is used to bond these sheets at the nodes, and, after expanding to the hexagonal or OX-Core® configuration, the block is dipped in phenolic resin. HRH-10® is available in Flex-Core®, a very flexible core material. It is formable, fire-resistant (self-extinguishing), water and fungus resistant, and has excellent dielectric properties.

HRP is a fiberglass fabric reinforced honeycomb dipped in a heat-resistant phenolic resin to achieve the final density. HRP has proven to be an excellent core material for radomes and aircraft structural parts, which provides good retention of strength for temperatures up 180°C.

HRH-327 is a fiberglass fabric, polyimide node adhesive, bias weave reinforced honeycomb dipped in a polyimide resin to achieve the final density. This material has been developed for extended service temperatures up to 260°C with short range capabilities up to 370°C.

Another nometal honeycomb used in the application of radome manufacturing is made by quartz fiber fabrics, which is a very low density honeycomb core for using in highly weight-sensitive radome structures. It is expected that will be used in airborne radome, spacefraft antenna reflectors, optical benches and other lightly loaded structures whtich must be positioned with accuracy and stability in order to properly perform their intended purpose. The dielectric constant and the loss tangent factor of quartz honeycomb are the most distinguished amounst all mineral fibers. Table 9.3 shows the dielectric properties of 48 kg/m^3 density quartz honeycomb tested at average frequencies of 9.5 GHz.

The honeycomb cores have open-cell structure which encourages moisture intrusion that can destroy the radome, and it has relatively poor impact resistance. Foams are also used for the core of sandwich radome constructions. Due to the closed cell formation of structural foam, the moisture absorption of the foams is minimal, and also it saturates at a very low value compared to aramid-based honeycombs.

The foam cores can be closed-cell rigid foam plastics based on polymethacrylimide (PMI), or they can be polyurethane (PUR) foams, polyethylene terephthalate (PET), polyetherimide (PEI), or they can be made of polyvinylchloride (PVC) cellular plastics. They all come in different densities, which may be suitable to different applications. Normally radomes with high loading, such as hail or bird strikes, require high density foams. The dielectric properties do change with the density of

TABLE 9.3
Dielectric Properties of a Quartz Honeycomb[69]

Density	Average Frequency	Dielectric Constant Along Ribbon	Loss Tangent Along Ribbon	Dielectric Constant Across Ribbon	Loss Tangent Across Ribbon
48 kg/m^3	9.5 GHz	1.042	0.00083	1.06	0.00036

the foam as well as the frequency. For X-band, the relative permittivity for foam would increase linearly according to

$$\varepsilon_r = 1.023 \times 10^{-3} d + 1.01931$$

where d is the density in kg/m³, and the range of densities is from 30 up to 250 kg/m³. The loss tangent would follow the law

$$\tan \delta = 1.545 \times 10^{-5} d + 1.3636 \times 10^{-4}$$

for the same range of values of density.

The dielectric constants of PET, PVC and PEI foams in different densities are shown in Figure 9.64. Table 9.4 shows the dielectric properties of PMI foam in three deferent densities.

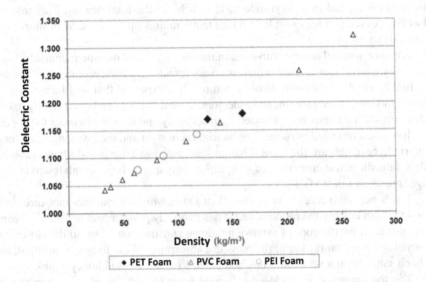

FIGURE 9.64 Dielectric constant of three foam core materials (tested under 12–13 MHz).

TABLE 9.4
Dielectric Properties of PMI Foams[70]

Properties	Frequency (GHz)	1 kg/m³	51 kg/m³	71 kg/m³
ε_r	2.5	1.05	1.057	1.075
	5.0	1.043	1.065	1.106
	10.0	1.046	1.067	1.093
$\tan \delta$	2.5	<0.0002	<0.0002	<0.0002
	5.0	0.0016	0.0008	0.0016
	10.0	0.0017	0.0041	0.0038

In radome manufacturing, prepregs (preimpregnated composites), which contain the required amount of resin (B-stage) in the fabric, are more popular, because they offer ease of handling, better resin control and dimensional control. For curing of prepregs, autoclave is recommended to obtain uniform resin flow and better compaction. Wet resin co-curing processes, such as resin infusion and RTM, also are used with foam core sandwich structures. However, due to certain reasons, use of an oven for curing of a high technology large size airborne radome was necessitated. Curing is a critical operation in obtaining the optimum properties in the material for a radome.

9.6.2 Medical Equipment

Traditionally, metals such as aluminum, stainless steel, and titanium have been used for structural components in the medical device industry. But these materials are radiopaque – that is, they obstruct X-rays. Accordingly, a metal device located in front of a trauma region would restrict X-ray visibility to the region. Carbon fiber has very low X-ray absorption characteristics, and low-density structural plastic foam and nonmetal honeycomb also have low radio absorption behavior, and the uses of carbon fiber composite sandwich panels in medical radiology equipment have become popular. The sandwich panels are also significantly better than solid plastics, such as polycarbonate, phenolic resin, and other traditional decks for improving overall performance of equipment. Therefore, medical device manufacturers have gradually replaced traditional materials bed panels with sandwich structural panels, which have broad market prospects and development in the medical device field in future.

Applications of the sandwich composite panels in medical devices include X-ray, CT, and MRI scanning bed boards/pallets; operating bed supports; various treatments, inspection, and transport bed supports. Figure 9.65 shows some applications of sandwich panels as patient support for inspections.

In medical applications such as CT tables, PMI foam is the most common choice. Among polymeric rigid foams, PMI foam is the core material with the highest strength and stiffness at minimum weight. That means couch and tabletops made with it are not only thinner than ones made with other core materials but also lighter in weight and therefore easier to handle in everyday use.

Their reduced mass means PMI foam cored tabletops absorb lower radiation levels than tables with other core materials. The radiation required for radioscopy can therefore be minimized, so the patient is exposed to lower levels and the health risk is reduced accordingly. In addition, thinner tabletops reduce scatter radiation and provide X-ray images of much higher quality. Components of X-ray diagnostic and therapy equipment, such as X-ray and CT tabletops, are increasingly being built and manufactured with novel and sometimes complex geometries. Rising requirements for geometrical freedom can be easily met using sandwich construction techniques. PMI foam can be precisely thermoformed with simple tools. In many cases, this saves considerable machining costs and minimizes costly material losses. Figure 9.66 shows the curved panels made by PMI foam core and carbon fiber composite facings.

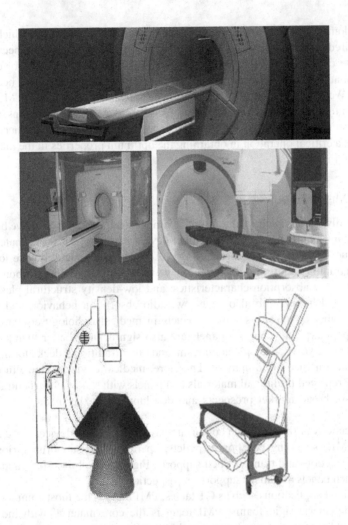

FIGURE 9.65 Applications of sandwich panels as patient support for inspections.[71]

FIGURE 9.66 Curved sandwich panels made by PMI foam and carbon fiber composite skins.[72]

The X-ray transmission ratios of face material, core material and sandwich panel are related to thickness. The relationship between the X-ray transmission rate and the dimension for thickness of facing, core and sandwich were derived by some research works. In an example research, using the optimization process, the thicknesses of face and core materials for sandwich cradles were determined to minimize the cost of used materials. They also met the criteria that the deflection should not be more than 20 mm, and the X-ray transmission rate of the cradle should be equal to or greater than that of aluminum at 1.5 mm thickness. To ensure medical accuracy, in many decks requiring a suitable aluminum equivalent, it is also required to have excellent rigidity and strength to ensure that the medical bed deformation at work is sufficiently small.[73]

In this research work, it was found that the X-ray transmission rate decreased rapidly from 96.8% to 74.7% as thickness increased from 0.7 to 4.5 mm. Regardless of the type of molding technique and material, the transmission rate was approximately the same for the same thickness. Therefore, the relation between the transmission rate and thickness of face material is expressed in the following equation.[73]

$$\text{Transmission rate of face material}(\%) = -5.7752 \times \text{thickness}(\text{mm}) + 100.89 \qquad (9.1)$$

The X-ray transmission rate of core materials change with the thicknesses and densities of foam. Thicknesses of 2, 11, 30, and 45 mm were considered for PMI 32 kg/m^3 foam, and an evaluation was performed for PVC 50 kg/m^3 foam at thicknesses of 45 and 60 mm. For the PMI 32 kg/m^3 foam, as the thickness of core increased, the transmission rate decreased slightly.[73] PVC foams showed a same transmission rate of 91.5% for a thickness of 60 mm when compared to the transmission rate of PMI 50 kg/m^3 with the same thickness. The relationship between the transmission rate and thickness of PMI 32 kg/m^3 core material is predicted by the following equation.

$$\text{Transmission rate of core material}(\%) = -0.0822 \times \text{thickness}(\text{mm}) + 99.659 \qquad (9.2)$$

The X-ray transmission rate of sandwich structure can be estimated by multiplying that of face material calculated by Equation 9.1 and that of core material calculated by Equation 9.2. Most cases agree with the following Equation 9.3 except for some structures made of PVC foam and a thick face.

$$\text{Transmission rate of sandwich structure}(\%)$$
$$= \text{Transmission rate of face material}(\%) \times \text{Transmission rate of core material}(\%)$$
$$(9.3)$$

Polyurethane (PUR) foam has a strong track record of use with medical imaging equipment. The high-density foam material is lightweight yet strong, while providing a homogeneous background that is transparent to X-rays. These characteristics make PUR foams ideal as core panels for X-ray and diagnostic tables, beds for CT scanners, and as calibration materials. And, thanks to its low Al equivalent, the special PMI foam supports high-quality X-ray and CT images with minimal patient radiation exposure. The PUR foam core and carbon fiber composite facing sandwich panels are used in the applications of radiological fixtures, radiological suite floors, tables, medical devices, radiological scanner covers, etc.

PUR foam core can provide a cost-effective option for specialized medical positioning table design. To meet customer requirements where the table must be lightweight, durable, transparent to X-ray radiation, easily machined to accommodate a specific type of insert for the attachment of medical instruments and clamps, as well as produced cost-effectively with short lead times. PMI foam, which is often used in these types of tables, is outside of the target price range and required expensive tooling to cure.[74]

Next generation radiolucent patient tables comprise carbon fiber-reinforced composite skins with a foam core. Carbon composite sandwich tabletops have twice as much X-ray permissibility as wooden tabletops and five times that of solid plastic tabletops. Improved radiolucency results in reducing a dosage of X-rays and associated health risks to a patient. The challenges associated with using a foam core in conventional radiolucent patient tabletops are described below.

Foam core sandwich structures can have less fatigue resistance and may structurally weaken over repeated loading and unloading of a patient table. Hence, manufacturers typically recommend a permissible working life period for these tables, after which they need to be replaced. The radiolucent property of a foam core is less than that of carbon, and there is a possibility of contamination in the foam core made of polyurethane that may show up as errors during X-ray examination. The volume and amount of material in the tabletop must be reduced to a minimum. Nonradiolucent foreign materials in the tabletop can result in a misdiagnosis of the patient. Foam core sandwich structures can have poorer relative strength and stiffness when compared to honeycomb sandwich structures. To improve stiffness, cantilevered foam core tabletops have to be made thick. However, such thick tabletops interfere with the easy maneuverability of the tabletops, and therefore structural and ergonomic design compromises need to be made.

One type of carbon fiber honeycomb core comprises unidirectional carbon fibers in a phenolic resin matrix, yielding a core density of 1,200 gsm for a 24 mm thick core (50 kg/m^3). A sandwich panel was made with carbon fiber biaxial face sheets and the carbon fiber honeycomb core by using a radiolucent epoxy adhesive film in a vacuum bag with a breather layer and heated in an oven over a 2-hour period ramping up to 80°C, holding for half an hour, ramping up to 120°C, holding for 45 minutes, and then cooling down to room temperature.[75]

The carbon fiber honeycomb core sandwich panel has improved radiolucency performance over existing foam core solutions. Hence, the X-ray dosage to the patient can be reduced as a lower X-ray intensity is sufficient for the X-ray scan. The strength-to-weight ratio of the face sheets and the carbon fiber honeycomb core is high due to the increased stiffness of the carbon fibers in the honeycomb core. The crush strength and fatigue performance of honeycomb sandwich structures are substantially higher than that of comparable foam core with similar density. Hence, the total weight for a given strength performance, that is, the material present in the radiolucent patient tabletop is lesser compared to that of conventional foam core solutions. Furthermore, less material consumed minimizes a risk of contamination by radiolucency of foreign materials.

The cradle panel designed as a sandwich structure is composed of the top face, bottom face, and the core between faces. One example product was made by the following procedure after FEA design works. The first layers of upper and lower face were formed by infusion and the remaining layers were formed using prepreg. The upper and lower faces were molded after stacking carbon UD or carbon fabric as per the thickness estimated by FEA process. After applying additional resin on the cured face, the prepared core was inserted between two molds, and it was assembled by clamping. Then, the assembly was cured in a dry oven. After curing, the assembly was demolded, and the completed cradle was trimmed at the bonded area.[76]

9.6.3 Acoustic Barriers

Sandwich composite structures are extensively used in constructional, naval, and aerospace structures due to the high stiffness and strength-to-weight ratios as what are presented in early sections. One critical shortcoming of these materials, however, is their suboptimal acoustical performance: they allow sound to pass through rather easily and therefore yield a low sound transmission loss (STL). This phenomenon can be in part explained by the mass law, which states that the transmission of the noise through the material is inversely proportional to the product of the thickness, the density of the material, and the frequency. In other words, a lightweight material in theory translates to a high sound transmission, particularly at low frequencies. The optimal design of lightweight and high STL of the structure would therefore usually require design trade-offs. However, advanced composite sandwich structures can be optimized for improved structural-acoustic performance. Recently, we have seen a growing interest in optimizing composite sandwich structures.

Sound insulation is a very important issue for ships, buildings, and transportation vehicles. Investigation showed the comfort of on-board large marine crafts where the stay on board for passengers ranged from several hours to several weeks.[77] The study surveyed 100 people during two sea trials, and it identified that acoustics was the area that required the greatest improvement. The marine craft was divided into sections where it was found that cabins were the most uncomfortable place acoustically and therefore required the most improvement. Sources of noise that provided the most discomfort fell into the three categories of engine noise, ventilation and whistles, and squeaking, clattering, clattering, cracking, and creaking.

9.6.3.1 Sound Transmission through Sandwich Structures

The problem of sound transmission through sandwich panels is schematically stated in Figure 9.67. Incident acoustic power P_i (sound waves) hits a sandwich panel. Some of P_i is reflected back, this is denoted by P_r, and some of it excites a wave motion in the panel. During this motion, some power is lost through dissipation in the material P_d and some causes radiation of power P_t in the form of transmitted sound from the opposite side of the panel.

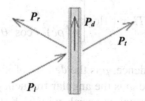

FIGURE 9.67 Schematic description of sound transmission through a sandwich panel.

The dominating wave type in this excitation-radiation system is the bending wave, since it has a relatively large velocity component perpendicular to the panel surface and thus interacts efficiently with the surrounding medium. The sound reduction index (R) can be defined in this most general form based on the incident and transmitted acoustic power as

$$R = 10\log(P_i / P_t) \tag{9.4}$$

The sound reduction index depends on the incidence angle and on the frequency of the sound waves. The sound field is usually variable both in frequency content and the incidence direction and it is difficult to predict the exact properties of the incident field. Thus the assumption of a diffuse field is often used to predict the sound reduction of a construction when there is no available information about the real conditions. In the diffuse field, sound waves arrive with equal probability from all directions and the reduction index is integrated for all incidence angles.

Sound transmission loss (STL), or referred to as the sound reduction index, of a sandwich panel is the difference between the incident sound power level and the transmitted sound power level for a specified frequency or frequency band. This number indicates the noise insulation capability of the panel. The sound transmission loss is often an important consideration in the analysis and design of partitions or panels separating adjoining spaces in industry, housing, and various types of vehicles. It can be described in terms of the panel impedance.

An intensity probe is used to measure the sound intensity in the anechoic room, L_{I2}. Then the sound transmission loss is the difference in the sound intensity levels incident on the panel and transmitted by it into anechoic room:

$$STL = L_{I1} - L_{I2} = \frac{\langle p_i^2 \rangle}{4\,\rho c} - L_{I2} \tag{9.5}$$

where $\langle p_i^2 \rangle$ is the spatial average of incident sound pressure square, and ρ and c are media density and sound velocity.

The mass law states that for any given frequency, the STL of a panel increases with increasing mass per unit area at a rate of 6 dB per octave. The mass law is effective over the entire frequency range but is the dominant factor in the mass law region. For a thin panel, neglecting stiffness and damping, the TL in this region is governed by Equation 9.6.

$$STL = 10\log\left[\frac{\omega^2 m^2}{4(\rho c)^2 / \cos^2\theta}\right] \tag{9.6}$$

where θ is the angle of incidence, ρ is the density of air, c is the speed of sound, m is the mass surface area, and ω is the angular frequency. Sound propagates through all materials but is faster through materials with high stiffness and low density. The sound energy vibrates particles doing work on the material.

The three essential parameters that determine the dynamic responses of a structure and its sound transmission characteristics are mass, stiffness and damping. Mass and stiffness are associated with storage of energy. Damping results in the dissipation of energy by a vibration system. For a linear system, if the forcing frequency is the same as the natural frequency of the system, the response is very large and can easily cause dangerous consequences. In the frequency domain, the response near the natural frequency is "damping controlled". Higher damping can help to reduce the amplitude at resonance of structures. Increased damping also results in faster decay of free vibration, reduced dynamic stresses, lower structural response to sound, and increased sound transmission loss above the critical frequency.

The behavior of a sandwich structure in the low frequency region is determined by pure bending of the entire structure. In the middle frequency region, the rotation and shear deformation of the core become important. At high frequencies, the bending of the face sheets is dominant. Therefore, if the damping in the core is higher than that in the face sheets, then the overall damping has a maximum value in the middle frequency range. On the other hand, if the damping in the core is less than that in the face sheets, then the total damping has a minimum value in the middle frequency range.

When the wavelength of sound in air is shorter than the wavelength of transverse waves propagating along a panel, the panel becomes an efficient sound radiator. This phenomenon is called coincidence because the trace wavelength of the radiated sound is equal to the transverse wavelength of the panel. At coincidence sound travels through the panel more easily for a particular angle of incidence and transmission. The critical frequency is the lowest frequency at which coincidence occurs. This happens when the longitudinal wavelength in air is equal to the bending wavelength in a finite panel. It is well known that the sound transmission through panels is primarily by bending waves. These waves are dependent on both the material and geometric properties.

The complex nature of the sound insulating properties of composite sandwiches was found to be dependent not only on the physical properties but also the frequency of incident noise. When building enclosure or wall systems, a number of considerations must be taken into account. When noise transmitted through air reaches a wall, it is reflected, absorbed, or transmitted through the wall. For attenuation of low frequencies (below 500 Hz), a stiff, thick panel should be used. At frequencies above 1 kHz, less stiff material should be used and/or viscoelastic material should be added to the panel.

9.6.3.2 Sound Transmission Reduction by Adding a Layer of Membrane-Type Acoustic Meta-Materials

There are a range of methods recognized in the literature for increasing the STL through sandwich composite structures. A very effective and simple way to do this is by increasing mass, although applications such as the aircraft and marine industries often have weight restrictions constrained by efficiency and running costs. Adding mass raises the STL curve in the mass law region extending from the resonance-controlled region to the coincidence-controlled region. Decreasing the bending stiffness of a panel shifts the coincidence effects to higher frequencies, thereby extending the mass law region. Increasing the internal damping reduces the effect of coincidence, therefore increasing the STL. This is achieved by the addition of layers to the core that are highly viscous but unfortunately are often too expensive to be economic to implement.

Noise insulation methods aiming at achieving noise reduction with minimum weight penalties were traditionally employed by using porous materials, perforated media, or acoustic blankets with mass inclusions. They have shown excellent performance in the high frequency region but fail at low frequencies.

Extensive efforts have been made to develop novel, lightweight materials that can achieve excellent noise reduction particularly at low frequencies. Recent studies on membrane-type acoustic meta-materials (AMs) appear to open up this possibility and have attracted much attention. By attaching a small mass onto the membrane with clamped boundaries, a narrow-band negative dynamic effective mass was tuned located between the first two resonance frequencies which resulted in total reflection, therefore breaking the mass law. Noise reduction in broadband can be achieved by stacking up several membrane panels and the total mass per unit area was only around 15 kg/m². Shortly after, other research results showed that low-mass-attached membranes or plates clamped at boundaries could also introduce high STL at low frequencies. Although the low-mass-attached AMs could be potentially lighter in weight, the mass-attached and locally resonant membrane-type AMs are the ones that have been predominately investigated by researchers.[78]

Inspired by the wide-spread use of honeycomb structures and recent development of low-mass-attached membrane-type AMs, a honeycomb acoustic meta-material was researched, designed, theoretically studied, and experimentally validated. Figure 9.68a shows a unit cell of the honeycomb acoustic meta-material, where an isotropic membrane is adhered on the top of the honeycomb structure. This material is termed as a lightweight yet sound-proof acoustic meta-material. The honeycomb core was made from aramid fiber sheets. The membrane material was latex rubber with a thickness $h_m = 0.25$ mm.

The static mass density of the honeycomb structure was measured to be $\rho^*_c = 32$ kg/m³. The Young's modulus, mass density, and Poisson's ratio for the latex rubber were $E_m = 7$ MPa, $\rho_m = 1,000$ kg/m³, and $\nu_m = 0.49$, respectively. The effective static density of the meta-material composite was 41.58 kg/m³. The meta-material composite thereby inherits the lightweight property of the honeycomb structure. The mass per unit area of the meta-material composite was 1.05 kg/m².

Figure 9.69 shows the experimental and simulation STL results for honeycomb cell structure only and the honeycomb meta-material. The STLs are shown in unit of decibel (dB). Since the honeycomb cell structure is hollow hole and lightweight, as expected it exhibited very poor acoustical performance. As shown by a solid line along the x-axis in the figure, the sound is almost 100% transmitted. With membranes added, the resulting structure showed distinct STL improvement over the entire frequency range under consideration (the solid line at top). The simulated STL results by considering different loss factors for the membrane captured the same trend as experiment and the predicted STL at the STL dip frequency increased as the loss factor increased.

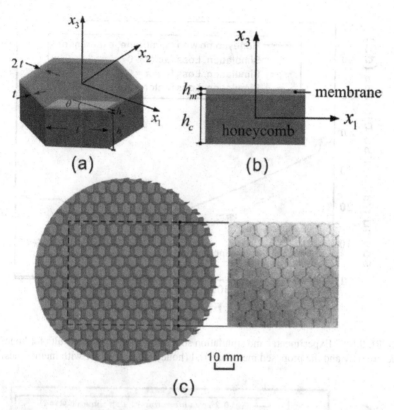

FIGURE 9.68 (a) Unit cell of the honeycomb acoustic meta-material, (b) side view of the acoustic meta-material, and (c) the meta-material prototype used for the acoustical test.[78]

To further improve the acoustical performance, the meta-material with membranes adhered on both the top and bottom sides of the honeycomb structure was also explored. With a membrane in a thickness of 0.25 mm, the effective static density of the meta-material composite was increased to $\rho = 50.98 \, kg/m^3$. Figure 9.70 shows that comparing with the sample with only a single membrane, two membranes combined can enhance the STL by as much as another 20 dB. The STL at low frequencies (<500 Hz) was consistently greater than 45 dB. The average STL was 37 dB over the frequency range from 50 to 1,600 Hz. This remarkable result was achieved at a very low mass per unit area (1.3 kg/m²).

A great potential of the proposed acoustic meta-material lies in that it can be readily modified to make honeycomb sandwiched structures which could potentially be simultaneously strong, lightweight, and soundproof. While conventional honeycomb structures comprise a honeycomb panel sandwiched between two face sheets and are well known to have extremely high stiffness to weight ratios, a different sandwich structure would consist of two face sheets with one or more membranes sandwiched between honeycomb panels. Such a structure with one membrane incorporated was measured in comparison with a conventional sandwich panel for their STL and the results are shown in Figure 9.71. The two face sheets are made of carbon fiber and had a thickness of 1 mm.

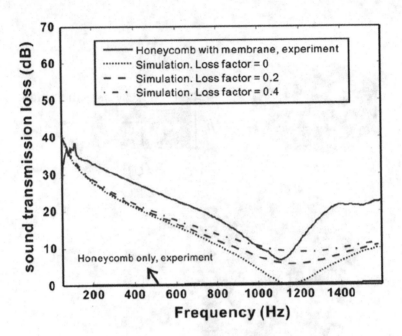

FIGURE 9.69 Experimental and simulation sound transmission loss results for honeycomb structure only and the proposed meta-material (honeycomb structure with membranes).[78]

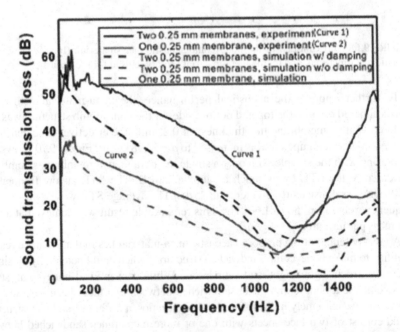

FIGURE 9.70 Comparison of the sound transmission loss for the meta-material sample with two membranes attached on the top and bottom surfaces of the honeycomb walls and the meta-material sample with a single membrane.[78]

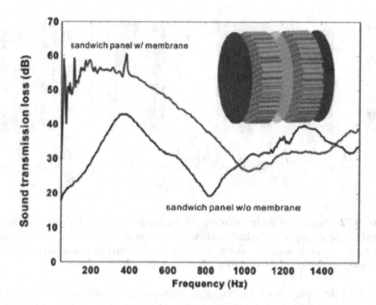

FIGURE 9.71 Comparison of the sound transmission loss for a sandwich panel with membrane and one without membrane.[78]

The presence of the membrane dramatically improved the STL, particularly in the low frequency region below the first resonant frequency of the membrane. The properties of the membrane seem to dominate the acoustical performance the sandwich panel. The STL at low frequencies (<500 Hz) was consistently greater than 50 dB. The average STL over the 50–1,600 Hz range was 40 dB (w/membrane) vs. 31 dB (w/o membrane). A higher STL is expected with more membranes added. However, it is noted that the addition of the membranes is not expected to degrade the mechanical property of the honeycomb, which is crucial for the future applications.

9.6.3.3 Reduce the Sound Transmission by Acoustic Separation of the Layers of Sandwich

A description of sound transmission via multilayer enclosures of finite dimensions was given by the theory of self-coordination of wave fields. In this case, the resonant transmission of sound and the inertial transmission of sound via enclosures are considered. The resonant transmission of sound takes place in the mode of self-induced vibrations depending on the degree of self-coordination of its own wave field with sound fields present in the air on both sides of the enclosure. The inertial transmission of sound takes place in the mode of forced vibrations depending only on the mass and the geometrical dimensions of the enclosure.

The sound transmission via building enclosures of finite dimensions has a resonant and inertial component. The resonant transmission of sound takes place in the mode of self-induced vibrations depending on the degree of self-coordination of its own wave field with sound fields present in the air on both sides of the enclosure. The most effective method of sound insulation enhancement of sandwich panels is to reduce the resonant transmission of sound by acoustic separation of the layers.

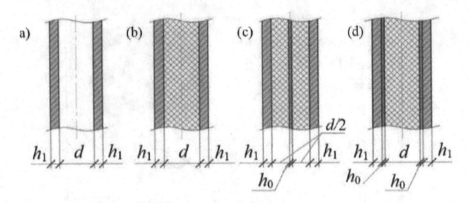

FIGURE 9.72 Structural design solutions of the sandwich panel: (a) two cladding sheets separated by the air gap, (b) standard sandwich panel (without acoustic layers separation), (c) sandwich panel with acoustic separation of the core in two equal parts, and (d) sandwich panel with acoustic separation of the cladding sheets from the core.[79]

Based on the analysis of the reference literature and papers, one can assert that three-layer sandwich panels with a core of rigid foam or mineral wool feature insufficient sound insulation properties stipulated by sound insulation reduction in the bandwidth close to the system frequency "mass – spring – mass" (f_{msm}). In subject enclosures with 30–150 mm thickness, an abrupt reduction of sound insulation is observed in the frequency band of 200–1,000 Hz. Acoustic separation is understood as introduction into the sandwich panel structure of one or several glued-in thin layers of elastic material splitting the core in several parts or between the core and the cladding sheets as shown in Figure 9.72.

Gypsum fiber boards were used as external cladding sheets ($h_1 = 12.5$ mm, 1,150 kg/m³ density). The core was made of mineral wool with elasticity module: $E = 0.8$ MPa, 25 kg/m³ density, as well as of polystyrene with elasticity module: $E = 8.5$ MPa, 15 kg/m³ density. The thickness of the core was $d = 50$ mm. The separation layers were made of roll material of a resilient polyether synthetic fiber ($h_0 = 4$ mm, $\rho = 75$ kg/m³ density, elasticity module $E = 0.3$ MPa). The claddings, the core, and the acoustic separation layers (samples in Figure 9.72b–d) were glued to one another with polymeric glue.[79]

Measured sound insulation curves of sandwich panels as displayed in Figure 9.72 with polystyrene foam core are shown in Figure 9.73. Due to the acoustic separation of the layers, there is a displacement of the resonant frequency of the system f_{msm} to the lower frequency band. Such a resonant displacement to the value of the resonant frequency of the sample with an air gap is observed for both sample types of the PS foam and mineral wool core. This value is a limit value for structures of such type since at equal limit conditions, in enclosures with an air gap, the cladding layers are connected only by the elasticity of the air enclosed between them.

The integration of a core with acoustic separation in the sandwich panel design extends the frequency band where the experimentally obtained data exceed the values of the mass law in the band above the resonance frequency f_{msm}. For sandwich panel with polystyrene foam core and acoustic separation dividing the core into two

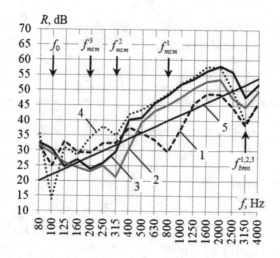

FIGURE 9.73 Measured sound insulation curves of sandwich panels with polystyrene foam core: 1 – panel b, 2 – panel c, 3 – panel d, 4 – panel a, and 5 – R from the mass law for panel b.[79]

equal parts, this band is 500–2,000 Hz; the separation of both the claddings and the core extends the band limits to 400–2,500 Hz accordingly, whereas for a standard sandwich panel, these limits are 1,250–2,000 Hz. It has been established that the most effective reduction in the resonant transmission of sound occurs when acoustic separation is introduced between the sheets cladding and the core. This design solution provides the greatest reduction of the resonance frequency f_{msm}.

The results obtained allow to develop rational constructive solutions for enclosures of civil and industrial buildings of sandwich panels, taking into account the spectrum of insulated noise. Improving the sound insulation of sandwich panels is provided without a significant increase of their mass and thickness.

9.6.3.4 Introduce Air or Sound Insolation Gap to Core or between Panels

The acoustic properties of composite sandwich panels can be improved by introducing air void inserts, glass wool inserts, and glass wool layers or by connecting two panels through a wool layer that may lift the sound insulation curve and shift the resonant frequencies. One example is a work to improve the sound insulation properties of CSIPs that consist of the glass fiber-magnesium-cement boards (face sheets) with the thickness of 11 mm and expanded polystyrene EPS core (thickness 150 mm) connected by thin adhesive layers. As compared to classical SIPs, they are significantly stronger and fire- and biological corrosion-resistant. Unfortunately, the high stiffness-to-mass ratio causes a low level of the sound insulation due to the appearance of resonant frequencies in the EPS core.[80]

The theoretical study indicated that as the core shear stiffness increased, the coincidence critical frequency moved to the lower frequency range and the sound insulation was improved in a medium and high frequency range. The study results showed that attaching the honeycomb core directly to the interior side of face sheets improved the sound insulation in a low frequency range.

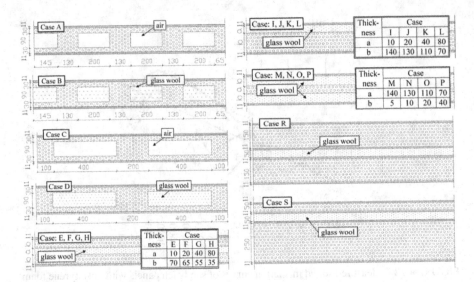

FIGURE 9.74 CSIP with different modifications (dimensions in millimeters).[80]

Several modifications of CSIPs were numerically tested in steady-state 2D FE analyses to improve its sound insulation. Two type modifications were considered: (a) glass wool layer, air void insert, and glass wool insert were put symmetrically against the horizontal axis or (b) two CSIPs were mutually connected through a glass wool layer as shown in Figure 9.74.

The geometry of air voids and the glass wool insert was the same. The following material parameters were assumed for the glass wool: $\rho = 101$ kg/m^3, $E = 303$ Kpa, $\nu = 0$ and loss factor $\eta = 0.05$, and for the air layer: $\rho = 1.2$ kg/m^3, and $K = 142$ Kpa. The second-order elements were used for the acoustic medium. The influence of the different CSIP modifications on the sound insulation curve is shown in Figure 9.75.

The suggested modifications of the EPS core are able to improve the sound insulation of composite panels, in particular with the aid of a glass wool layer. For the modified CSIP with the air void inserts 20cm wide (case "A"), the first resonant frequency was shifted from 630Hz down to 500Hz and the sound reduction index R increased from 15.7 dB up to 29.2 dB in the area of the mass-spring-mass resonance. However, this modification decreased the sound insulation at the low frequency range (250–315 Hz). On the other hand, the 40cm wide air void inserts (case "C") allowed for a sound insulation growth in a broad frequency range (from 160Hz up to 1,000 Hz). The index R increased from 15.7 dB up to 23.6 dB in the area of a symmetric coincidence frequency. The low frequency acoustic insulation was smaller.

A glass wool (cases "B" and "D"), the sound insulation increased obviously with the growing width of wool inserts. An increase of the sound insulation at the broad frequency range was obtained in the cases "B" and "D" due to the weight augmentation. Thus, a noticeable increase by 2.5–6.1 dB was obtained. A significant positive effect of an additional damping layer (glass wool) on the shift of the resonant frequency is showed in Figure 9.75.

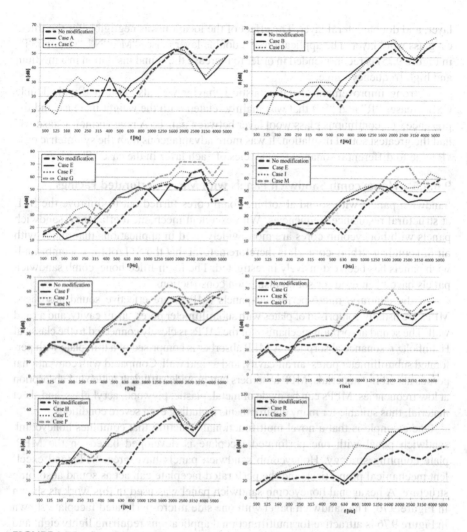

FIGURE 9.75 Sound insulation curve for modified CSIP influence of air gap and glass wool layer position as shown in Figure 9.120.[80]

All modified panels (cases "E", "F", "'G", and "H") had always a greater sound insulation in a middle region of the frequency band. The addition of the 1 cm wide damping layer (case "E") shifted the low resonant frequency to 315 Hz and increased R up to 43.8 dB in the area of resonant frequency ($f = 630$ Hz). This is related to the smaller sound insulation in a low frequency range. The panel with the additional 1 cm thick damping layer (case "E") was the most advantageous because it was characterized by the greatest growth of sound insulation. The greatest increase of the R value was occurred in the case "H" (8 cm thick additional glass wool layer).

The influence of an additional layer location on sound insulation curves is showed in from laminate I to P. The overall thickness of additional glass wool layers was the same. Three-layer types were taken into account: single mid-layer, single lateral layer, and double lateral layer. The effect of the location was negligible for the thickest glass wool layer. The application of a double layer (cases "M", "N", "O", and "P" in Figure 9.75) is recommended in order to increase the sound insulation in a medium and high frequency range.

A strong improvement of the acoustic behavior was also obtained with double CSIPs (cases "R" and "S") that was mainly related with the weight increase and the presence of a continuous glass wool layer between panels. A nonsymmetric case "S" had the greatest sound insulation. It was more advantageous than the symmetric case "R" due to different resonance frequencies of each layer in the case "S".

9.6.3.5 Honeycomb Sandwich Panels with Micro-perforated Facings

Sandwich panels with sound absorbing foam cores turned out to improve the STL at structural resonance frequencies. With excellent mechanical efficiency, sandwich panels with honeycomb cores are more widely used in applications than those with air or sound absorbing cores. It is therefore natural that the STL of honeycomb sandwich panels has been extensively investigated. The foam-filled honeycomb sandwich panels have been used for sound insulation constructions.

On the contrary, micro-perforated panels (MPPs) are effective sound absorbers. MPPs are usually comprised of plates with submillimeter pores, an air cavity, and a rigid wall. The sound absorption mechanism of the MPPs is closely connected to the classical Helmholtz resonance absorption. A MPP absorber is comprised of a thin plate with perforated submillimeter pores, an air cavity, and a rigid wall. Compared with conventional porous absorbing materials, MPP absorbers can provide sufficient wideband absorption at low frequencies. MPPs can be made of metal, plastic, plywood, acryl glass, and sheet material, thus suitable for many environments including even severe conditions.

One example is that a novel multifunctional structure that combines honeycomb sandwich panel with one perforated faceplate is developed based on the MPP-plate coupling strategy. Honeycomb sandwich panels have great STL (and excellent mechanical properties), and the perforated faceplate can act as sound absorbing structure. A hexagonal honeycomb sandwich panel was used in this work as shown in Figure 9.76. The sandwich panel with one side micro-perforated faceplate shown in Figure 9.76 is attractive for multifunctional applications requiring lightweight and simultaneous load carrying, sound insulation as well as sound absorption capabilities.

Figure 9.77 shows perforation-induced increment of STL within the frequency range of 700–1,200 Hz. The peak frequency in the STL curve is identical to the sound absorption coefficient. For the sandwich without perforation, since no acoustic energy can be consumed during sound propagation, the STL is decided by the reflection of sound wave. In the presence of perforation, sound wave enters the sandwich via the perforated pores and the acoustic energy is consumed due to viscous and thermal losses inside the pores.

Figure 9.78a shows three hexagonal sandwich panels having identical geometrical parameters but different perforation ratios. These sandwich panels have one, two and three pores in each unit cell of the faceplate as shown in the figure. All the pores have the same diameter of 0.5 mm, and the other geometrical parameters of the sandwich panels are the identical. Accordingly, the perforation ratios are 0.39%, 0.79%, and 1.2%, respectively.

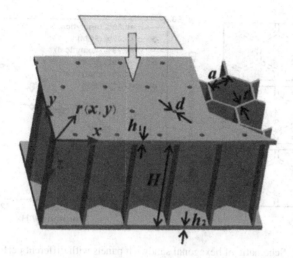

FIGURE 9.76 Schematic of one surface-perforated sandwich panel with hexagonal honeycomb core.[81]

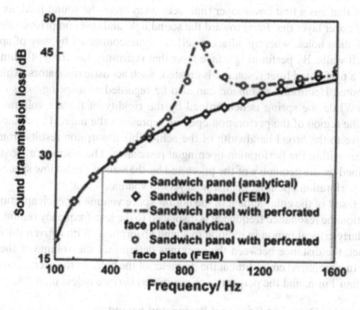

FIGURE 9.77 Comparison of sound transmission loss between sandwich panels with non-perforated and perforated faceplate.[81]

The influence of perforation ratio on the STL of hexagonal sandwich panels is displayed in Figure 9.78b. It can be seen from the figure that with the increase of perforation ratio, the peak frequency of the STL increases. As the perforation ratio is increased, the viscous and thermal losses inside the perforated pores are enhanced as a result of the increased contact area between air and solid frame, thus enlarging the resistance of the perforated faceplate.

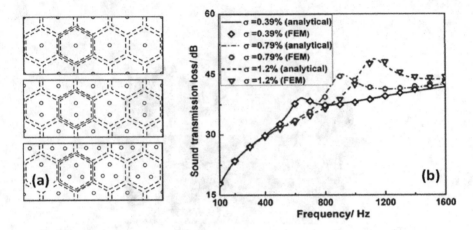

FIGURE 9.78 Schematic of hexagonal sandwich panels with different perforation ratios (a) and influence of perforation ratio on sound transmission loss (b).[81]

Another acoustic sandwich construction comprises a core layer with a honeycomb-like cells that has a first cover layer that faces away from the sound field, as well as a second cover layer that faces toward the sound field and that comprises a multitude of perforation holes, wherein adjacent cells are interconnected by way of apertures in the cell walls. By perforating a face layer that is arranged in front of a limited air volume, a perforated liner resonator is created. Such acoustic resonators, which form a distributed Helmholtz resonator, can also be regarded as a spring-mass damping system. While the spring is determined by the rigidity of the air volume, the air plug in the region of the perforation opening represents the mass. The damping that is decisive to the broad bandwidth of the achievable absorption results from losses that occur within the perforation opening at resonance. The resonance frequency is determined by the geometry of the pores and by the size of the hollow space volume of each perforation opening, i.e. the honeycomb volume.

As a result of the enlargement of the hollow space volume for each aperture of the perforation the resonance frequency is displaced to the low-frequency region, i.e., the particularly critical region in relation to sound absorption. With a given thickness of the panel, the distance between the openings determines the volume of the hollow spaces. In a preferred embodiment the diameter of the holes of the second cover layer is less than 1 mm, and the percentile of perforated surface is less than 1%.

9.6.3.6 Use Damping Core and Perforated Facing

A series of soundproof panels have been developed, which provides great acoustic comfort in premises that require it, as well as high levels of sound and heat insulation and excellent levels of noise absorption. The perforated design of one of the surfaces ensures that the panel gives excellent noise absorption and highly effective soundproofing. The design and development of the rock wool acoustic core with damping property represent an important step forward in the sectorization of buildings, for either industrial or leisure use, offering a series of soundproofing and noise-absorption advantages rarely found in a single product. The products include the

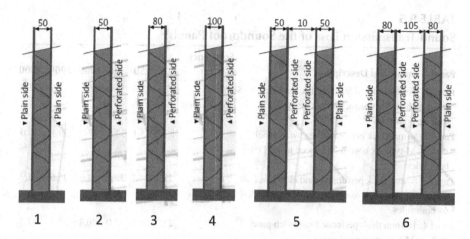

FIGURE 9.79 Configurations of rock wool core panel with one perforated facing.[82]

rock wool core and perforated facing on one side, which also has thermal insulation properties, offers a high level of noise absorption. The double panel systems consist of two panels and an air gap between, which offer very high levels of soundproofing property. Figure 9.79 shows a few of the soundproof panel configurations. Table 9.5 reports the STL values of the individual panels at different sound frequency. The results show that the double-panel system has 60–70 STL level at the certain frequency. The most common applications of the soundproof panels are transport, railways, roads and airports, power stations, electrical transformers, gas control centers, and water pumping stations.[82]

9.6.4 SPORTS AND LEISURE

With a growing economy, people's living standards improve, and people are afforded leisure and all kinds of sports venues. Particularly at the professional level, experts focus on scientific training and give great importance to the improvement and development of sports equipment. Because of their light weight, high strength, large degrees of freedom of design, easy processing, and forming characteristics, sandwich composites are obtaining widespread application in sports equipment.

Before the advent of composites, wood, steel, and aluminum alloy were mainly used for making sports equipment. Compared with these materials, sandwich composite materials have obvious advantages.

There are many different kinds of sports equipment, the following plate-like structure, such as skis, surfboards, windsurfing, table tennis paddles, slats and gliding wing spar etc. are commonly made by sandwich.

9.6.4.1 Sporting Boards on the Water

Boarding on the water includes surfing, kiteboarding, skimboarding, water skiing, and wakeboarding, which is an action sport in which the sportsman utilizes a special board to ride across the water. In all of these waterboarding sports, the board

TABLE 9.5
Sound Transmission Loss of the Soundproof Panels[82]

Panel Number and Description	Frequency (Hz)	125	250	500	1000	2000	4000
Panel 1: Non-perforated 50 mm thick sandwich panel with an M-type rock wool core, faced in steel on both sides	Sound Transmission Loss (dB)	22.3	27.0	31.0	35.1	33.6	32.8
Panel 2: 50 mm thick perforated sandwich panel with an M-type rock wool core faced in steel on both sides		20.2	25.1	32.7	37.0	39.1	36.0
Panel 3: 80 mm thick perforated sandwich panel with an M-type rock wool core, faced in steel on both sides		23.2	24.7	32.9	39.6	35.6	44.9
Panel 4: 100 mm thick perforated sandwich panel with an M-type rock wool core		25.7	29.8	35.0	40.8	32.4	47.0
Panel 5: Double panel. Consisting of two 50 mm thick perforated sandwich panels with an M-type rock wool core, faced in steel on both sides. Between the two panels there is a 10 mm air chamber. The perforated sides of the panel are positioned facing into the air chamber		31.6	42.8	49.5	52.0	58.0	60.4
Panel 6: Double panel consisting of two 80 mm thick perforated sandwich panels with an M-type rock wool core, faced in steel on both sides. Between the two panels, there is a 105 mm air chamber. The perforated sides of the panel are positioned facing into the air chamber		36.7	47.1	56.1	71.0	63.5	74.4

is the main instrument that keeps the rider on the water; however, some differences exist from one to another. Modern boards are mostly manufactured by the sandwich structures with foam or a lightweight core covered by the top and bottom shells. This creates boards with not only a high resistance to bending, preventing boards from breaking, but also an increase in buoyancy, stability, and improved user experience.

In 1920s, a surfboard was made by hollow wood that was a redwood board with hundreds of drilled holes into it and encased with thin layers of solid wood on the top and bottom. It was about 4.6 m long and weighed 45.4 kg, but the design made it much faster than traditional solid wood surfboards. It became the first mass-produced surfboard in 1930. The first rudder, or "fixed fin," was installed on surfboards in 1935. Similar to the rudder on a boat, the component stabilized the board's position in the water.

By 1932, redwood was out, and balsa wood was in. It was a breakthrough for fast, lightweight boards, since a balsa board weighed around 16 kg, or a little heavier if it was made of a balsa/redwood composite. These boards were then coated with layers of resin and fiberglass to make them waterproof.

Developments in industrial and manufacturing technology made during World War II extended to the water board world in the late 1940s and early 1950s. The most lasting and important development in surfboard design occurred in the late 1950s with the phasing out of wood in favor of fiberglass and plastic foams. The selling point of foam is that it's incredibly lightweight, making boards easier to control. Foam is also much easier to shape and cut than wood, allowing for quick mass production.

The major development in water boards over the last 20 years has been the many different kinds of water boards available, each suited to different tastes or wave-tackling abilities, the result of tinkering with heavily entrenched board designs. Developments have focused around the rail curve and what can be done with the board's underside.

Symmetric sandwich laminates are mostly used for making the water boards because of the complex deformation effects of asymmetric multilayer structures, which are also hard to handle in the complete production process. In the boarding sports industry almost all facing sheets of the products are made out of symmetric biaxial or triaxial glass fiber fabrics to increase the stiffness and strength of the sandwich composites. Carbon fiber-reinforced plastics are used to reduce the high-performance products' weight. Fiberglass face sheets are also well suited to integrate additional functions.

Modern water boards are made by using polyurethane or polystyrene foam as core materials covered with layers of carbon or fiberglass cloth, and polyester or epoxy resin. Recent developments in water board technology have included the use of carbon fiber and Kevlar composites, as well as experimentation in biodegradable and ecologically friendly resins made from organic sources. Higher-end products will often replace the PU or PS with better performing foam core such as PVC or SAN. Each year, approximately millions of boards are manufactured.

Surfboards have traditionally been constructed using polyurethane foam, and it remains a popular choice. They are made stronger with one or more stringers going down the middle of the board. The foam is molded into a "blank", in the rough shape of a surfboard. Once the blanks have been made, they are given to shapers. Shapers then cut, plane, and sand the board to its specifications. Finally, the board is covered in one or more layers of fiberglass cloth and resin. Vacuum forming and modern sandwich construction techniques borrowed from other industries have also become common. It is during this stage that the fins or boxes for removable fins are attached and the leash plug installed.

9.6.4.2 3D Printing Sport Boards

Recently, 3D printing has been introduced for making sport boards. The main benefits of 3D printing could be listed as design freedom, waste diminution as well as the ability to fabricate complex structures with sufficient geometrical precision. The literature shows that the board geometry and different patterns of the cores were studied for making the sport boards.

In one example, the board was made by printing two primary parts using polylactic acid (PLA), a top shell, and a merged bottom shell and lightweight honeycomb core that can be designed and 3D-printed with different patterns. The two parts glued together with a strong adhesive after the 3D printing process. Furthermore, the boards are manufactured with a bottom curvature, which aims at better edging and upwind ability, and is significant for beginners, providing more grip and stability when compared to flat boards.[83]

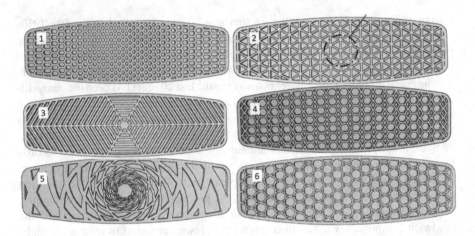

FIGURE 9.80 3D printed core pattern for sandwich sport boards: (a) bamboo-inspired graded honeycomb, (b) triangle honeycomb, (c) Spiderweb-inspired core, (d) hexagonal-rhombic structure, (e) pinecone and sunflower-inspired patterns, and (f) hexagonal carbon lattice.[83]

A few of different structure cores also were printed in the study, which are tested by three-point bending and compared to each other by the curves of loading force versus deflection. The printed and tested pattern included bamboo-inspired graded honeycomb, triangle honeycomb, spiderweb-inspired core, uniform honeycomb, hexagonal-rhombic structure core, pinecone and sunflower-inspired pattern core, hexagonal carbon lattice, and solid core as shown in Figure 9.80.

After designing and printing different patterns, every board with the previously mentioned core structures was applied to a three-point bending loading tests. The constant displacement of 4 mm in the z-direction was executed by means of the loading nose, and the reaction forces for each designed board was determined. Figure 9.81 compares reaction force–displacement curves of different core structures. The preliminary conclusion drawn from this figure is the fact that the graded honeycomb structure and fully filled board can tolerate maximum and minimum forces, respectively, while the rest of the patterns experienced an intermediate force. As the maximum stress occurs in the middle of the board right below the load application area, it can be found that the graded honeycomb pattern showed the best bending resistance when compared to other non-uniform patterns.[83]

9.6.4.3 Skis and Snowboard

Generally, snow ski boards have been made by wood, metal, and fiber composites. Wood is light and inexpensive, but easy to be affected with damp and out of shape. Aluminum alloy metal skis price is higher, the requirement to the snow, high adaptability is poor. Fiber-reinforced sandwich composites of ski board is suitable for any types of snow, and easy in maintenance. 99% of all skis on the market today are made of composite or sandwich composite. The core material for ski board is made from light wood, or polymer foam, the specific product chosen for the performance

Sandwich Structural Composite Applications

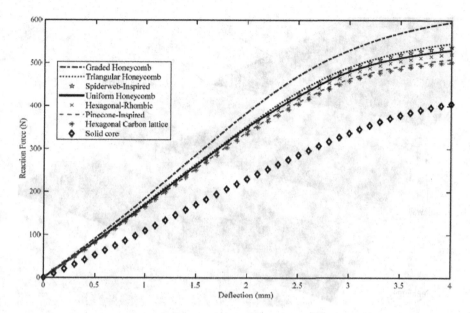

FIGURE 9.81 Reaction force-deflection curves for all core structures.[83]

vs. price target. The skins are often a combination of fiber types. The designers may place specific materials along different sections of the ski (not just top vs. bottom skin) to obtain a specific flexural and damping response.

Sporting goods require something that can last for many years and take a lot of abuse without braking. This is why most people use composites in combination with wood or another sandwich core. The core is what defines the size and shape, which is then wrapped with composite materials.

In one example, a core is made of paulownia or other hardwood that is very light and gives a light ride. In general for skis, poplar, birch, or bamboo or any local wood can be used. They are hard woods that give the durability you need in this kind of sport and also take well the screws for the bindings. The sport boards can be made by using hand layup with vacuum bagging or compression molding.

Each manufacturer has slightly different methods to making skis or snowboards, but the basic structure stays very similar. The basic structures of ski and snowboard, for example, with a wood core are shown in the graphic in Figure 9.82, with each part explained in the following section.

Wood is used both as skin and core, albeit as "skin", it is not on the outer surface. Wood typically gives a lively feel with good vibration damping; it keeps its shape well and has less resonance than foam or plastic. Carbon fiber fabric seems to be preferred for skis and glass fabric for snowboard. Wood cores are made of strips of laminated hardwood that run along the length of the board. The strips used can be made of different woods, arranged in different patterns (both flat-grain & end-grain) and have areas that incorporate other materials like foam. This is done to give different levels of strength, flex, and weight to different areas of a board, creating different behaviors and responses. The laminated wood strips are glued together and

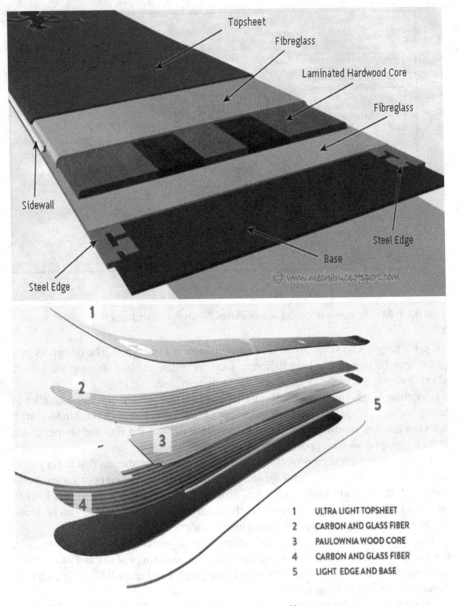

FIGURE 9.82 Structure of snowboard[84] (a) and skiboard[85] (b).

then precisely are cut by CNC machine into shape. Carbon fiber rods, and many other materials/structures are often applied to the core as well to further enhance the desired board characteristics.

Fabric orientation of boards is generally one of two routes: bi-axial wrap – the fiberglass strands are weaved together at 90° to each other; this produces a lightweight, dependable, and forgiving layer; and tri-axial wrap – weaved at +45°, 0°, and

Sandwich Structural Composite Applications

−45° or +60°. This provides increased torsional stiffness and response compared to the bi-axial wrap.

Base is the bottom surface always in contact with the snow. There are two types of bases used for making boards: extruded and sintered: extruded plastics. This is where the base material is melted and then cut into shape. Extruded bases are cheap to make and low maintenance but are less durable and slower. They are smoother and less porous so don't absorb so much wax, but if the base is left unwaxed, the overall performance is not affected so much. Sintered composite, another base material, is ground into powder, heated, pressed, and sliced into shape. Sintered bases are more expensive to make but are more durable and faster. They are very porous and absorb wax well but will lose performance if they are left unwaxed and are more difficult to repair.

9.6.4.4 Sandwich Construction for Making Canoes, Kayaks and Paddleboards

Canoes and kayaks are constructed of wood, aluminum, FRP, and thermoplastic. When the consumer desires low weight or high performance, sandwich construction dominates. Lots of kayaks have some sandwich construction to create shallow ridges where extra stiffness is useful, such as on the hull beneath the seat and on the underside of the foredeck. A few composite manufacturers use it extensively. Like other sport and recreational goods, the choice of face sheet and core material will depend on the level of performance vs. the price. Laminates of E-glass and bulker mats are common. Today lamination process is often infusion, with some prepreg. Aramid or carbon and PVC or SAN foam core is customary for higher performance, generally with core thickness ~6mm. Recent developments include polymer fiber such as ultra-high-molecular-weight polyethylene (Dyneema®) or polyolefin polypropylene (Innegra™) in combination with carbon. Some racing kayaks are made with honeycomb core.

Obviously, covering a mold with a thin layer of wood takes a lot of time so manufacturers prefer flexible core materials such as polyester felt. This looks and feels like thick blotting paper, and drapes quite nicely onto a mold. To make it absorb less resin, it may contain tiny plastic bubbles (microspheres). Examples are the core mats. They add stiffness but not much strength. Racing kayaks are sometimes made with a synthetic honeycomb core which adds both stiffness and strength. Figure 9.83 shows the construction of a kayak with core mat sandwich laminate, and a kayak product.

Company-built kayaks are equipped with sandwich-reinforced keel area ensuring scratch and impact strength. Hull and deck are joined using internally both fiberglass and aramid taping and externally a very robust aramid taping. All decks are built in sandwich as shown in a figure above. A few kayak manufacturers go even further down the high-tech road and use vacuum resin infusion, putting the reinforcement in the mold, putting the mold in a vacuum bag, sucking out the air with a powerful pump, and then allowing the resin to infuse through tubes. A manufacturer may save more resin by buying the reinforcement which is already saturated ("preimpregnated") with the right amount of resin. Prepreg has to be stored in a fridge and may have to be baked in an oven. It is more often used in the aerospace industry than by kayak builders.

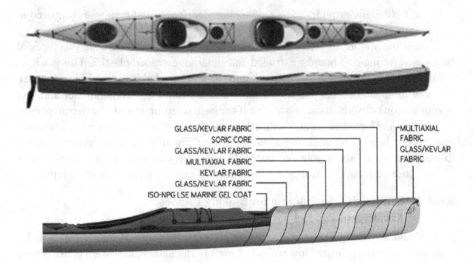

FIGURE 9.83 Sandwich construction and product of a kayak.[86]

REFERENCES

1. E. Greene, *Marine Composites* (2nd ed.). Annapolis, MD: Eric Green Associates Inc., 1999.
2. "Quality of Manufacturing" – Part 3 of the Best Center Console Make or Break Features, http://bahamaboatworks.com/news/quality-of-best-center-console-boats-manufacturing-part-3/.
3. "Foam Sandwich Construction," https://boatbuildercentral.com/support-tutorials/Tutorials/foam-sandwich-how-to.pdf.
4. "40M Resin Infused Yacht," http://flink.online/projects/40m-resin-infused-yacht/.
5. B. Veritas, J. L. Meyer and S. Marinebouw. "Best Practice Guide for Sandwich Structures in Marine Applications," https://trimis.ec.europa.eu/sites/default/files/project/documents/20130201_105501_16344_RevisedBPGv1-00---Report-Format.pdf.
6. S. Bartlett and B. Jones, "Composite Ship Structures," NSWC Carderock Division, Feb 19, 2013.
7. "River cargo barges," https://maritime-executive.com/
8. "Details of a wind power station," https://www.energy.gov/eere/wind/wind-energy-technologies-office.
9. O.T. Thomsen, "Sandwich materials for wind turbine blades – present and future," *Journal of Sandwich Structures and Materials, MuPAD Tutorial*, vol. 11, no. 1, pp. 7–26, 2009, and L. Kühlmeier, "Buckling of wind turbine rotor blades: Analysis, design and experimental validation," Ph. D. Thesis, Department of Mechanical Engineering, Aalborg University, 2007.
10. L. Mishnaevsky Jr., K. Branner, H. N. Petersen, J. Beauson, M. McGugan and B. F. Sørensen, "Materials for wind turbine blades: An overview," *Materials*, vol. 10, p. 1285, 2017.
11. T. K. O'Brien and I. L. Paris, "Exploratory investigation of failure mechanisms in transition regions between solid laminates and X-Cor® truss sandwich," *Composite Structures*, vol. 57, no. 1–4, pp. 189–204, 2002.
12. B. Castanie, C. Bouvet and M. Ginot, Review of composite sandwich structure in aeronautic applications, Composites Part C: Open Access, 2020.

13. B. Hashemi, Aircraft Structures, Chapter 1, A Brief History of Aircraft Structures, https://www.academia.edu/37444717/Aircraft_Structures_A_Brief_History_of_Aircraft_Structures.
14. A.S. Herrmann, P. C. Zahlen and I. Zuardy, "Sandwich structures technology in commercial aviation, present applications and future trends," In *Sandwich Structures 7: Advancing with Sandwich Structures and Materials*, O.T. Thomsen et al. eds., pp. 13–26. Dordrecht: Springer, 2005.
15. S. Black, "Fuselage skins redesign streamlines production" *Composites World*, 12/18/2009, https://www.compositesworld.com/articles/fuselage-skins-redesign-streamlines-production.
16. A. Airoldi, Aerospace Structures, Politecnico di Milano, https://nanopdf.com/download/introduction-to-the-course-2_pdf.
17. C. Le, "New Developments in Honeycomb Core Materials", www.ultracorinc.com.
18. H.R. Hull, "Vented Flexible Honeycomb," US Patent 6,003,283, Dec. 21, 1999.
19. Vented Aluminum Honeycomb, Argosy International, https://www.argosyinternational.com/products/honeycomb/vented-aluminum-honeycomb/.
20. B. Morey, "Sandwich Cores for the Future," https://www.aerodefensetech.com/component/content/article/adt/features/articles/23568.
21. S. Francis, "Tooling, precision enable composites in satellite subsystems," *Composites World*, 4/27/2020.
22. "Sandwich Construction for Public Transportation," Presented in Troy, MI, Dec. 2002, https://www.diabgroup.com/en-GB/Markets/Transport.
23. A. Starlinger, Sandwich Structures, lecture ETH Zurich sandwich rail cart 2013. https://www.stadlerrail.com/en/.
24. D. Wennberg, "Light-Weighting Methodology in Rail Vehicle Design through Introduction of Load Carrying Sandwich Panels," 2011. http://www.diva-portal.org/smash/get/diva2:416836/FULLTEXT01.pdf.
25. M. Wilson and J. Roberts, "Composite Materials for Railway Applications," Bombardier Transportation, Genoa, June 2002. https://www.bombardier.com/en/transportation.html.
26. J. Carruthers, "Composites in the Rail Industry: Overview and Future Development," In *10th Technical Forum of the Composites Processing Association*, Derby, UK, November 2005.
27. D. Singh, "Comparison of Carbon Steel and Composite Side Wall of Light Rail Vehicle by Finite Element Analysis," of M. S. Thesis, the University of Texas at Arlington, TX, May 2017.
28. Sandwich structural components in passenger rail cars, https://www.hexcel.com/site/search/?search=passenger+rail+cars.
29. M. Robinson, E. Matsika and Q. Peng, "Application of composites in rail vehicles," In *21st International Conference on Composite Materials*, Xi'an, 20–25th August 2017.
30. D. Y. Jeong, D. C. Tyrell, M. E. Carolan and A. B. Perlman, "Improved tank car design development: Ongoing studies on sandwich structures," In *Proceedings of 2009 ASME Joint Rail Conference*, Pueblo, Colorado, USA, March 3–5, 2009.
31. M. E. Carolan, D. Y. Jeong, A. B. Perlman and Y. H. Tang, "deformation behavior of welded steel sandwich. panels under quasi-static loading," In *Proceedings of the ASME/ASCE/IEEE 2011 Joint Rail Conference*, Pueblo, Colorado, USA, March 16–18, 2011.
32. M. Carolan, B. Talamini and D. Tyrell, "Update on ongoing tank car crashworthiness research: Predicted performance and fabrication approach," In *Proceedings of 2008 IEEE/ASME Joint Rail Conference*, Wilmington, Delaware, USA April 22–23, 2008.
33. H. Ko, K. Shin, K. Jeon and S. Cho, "A study on the crashworthiness and rollover characteristics of low-floor bus made of sandwich composites," *Journal of Mechanical Science and Technology*, vol. 23, no. 10, pp. 2686–2693, 2009.

34. Hybrid and Electric Vehicles the Electric Drive Delivers, April 2015. http://www.ieahev.org/assets/1/7/Report2015_WEB.pdf.
35. https://www.newenglandwheels.com/.
36. H. Ning, G. M. Janowski, U. K. Vaidya, and G. Husman, "Thermoplastic sandwich structure design and manufacturing for the body panel of mass transit vehicle," *Composite Structures*, vol. 80, no. 1, pp. 82–91, 2007.
37. D. Dawson, "A clean technology for clean, zero-emissions buses, Monocoque composite body designed to support all bus loads," *Composites World*, 10/1/2018.
38. https://www.metyx.com/automotive/.
39. "Design composite TOP-series for commercial vehicles," https://www.design-composite.com/en/lightweight-construction/commercial-vehicles.
40. "Light Weight Freight Truck Body," https://www.holycorepanel.com/truck-body/freight-truck-body/light-weight-freight-truck-bodys.html.
41. "Vixenite™ Composite Panels," https://vixencomposites.com/products/.
42. "Strick® Heads To NPTC…New FRC Trailers In Tow," https://www.stricktrailers.com/.
43. "Advanced Molded Structural Composite (MSC) Technology," https://wabashnational.com/brands/wabash-national/wabash-product-portfolio/wabash-refrigerated-vans/msc-reefer.
44. "Lightweight honeycomb production technology for transportation," https://www.engineerlive.com/content/econcore-0.
45. "Fiberglass to Carbon Fiber: Corvette's Lightweight Legacy," https://media.gm.com/media/us/en/gm/news.detail.html/content/Pages/news/us/en/2012/Aug/0816_corvette.html.
46. C. J. Cameron, "Design of Multifunctional Body Panels in Automotive Applications," Licentiate Thesis, Stockholm, Sweden, 2009.
47. S. Milton and S.M. Grove, "Composite Sandwich Panel Manufacturing Concepts for a Lightweight Vehicle Chassis," ACMC, University of Plymouth, https://citeseerx.ist.psu.edu/viewdoc/download?doi=10.1.1.200.8212&rep=rep1&type=pdf.
48. M. Kriescher, S. Brückmann and G. Kopp, "Development of a lightweight car body, using sandwich-design," *Transport Research Arena*, Paris, 2014.
49. H.C. Davies, M. Bryant, M. Hope, and C. Meiller, "Design, development, and manufacture of an aluminum honeycomb sandwich panel monocoque chassis for Formula Student competition," *Proceedings of the IMechE Part D: Journal of Automobile Engineering*, vol. 226, pp. 325–337, 2011.
50. G. Savage, "Composite Materials Technology in Formula 1 Motor Racing, Honda Racing F1," Composite Materials Technology in Formula 1 Motor Racing, Honda Racing F1, July 2008.
51. "RAY-CORE SIPs ™ - The Better Structural Insulated Panels," https://raycore.com/sips-structural-insulated-panels.
52. "KS1000RW Kingspan Insulated Composite Panels," https://www.steelroofsheets.co.uk/products/ks1000rw-composite-panels/.
53. K. Simon, M. Weinfeld, T. Moore and C. Weincek, "Structural Insulated Panels (SIPs)," https://www.wbdg.org/resources/structural-insulated-panels-sips.
54. M.R. Legault, "Composites for builders: Establishing structural foundations," *Composites World*, 3/10/2017.
55. S. Bida, F. Aznieta, A. Aziz, M. Jaafar, F. Hejazi, A. Nabilah, "Advances in Precast Concrete Sandwich Panels toward Energy Efficient Structural Buildings," www.preprints.org Posted: 8 October 2018.
56. "ArmaPET™ boosts sustainable building, providing a higher degree of freedom in design, while contributing to improved safety and comfort," https://local.armacell.com/en/armapet/markets/infrastructure/.

57. T. Keller, C. Haas and T. Vallée, "Function-Integrated GFRP Sandwich Roof Structure," Asia-Pacific Conference on FRP in Structures (APFIS), 2007.
58. K. Jimison, R. Moreau, "SPS Bridge Decks," https://www.spstechnology.com/.
59. C.S. Cai, A. Nair, S. Hou and M. Xia, "Development and Performance Evaluation of Fiber Reinforced Polymer Bridge," https://www.ltrc.lsu.edu/pdf/2014/fr_472.pdf.
60. J. Cotter, Balsa cored bridge, https://www.materialstoday.com/composite-applications/news, November 6, 2009.
61. "Bridge Decking Product Lines," www.compositeadvantage.com.
62. A. Manalo and S. Douglas, "Fibre composite sandwich beam: An alternative to railway turnout sleeper," In *Southern Region Engineering Conference*, Toowoomba, Australia, Nov. 2010.
63. A. Manalo, "Fibre reinforced polymer composites sandwich structure: Recent developments and applications in civil infrastructure," *Materials Science*, 2013. https://eprints.usq.edu.au/24549/7/Manalo_%20IRCIEST_2013_PV.pdf.
64. H. Fang, Y. Mao, W. Liu, L. Zhu and B. Zhang, "Manufacturing and evaluation of Large-scale composite bumper system for bridge pier protection against ship collision," *Composite Structures*, vol. 158, pp. 187–198, Dec 2016.
65. P. Qiao, M. Yang, A. Mosallam and G. Song, "An Over-Height Collision Protection System of Sandwich Polymer Composites Integrated with Remote, Monitoring, for Concrete Bridge Girders," June 30, 2008, http://worldcat.org/arcviewer/1/OHI/2009/04/30/H1241104309097/viewer/file1.pdf.
66. "Radome framework with orange different shadow," http://www.radome.net/tl.html.
67. S. Papadopoulos, "Electromagnetic Modelling of Dielectric Geodesic Radomes using the Finite Difference-Time Domain Method," Ph. D. Thesis, University of London, London, UK. https://ethos.bl.uk/OrderDetails.do?uin=uk.bl.ethos.271150.
68. "Dielectric constant as a function of core density for several honeycomb types," http://www.hexcel.com/.
69. "Dielectric properties of a quartz honeycomb," http://ultracorinc.com/.
70. "Dielectric properties of PMI Foams," https://www.rohacell.com/product/rohacell/en/.
71. "Applications of sandwich panels as patient support for inspections," https://www.rohacell.com/product/rohacell/en/.
72. "Curved sandwich panels made by PMI foam and carbon fiber composite skins," https://www.rohacell.com/product/rohacell/en/.
73. S. J. Lee and I. S. Chung, "Optimal design of sandwich composite cradle for computed tomography instrument by analyzing the structural performance and X-ray transmission rate," *Materials*, vol.12, no. 2, p. 286, 2019.
74. H. Mason, "Composite sandwich panels enable flexibility in medical table design," *Composites World*, Sept 2020. https://www.compositesworld.com/articles/composite-sandwich-panels-enable-flexibility-in-medical-table-design.
75. G. Aravamudan, "Radiolucent Patient Table," US Patent 9282938, March 15, 2016.
76. S. J. Lee, M. W. Kim and J. C. Kim, "Manufacturing process of sandwich cradle for the computed tomography medical instrument," In *18th International Conference on Composite Materials*, Jeju Island, S. Korea, August 21, 2011.
77. A. Cowan, "Sound Transmission Loss of Composite Sandwich Panels," M. S. Thesis University of Canterbury, New Zealand, 2013.
78. N. Sui, X. Yan, T. Huang, J. Xu, F. Yuan, and Y. Jing, "A lightweight yet sound-proof honeycomb acoustic meta-material," *Applied Physics Letters*, vol. 106, no. 17, p. 171905, 2015.
79. V. I. Erofeev and D. V. Monich, "Sound insulation properties of sandwich panels," *Materials Science and Engineering*, vol. 896, p. 012005, 2020.

80. A. Wawrzynowicz, M. Krzaczek and J. Tejchman, "Experiments and FE analyses on airborne sound properties of composite structural insulated panels," *Archives of Acoustics*, vol. 39, no. 3, pp. 351–364, Mar 2015.
81. H. Meng, M.A. Galland, M. Ichchou, F.X. Xin and T.J. Lu, "On the low frequency acoustic properties of novel multifunctional honeycomb sandwich panels with microperforated faceplates," *Applied Acoustics*, vol. 152, pp. 31–40, 2019.
82. "ACH Noise Solutions," http://www.panelesach.co.uk/assets/documentacion/C-PA-V5.pdf.
83. A. Soltani, R. Noroozi, M. Bodaghi, A. Zolfagharian and R. Hedayati, "3D printing on-water sports boards with bio-inspired core designs," *Polymers*, vol. 12, no. 1, p. 250, 2020.
84. "Snowboard Construction," http://www.mechanicsofsport.com/snowboarding/equipment/snowboards/snowboard _ construction.html.
85. "The Ultra-Light Free-Touring Performance package," https://www.dynastar.com/mythic-skis.
86. "Production of Kayaks and Canoes," https://www.taheoutdoors.eu/index.php/ro_en/production-of-kayaks-and-canoes/.

10 Sandwich Composite Damage Assessment and Repairing

Wenguang Ma

Russell Elkin

CONTENTS

10.1 Detection and Estimation of Defects and Damages 428
    10.1.1 Personal and Visual Inspections 428
    10.1.2 Inspect by Instruments 429
    10.1.3 Damage Estimation 431
10.2 Different Type of Damages 432
    10.2.1 General Clarifications of the Damages 432
    10.2.2 Detail Damages of Special Products 433
10.3 Repairing Procedures and Methods 433
    10.3.1 Repair Plan Design and Option Selection 433
    10.3.2 Damage Cleaning 434
    10.3.3 Repairing Materials and Preparation 435
    10.3.4 General Repair Processes 437
        10.3.4.1 Repair of Type A Damage 437
        10.3.4.2 Repair of Type B Damage 438
        10.3.4.3 Repair of Type C Damage 439
    10.3.5 Special Repair Processes 441
        10.3.5.1 Surface Repair 441
        10.3.5.2 Brief or Short-Term Repair 441
        10.3.5.3 Structural Repair 441
        10.3.5.4 Vacuum Bagging Pressure Curing 443
        10.3.5.5 Repair of Structure Face with Cracks or Surface Defects ... 443
        10.3.5.6 Re-bonding Delaminated Skin to a Core by Resin Injection 443
        10.3.5.7 Vacuum Infusion Bag Pressure Curing Repair Technology 444
        10.3.5.8 Temperature Control During Repair Curing 445
10.4 Quality Inspection During and After Repairing 446
References 446

DOI: 10.1201/9781003035374-10

Fiber-reinforced composite sandwich structures have been increasingly adopted for their high stiffness-to-weight and strength-to-weight ratios. These structures can be optimized so that each element operates near its material limit, which results in a structure with a very high ratio of bending stiffness to weight. However, these structures undergo various damages during service, and depending upon the damage type, location, size, and severity, appropriate repair techniques need to be performed. Any permanent repair technique adopted has to restore the full integrity of the damaged composite sandwich structure. The damage can be limited to one skin (Type A), to one skin and the core (Type B), or to both skins and the core (Type C). The approach used in performing the repair is critical to ensure that the strength and stiffness of the structure are restored.

Various repair configurations such as overlap patch, scarf repair, and step repair are adopted for the repair of the sandwich structures, and among them, the scarf repair technique has provided greater efficiency. Apart from repair configurations, the production technique also plays an important role in improving the efficiency. Hand layup with the vacuum consolidated scarf repair technique is widely adopted for marine structures. The hand layup alone generally results in an inconsistent repair, low volume fraction, and high percentage of voids within the bond line that lead to reduced load-carrying capabilities of the repaired component. Greater care has to be taken to produce a good quality repair. Combining scarf repair configuration and the modern lamination technique of vacuum-assisted infusion, vacuum-assisted resin-infused scarf repair has been adopted to repair the Type B and C damages of the sandwich structure.

The overall processes of sandwich composite repairs are displayed in Figure 10.1. The first step is damage assessment. Some damage to composites is obvious and easily assessed, but in some cases, the damage may first appear quite small, although the real damage is very much greater. Impact damage to a fiber can appear as a small dent on the composite surface, but the underlying damage can be much more extensive. The decision to repair or scrap is determined by considering the extent of repair needed to replace the original structural performance of the composite. Other considerations are the repair costs, the position and accessibility of the damage, and the availability of suitable repair materials.

Easy repairs are usually small or do not greatly affect the structural integrity of the component. These repairs are made by the simple guidelines indicated for sandwich panels. Complex repairs are needed when the damage is extensive and needs to replace the structural performance of the component. The best choice of materials would be to use the original fibers, fabrics, and matrix resin. Any alternative would need careful consideration of the service environment of the repaired composite, i.e., hot, wet, and mechanical performance. The proposed repair scheme should meet all the original design requirements for the structure.

Temporary repair can be performed when a composite repair is needed for components in use. Some repairs need the special equipment of the workshop, and some form of improvised repair is needed to return the component to a suitable repair workshop. A temporary repair, usually in the form of a patch, can be fixed to the component. Usually, a "belt and braces" approach is taken to ensure safety until the component can be repaired at a later date.

Sandwich Composite Damage Assessment

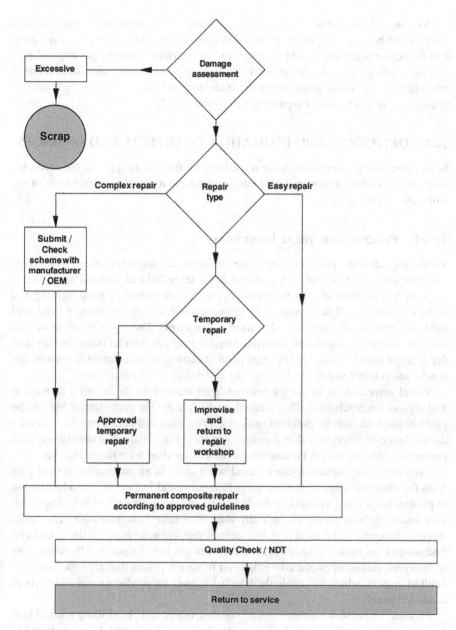

FIGURE 10.1 Overall processes of sandwich composite repair.[1]

After thoroughly inspecting and evaluating the damage, a repairing plan should be proposed by following a theoretical analysis and general guidelines. Repair operations should be carried out in controlled workshop areas to ensure high-quality repairs. Temporary enclosures are used for field repairs. Good housekeeping and attention to repair detail will ensure success.

For comprehensive inspection of repaired parts, a number of nondestructive tests (NDTs) can be used. The inspector should examine the quality of the repaired area, and particular attention should be given to the interface between the original part and the repaired area. In this chapter, damage inspection and estimation, damage cleaning and preparing, different repairing techniques based on damaging types and grades, and inspection after repairing will be presented.

10.1 DETECTION AND ESTIMATION OF DEFECTS AND DAMAGES

Before discussing damage assessment methods for the failure types of the sandwich structures, let's take a look at what tools are available to determine the extent of the damages or defects.

10.1.1 PERSONAL AND VISUAL INSPECTIONS

Visual inspection is a primary method for the in-service inspection of sandwich composite structures. It is relatively fast and has a large field of view. General visual inspection is performed with the naked eye under conditions of good lighting and surface cleanliness. The sensitivity of inspection is enhanced by using a hand-held light and viewing the surfaces also from a low angle. The experiences show that visual inspection is capable of detecting impact damages with an initial impact dent depth larger than 0.5 mm. On the other hand, it cannot detect delaminations and disbonds, and it is not suited for defect sizing and defect depth estimation.

Visual inspection is by far the most utilized method for discovering damage to a fiberglass sandwich boat. This method works fine for the part exterior but can be problematic with heavily outfitted interiors. The visual inspection can be tricky for the beginner, as fiberglass skin damages can look worse than it is structurally, and conversely, the extent of delamination can be greater than what meets the eye.

Remote visual inspection can be used when there is an interrupted optical path from the observer's eye to the test area. Remote visual inspection includes the use of photography, video systems, and robots. The endoscope is a small instrument that uses visible light to create an optically magnified image on a monitor. The endoscope technique can be used to detect defects that are not visible to the naked eye. Endoscopes are optically sensitive, and their images are obvious and therefore easy to interpret. Endoscopes are also employed to access places that are otherwise difficult to inspect, which has made them valuable tools for producers and end-users of sandwich panels.[2]

Personal inspection also includes minor destructive tests by drilling a small hole or cutting out a small area to determine the damages. Suspected delaminations and water intrusion can be confirmed. In some cases, it may be necessary to take hole-saw samples for resin content tests or even larger samples to test mechanical properties.

Tap testing is a common procedure for quantifying suspected delaminations not obvious by visual inspection. It includes tapping on the sandwich structure with a small metal object or plastic hammer and listening for changes in the sound. Not any object or hammer will be suitable. Typically, areas of delamination will have a "lower-pitched" response, corresponding to a lower natural frequency of the local structure. This technique can also reveal disbonds and/or delaminations between sandwich skins and cores, although only on the side being tapped. Tap testing is a subjective technique that relies on a trained surveyor's ear to differentiate between actual damage, delaminations, and inherent structural discontinuities.[3]

To apply a more scientific approach to the ancient art of "sounding," battery-driven electromagnetic hammer has been developed to tap test a composite or sandwich structure. The instrumented tap hammer is a light-weight handheld device (about 0.5 kg) that uses a battery-driven solenoid hammer with a force sensor (accelerometer) built in the hammer tip.[4] Practically, the time during which the hammer is in contact with the surface of the test part is measured. This contact time will increase in areas with defects such as disbonds that lower the local contact stiffness of the part. The instrument is a low-cost, couplant-free inspection unit for smaller areas where damage is suspected; for global inspection of large surface areas, it is less suited because of its spot measurement performance. Impact damages are generally well detectable; only a few minor impacts were missed in the evaluation. The detectability for delaminations and disbonds, on the other hand, was varying and not always consistent. Furthermore, a varying skin thickness (e.g. lay-up differences) or the presence of back-up structure can influence the tap tester response. The capability for defect sizing is limited, and the technique is not suited for defect depth estimation. The newer model with LCD display that shows the measured contact times together with a separate monitoring unit may have an increased detection performance for in-service defects.

10.1.2 INSPECT BY INSTRUMENTS

(Refer to Section 6.8 No Destructive Tests for more details.)

Pitch-catch method is for detecting the damages of the sandwich structures because of its relatively low cost and its multimode inspection capabilities (pitch-catch, mechanical impedance, resonance). The couplant-free pitch-catch (PC) technique is proved to be the most promising inspection mode. In the evaluation, however, the method showed a limited detection performance for in-service defects; the detectability was also quite variable. Impact damage is the defect type best detectable with the method, but not, as might be expected, the different disbond types. Although couplant-free, the pitch-catch method is not well suited for global inspection of large surface areas (because of its spot measurement performance), but more for the verification of suspect areas. More detailed information can be found in Section 6.8 No Destructive Tests.

Handheld UT camera is a handheld ultrasonic imaging camera for fast and real-time UT inspection. It produces a high-resolution C-scan image over an area of the specimen utilizing an array of 120×120 piezoelectric sensing elements. The handheld UT camera can indeed be used for relatively fast and real-time UT inspection of smaller areas where damage is suspected. For global inspection of large surface areas, however, the camera is less suited because of the limited field of view (about 1 square inch). A limitation, inherent to ultrasonic testing, is that a couplant is required between the camera and test part. The detectability for defects is generally good, especially when scanning the camera over the test specimen. For honeycomb structures, the camera was somewhat less successful in the evaluation: some disbonds and impacts were only detectable with limitation.

The ultrasonic phased array (UT-PA) method is a special UT method that makes use of transducers consisting of multiple ultrasonic elements that can each be driven independently. The PA transducers can have a different geometry (e.g. linear, matrix, and annular), and the PA beams can be steered, scanned, swept, and focused electronically. Ultrasonic coupling between transducer and test part was accomplished by applying a water film between the wedge and test part; the water is herewith guided through two small-diameter holes machined in the wedge. UT-PA is very suitable for the in-service inspection of the sandwich composites. The handheld probe can be used for relatively fast and real-time UT inspection of larger areas. The detectability for defects such as delaminations, disbonds, and impact damages is excellent. The UT-PA technique is however not couplant-free and that can be a limitation for in-service use. Also, careful scanning is necessary in order not to damage any surface protection system present (scratches can occur).

UT-PA dry-coupling roller probe employs a UT-phased array probe housed within a rubber-coupled and water-filled wheel probe. The probe can be used without a couplant, but generally, a fine water spray on the test part is used for optimum coupling. The larger probe is meant for flat surfaces, the smaller one can also be used on slightly curved parts. The handheld roller probe together with a 7-axis scanning arm can be used for fast and real-time UT inspection of relatively large areas. The dry-coupling roller probe is very suitable for the in-service inspection of composites, being a special UT-PA technique. The detectability for defects such as delaminations, disbonds, and impact damages is excellent.

Shearography is an optical method, based on speckle interferometry, for the noncontact measurement of out-of-plane deformations of a material surface. Digital image processing of the data is further done to enhance the defect presentation. Thermal load can be applied by heating lamps, and vacuum load can be applied by connecting a vacuum hood which sucks directly to the surface to be inspected. Shearography is a relatively fast, noncontact technique that requires no coupling or complex scanning equipment. Because of the optical technique, the specimen should not have a shiny surface. The inspection time is largely determined by the limitations of the field of view. The impact damages were readily detectable (including the nonvisible impacts) but generally undersized when compared to the baseline UT results. The detectable defect size decreases with increasing defect. Shearography can be used with limitation for defect sizing, but the technique is not suited for defect depth estimation. Shearography inspection seems most promising for the inspection of sandwich structures.

Thermography is a relatively new technique for evaluating the integrity of sandwich composite laminates over large areas and uses a high-precision infrared camera to measure small temperature differences over the area of interest. Typically, the area of concern is heated locally, and voids or trapped moisture show up as temperature differences as the subject area cools down. It's very easy to mark suspected areas on the laminate surface based on the image seen on the camera. Like other investigative tools, it should not be used as the sole source for identifying a problem. Thermographic techniques are well applicable to composite materials because of their relatively low thermal conductivity, which implies a slow lateral heat flow with closely spaced isotherms, resulting in a good defect resolution. The detectable flaw size is in general larger than the depth of the flaw. The thermography is a fast, global, and noncontact method that requires no coupling or complex scanning equipment. Most impact damages were readily detectable, except for some smaller and nonvisible impacts. The detectability for disbonds was somewhat better, such as skin-to-core disbonds were readily detected in the evaluation. Thermography can be used for defect sizing, but the technique is not suited for defect depth estimation. The thermography seems promising also for the inspection of water ingress in composite structures.

Moisture meter is the most commonly used tool to detect damage to sandwich composites. A moisture meter measures the radio frequency (RF) power loss caused by the presence of water. The moisture meter is often used to detect moisture, which provides a comparison of NDI testing equipment. With fiberglass, meter readings will always be on a "relative" scale, and destructive testing (i.e., cutting open the laminate) is required to get an absolute reading. It should be noted that laminates with wood core will often read as "wet" when using moisture meters designed to inspect composites. This is due to the fact that dry wood still has a moisture content of 6%–14%. The nearby presence of metal can also skew the result.

Radiography, often referred to as X-ray, is a very useful NDI method because it essentially allows a view into the interior of the part. This inspection method is accomplished by passing X-rays through the part or assembly being tested while recording the absorption of the rays onto a film sensitive to X-rays. Since the method records changes in total density through its thickness, it is not a preferred method for detecting defects such as delaminations that is normal to the ray direction. It is a most effective method, however, for detecting flaws parallel to the X-ray beam's centerline. Internal anomalies, such as delaminations in the corners, crushed core, blown core, water in core cells, voids in foam adhesive joints, and relative position of internal details, can readily be seen via radiography. Most composites are nearly transparent to X-rays, so low-energy rays must be used. Because of safety concerns, it is impractical to use around aircraft. Operators should always be protected by sufficient lead shields, as the possibility of exposure exists either from the X-ray tube or from scattered radiation. Maintaining a minimum safe distance from the X-ray source is always essential.

10.1.3 Damage Estimation

The defects in sandwich composites could come from production, transportation, and installation, or during use and maintenance, and by incidence. The damages in production include product design errors, process errors, human factors, core degassing or core defects, curing problems, and repairing errors.

Damages during operation include overload, external shocks, fatigue, lightning strikes, effects of environmental factors (temperature, humidity, air pressure, and changes in phase state of matter (freeze/thaw)), creep, and accident. The damages caused by the above reasons include delamination, peeling of the skin-core bond, skin cracking, water vapor entry, core cracking, fiber and resin matrix peeling, etc.

The damage could be very light so cannot be detected visually on both surfaces, invisible from the front surface, and visible on both sides. If the damage is very light, so it could be in allowable damage limit and not affect the product function. However, a product should have a critical damage point (CDP). It has to be repaired if the damage is over the CDP, otherwise, just continue using it. Repairing on time is important if the damage is over the CDP because the repairing can guarantee the product quality, prevent accident, improve service life, and increase market credibility.

10.2 DIFFERENT TYPE OF DAMAGES

10.2.1 GENERAL CLARIFICATIONS OF THE DAMAGES

Damage to sandwich composite structures can be assigned to the three groups generally defined and can involve various mechanisms. Type A damage generally involves matrix cracking, fiber breakage, and delaminations in the skin. The damage may or may not extend through the full thickness of the skin. Type A damage can also include debonding of the skin from the core. Type B damage involves Type A damage to one skin combined with crushing or shear cracking of the core. Type C damage involves the same damage mechanisms as Type B except that both skins are affected. Type C damage can fully penetrate the sandwich structure as shown in Figure 10.2.

While some instances of delamination and core debonding may be repaired through simple methods such as injecting a suitable adhesive or resin into the damaged area, damage to sandwich structures usually requires removal and replacement of the affected material. One of the primary reasons for removal of all damaged material is that damage will tend to grow under subsequent loading. Also, the detrimental effects of water ingress can be a particular concern and cause additional damage growth. Before any material is removed, it is necessary to know the extent of the damage, determined by the NDT techniques described above. In many situations, the most reliable method is removal of all damaged material starting at the center of the damaged region, working outwards until sound material is encountered.

Most damage to fiber-reinforced composites is a result of low velocity and sometimes high velocity impact. In metals, the energy is dissipated through elastic and plastic deformations and still retains a good deal of structural integrity. While in fiber-reinforced material, the damage is usually more extensive than that seen on the surface.

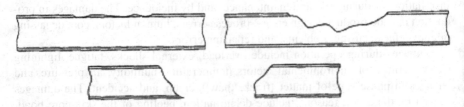

FIGURE 10.2 Damage Type A (a), Type B (b), and Type C.

Sandwich Composite Damage Assessment

10.2.2 DETAIL DAMAGES OF SPECIAL PRODUCTS

Punctures are the least mysterious types of damage we see in sandwich structures. Since punctures are clearly visible, the next question is how much the delamination extends beyond what is visible on the exterior or to the interior that may not be accessible for inspection. This is a good time to use the laminate tapping technique, marking suspected damaged areas as you go. The boundary of the damaged area should be confirmed by drilling a small hole since this area is going to be repaired anyway.

Delaminations within the laminate typically can't be diagnosed visually, although sometimes the panel will be "softer" when loaded. Tap testing or ultrasonic evaluation is required to quantify suspected delamination damage. Knowledge of laminate construction and interior structure enhances the value of NDT for laminate delamination detection. Before investigating the extent of delaminations or voids in a laminate, it is necessary to establish what we consider to be an "allowable" delamination or void size.

If the structure has internal stiffeners, fracture or debonding away from the primary laminate must be considered. Internal stiffeners "attract" loads and therefore are a location for stress concentrations. The bad news is that these stiffeners are not always accessible for visual inspection. Often, stiffener failure is revealed by stress cracks seen on the opposite side of the supported panel.

Water intrusion into the core of a sandwich laminate is difficult to casually observe. In marine, boats with severe enough water intrusion affect performance and the boat floating. A lot of times, water intrusion is discovered when holes are drilled to install new hardware. Indeed, poorly bedded hardware installation or panel edge close out is very often the cause of water intrusion. Saturated decks may be "spongy" to walk on and hull panels may flex more than usual. If the laminate is finished with gel coat, stress or "spider" cracks can be a sign of damage but may also be limited only to the gel coat itself. Gel coats fail at strain levels lower than the structural portion of the laminate, so the stress cracks serve as an indicator of potential future problems. Because gel coat stress cracks are by definition on the laminate surface, visual inspection is the most effective technique. Sometimes, a dye penetrant can be used to enhance visual inspection.

Most manufacturing defects can be uncovered through careful visual inspection. This is best done during key stages of construction before structural elements are hidden by outfitting. Conscientious builders understand that standardized, documented inspections can minimize future warranty claims and enhance brand identity. An infrared inspection of newly molded hulls can highlight areas where there are voids under the gel coat, which is a potential warranty item once the boat leaves the factory.

Heat damage is a local fracture with separation of surface plies. Its effect on the mechanical performance depends on the thickness of the part.

10.3 REPAIRING PROCEDURES AND METHODS

10.3.1 REPAIR PLAN DESIGN AND OPTION SELECTION

The repair philosophy for damaged sandwich composite structures depends greatly upon the particular component and the extent of the damaged incurred (for example, whether just one facing has been punctured or both). Since composite structures are employed in different industries with different design philosophies, the repair concepts include a wide range of approaches, ranging from highly refined and structurally efficient flush patch repairs to an external attached metal or composite patches.

One should consider damage type, allowable damage limits, repairable limit, repair type, access to the repair, surface appearance requirement after repairing, repair proximity limitations, and service life after repairing. It is important to know the product's basic design and manufacturing details, as well as what materials, tools, equipment, process guidance, and assistance are available to complete the repair, what kind of technical personnel, quality inspection, and nondestructive testing techniques are available, and what quality assurance and testing procedures should be followed.

Repair techniques used in the aerospace industry undergo extensive development and testing before implementation. Some of the approaches used in the aerospace industry can be adapted to other markets, but the materials used differ significantly. Aerospace composite laminates are usually autoclave-cured and, as a result, are of high quality (low porosity) having a high fiber volume fraction (>60%) and very low void content (<2%). Nonaerospace structures are usually made from E-glass fibers with polyester or vinyl ester resin. The lamination techniques are hand lay-up or vacuum bagging, and the resulting fiber volume fraction could be as low as 20%, and void content could exceed 10%. Core materials are different. Inferior laminate quality may affect the repair integrity. For example, adhesive shear strength, an important property in the bonded repair of composites, is dramatically reduced by the presence of porosity. A 5% void content reduces the shear strength by 20%.

After establishing a preliminary repair plan based on the above principles, some works should be done to confirm it. These works include making samples for static tests, such as bonding strength, impact, and flexural test; using test results in theoretical analysis to verify the allowable performance the structure; predicting stress/strain distribution by using finite element analysis software, and finally verify and approve the repairing plan.

Before starting a repairing work, one should consider about quality assurance factors, which are the performance of the materials selected for repairing, repairing layer thickness, patching procedure, working environment, technical staff training, and qualification assessment that include repeated training to maintain their technical level and ensure the quality of repaired products through nondestructive testing methods.

10.3.2 DAMAGE CLEANING

Before repairing, the damaged skin and core should be cleaned, and all contaminated materials should be removed. Excessive moisture levels should be dried. The damaged skin and core can be removed with a router cutting slightly deeper than the face sheet thickness. The skin can be pulled away from the core with a plier or cut loose with a knife. The area of the core to be removed is then trimmed as shown in Figure 10.3. The section of the core to be removed is pulled away from the opposite side skin by pliers, and the surface is made smooth with abrasive paper. A router may also be used to remove damaged core. After removing the core, the area is

Sandwich Composite Damage Assessment

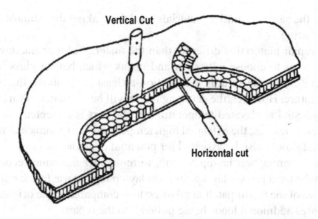

FIGURE 10.3 Removing damaged core with a knife.[5]

vacuumed and wiped with acetone or other chemicals. But the repair work must be done after the acetone is completely dry (at least 1–3 minutes). Acetone and other solvents should not be applied directly to core materials as they will readily absorb the chemicals, which are not as easily released, until such time as to interfere with the bond.

Liquids should be removed via absorption. When the liquid is not water, contamination must be removed from repair area, and then the repairing area must be cleaned and dried. Moisture and water usually exist in porosity cores, voids in skin and interfacing layer, delaminations, or honeycomb cells. Epoxy resin and adhesives can absorb water. Water also condenses on the surface of repair materials if improperly thawed after removal from freezer storage. Water and moisture present during the cure cycle can have negative consequences that include the disbonded face sheets, so the sandwich parts end up with larger damage than they started with, core tension failure, and porous bond lines to reduce strength and future potential for water ingress in service.

If the damaged product is contaminated by salt water, they must be thoroughly cleaned with fresh water and then given adequate natural drying time. Drying can be accelerated with a hot air blower, or a vacuum bag can be used to assist with exhaust drying. After drying, take a sample to determine the humidity. Generally, the humidity of the saturated composite material layer can reach 3%, and the repair work should be carried out when the humidity is equal to or less than 0.5%.

10.3.3 Repairing Materials and Preparation

The following items are important for choice of repair material when designing a composite repair:

1. To understand the original design criteria, operating temperature environment, material properties, and loads for the part. Reverse engineering a repair in composites is not as practical as it is in metals.

2. To use the same or similar materials used for making the damaged product for repairing.
3. If the repair material is different than the material of manufacture, a good guideline is to choose adhesives and resins which have a glass transition temperature in the saturated condition at least 50°F above the operating temperature. However, the deciding factor will be the material's mechanical ability tested at elevated temperature. If the part is subjected to sustained high temperatures, the extended high temperature performance of the repair material should also be assessed for potential degradation.
4. When designing wet lay-up repairs, to remember that unidirectional tape cannot be used for wet lay-up, and wet lay-up repairs use fabric only. Check stiffness of the repair patch in all directions compared to the original part to minimize additional loads being pulled into the repair.
5. Isophthalic unsaturated polyester and vinyl resin are suitable for repairing. Ortho-benzene type (universal type) is generally not suitable for repair. Epoxy resin is suitable for repairing the original parts that are epoxy resin products.
6. If possible, the reinforcement material used in the original product should be selected, especially when the product is under heavy load and is close to the design limit. If you cannot choose the same one, you can choose a similar fabric. When there is no similar material, you can choose multiple thin fabrics instead of heavy fabrics,

Pay special attention to the handling of the materials. Prepregs, film adhesive, and foam adhesives require refrigerated storage, need thawing prior to use, and have limited shelf life. Airline base maintenance facilities may have some commonly used prepreg materials available, but they are expensive to maintain inventory. Proper record keeping is essential to track out time to ensure the materials are all still within their shelf life. Resins and adhesives have a limited shelf life but are relatively cheap and readily available in small quantities. The original material of manufacture will usually offer the largest repair size. Wet layup repairs are usually heavier than prepreg repairs due to a higher resin content and fabric for tape substitution. This can be an important consideration for repair of balanced control surfaces.

Damaged core can be filled either with a foaming adhesive, a laminate, or a new core section bonded in place. The latter method is usually adopted as it best restores the properties of the sandwich structure. The first two approaches may be used where the damage is shallow and covers only a small area. Different approaches to this repair are required for Type B or C damage. The repair of Type C damage also depends on whether access can be gained from both sides. Differences arise in the method used to bond in replacement core. Air can escape through the open cells of the honeycomb or balsa as it is bonded in place, while with rigid foams, air can be trapped in the bond line leading to defects and a poor-quality repair. If the repair size is large enough, perforated sheets must be used.[5]

Sandwich Composite Damage Assessment

10.3.4 General Repair Processes

10.3.4.1 Repair of Type A Damage

The replacement of a skin for Type A damage is a straightforward procedure which is best accomplished using a scarf repair. An important aspect of such a repair is the scarf ratio, the angle of length vs. thickness to ensure good transfer of shear load in the bondline. The angle should be as low as possible; a ratio at least 10:1. Surface preparation is also critical. Provided sufficient care is taken, such a repair should have adequate strength and durability. Type A repair procedure is shown in Figure 10.4 and detailed below. In the following procedures, a layer is defined as one ply chopped strand mat (CSM) and one ply of woven roving.

1. Remove Damaged Material: The damaged skin is removed starting at the center, working downwards and outwards until the sound material is encountered.
2. Repair Preparation: If the damage is deeper than one layer, taper sand the surrounding skin.
3. Replace Skin: Replace the skin using the number of layers removed. Each successive layer is to be longer and wider than the previous layer. The amount of extra length will depend on the ply thickness. Apply one extra layer extending beyond all damage. A good rule is to increase this extension 2.5 times the distance used for the other plies.
4. Apply the resin, squeeze all air out for glass fabric, and make sure all fabric is saturated by resin.
5. Make sure resin is properly and fully cured before sanding and cleaning the repaired surface.

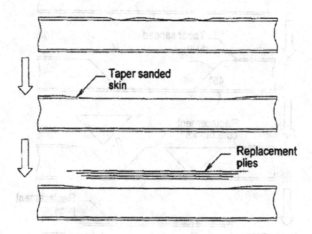

FIGURE 10.4 Method for the repair of Type A damage of foam core sandwich panel.[5]

10.3.4.2 Repair of Type B Damage

The repair of Type B damage of sandwich structures requires the replacement of one skin and the core. The repair starts from removing damaged skin from the center of the damaged region outwards until the sound material is encountered, then removing the exposed damaged core leaving the other skin intact.

Another modification is when installing the replacement core, use a paste adhesive designed to bond the core in the minimum amount of adhesive (bond line thickness 3 mm maximum). Make sure no voids should exist between the undamaged skin and the replacement core. The important procedures are shown in Figure 10.5. Applying resin and surface cleaning after curing are similar to the last section for the repair of the outer skin.

Two primary deficiencies were noted with the Type B repair technique introduced above. The first concerned preparing the core at a 45° angle which proved to be difficult. The second deficiency was entrapment of air between the replacement core and the existing skin during the bonding process. The modified techniques can be used to overcome these problems. In the modified Type B repair technique, emphasis was placed on simplifying the procedure. This was achieved by replacing the core in one section whenever practical and using 90° butt joints. To avoid entrapment of air when the replacement core was positioned, holes were drilled through the core at a spacing of between 50 and 100 mm. A problem associated with the 90° joins in the core was the difficultly in filling the bondline between the replacement and existing core. To overcome this, the repair was placed under a vacuum bag to draw adhesive up around the edges. A vacuum pressure of around 70 KPa (20 in Hg) was found to be sufficient, and the adhesive should have a reasonably short gel time (30–50 min). Also, the

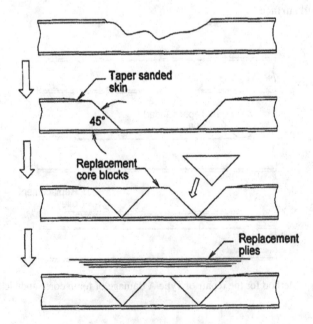

FIGURE 10.5 Method for repairing Type B damage to the sandwich panel.[5]

Sandwich Composite Damage Assessment

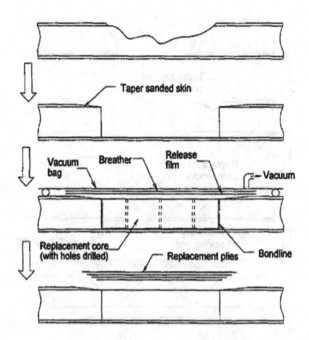

FIGURE 10.6 Modified method for the repair of Type B damage to the sandwich panel.[5]

correct amount of adhesive should be applied since significant bleed does not occur. Additionally, the adhesive should be applied only to the existing skin and core, not to the replacement core, as this prevents blockage of the holes drilled through the replacement core.

The procedures of the modified Type B repair method are shown in Figure 10.6. The major procedures are to prepare the foam core to an angle of 90°, sand the edge of the laminate to the proper taper, use a paste adhesive designed to bond the core (leave about 1 mm all round for the bondline), drill 3-mm-diameter holes through the core if necessary at 50–100 mm centers approximately, apply the correct amount of adhesive (calculated from the volume of the bondline) to the existing skin and core, apply a layer of perforated release film and breather over the repair area, position the vacuum bag over the repair area, sealing it to the surrounding structure, apply suitable vacuum until the adhesive has cured, and lean up area for laminating top skin. The remaining steps are the same as the procedures in the last section.

10.3.4.3 Repair of Type C Damage

The repair of Type C damage to sandwich structures requires the replacement of both skins and the core. The special repair procedure for Type C damage is shown in Figure 10.7. The other procedures are the same as the steps described above with the addition of laminating a backing plate where required. This will create a temporary mold surface and a proper seal against vacuum leaks.

The most difficult part of the repair was bonding the replacement core in position accurately without creating voids in the bondline. To avoid the requirement of a

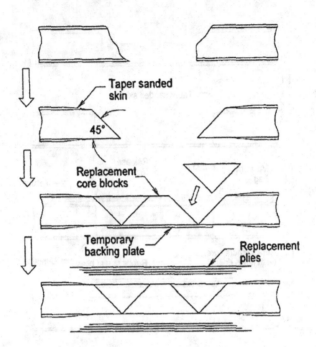

FIGURE 10.7 Method for the repair of Type C damage to the sandwich panel.[5]

backing plate, a lip was left in one skin against which the replacement core could rest. Holes were drilled through the replacement core into the bond line gap.[5] The holes should emerge near the bottom of the bond line gap, so air does not become trapped when the adhesive is injected. The spacing between the holes should be twice the core thickness. Bonding the core in place was then conducted in two stages using a caulking gun. First, a bead of adhesive was placed around the lip and the core placed in position. The adhesive was then allowed to cure to prevent the core from moving and excessive adhesive leaking during the next stage. The gap between the existing and replacement core was then filled with an adhesive by injecting it through the holes using a caulking gun. Following cure of the adhesive, the replacement skins can be laminated.

The modified Type C repair technique is shown in Figure 10.8, which is similar to the modified Type B repair techniques except for installing the core as described below:

1. Cut a piece of foam, allowing 1 mm all round for the bondline,
2. Drill 3 mm diameter holes through the core into the bondline gap. The spacing of the holes should be twice the core thickness,
3. Place a bead of adhesive around the lip and position the foam, forcing it down lightly,
4. After the adhesive has cured, inject adhesive into the bondline through the holes using a caulking gun. Clean up the area before the adhesive cures,
5. Laminate the skins on both sides, and then clean after curing.

Sandwich Composite Damage Assessment

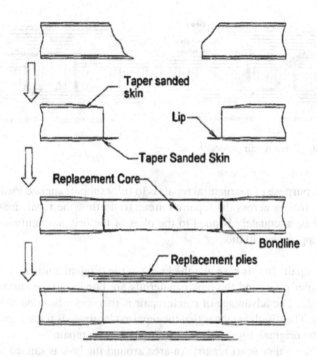

FIGURE 10.8 Modified method for the repair of Type C damage to the sandwich panel.[5]

10.3.5 SPECIAL REPAIR PROCESSES

10.3.5.1 Surface Repair
In this case, inspection has determined that the damage has not affected the structural integrity of the component. A cosmetic repair is carried out to protect and decorate the surface. This will not involve the use of reinforcing materials.

10.3.5.2 Brief or Short-Term Repair
It is often the case in service that small areas of damage are detected which in themselves do not threaten the integrity or mechanical properties of the component as a whole. However, if left unrepaired, they may lead to further rapid propagation of the damage through moisture ingress and fatigue. Simple patch-type repairs can be carried out, with the minimum of preparation, to protect the component until it can be taken out of service for a proper structural repair. Temporary repairs should be subject to regular inspection.

10.3.5.3 Structural Repair
If the damage has weakened the structure through fiber fracture, delamination, or disbonding, the repair will involve replacement of the damaged fiber reinforcement, and core in sandwich structures, to restore the original mechanical properties. Since a bonded-on repair constitutes a discontinuity of the original plies, and therefore a stress raiser, structural repair schemes normally require extra plies to be provided in the repair area. Structural repairs can be done as described in the last sections.[1]

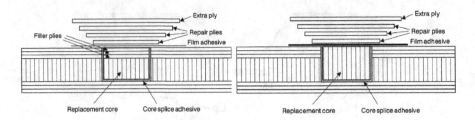

FIGURE 10.9 Patch repair process.[1]

The main purpose of a structural repair is to fully support applied loads and transmit applied stresses across the repaired area. To do this, the repair materials must overlap, and be adequately bonded to the plies of the original laminate. There are three basic approaches to this.

1. Patch repair: In this case, the thickness of the original laminate is made up with filler plies, and the repair materials are bonded to the surface of the laminate. The advantage of patch repair is that it can be done quickly and simply. The disadvantage is that the repaired laminate is thicker and heavier than the original. Figure 10.9 illustrates the patch repair.
2. Taper sanded or scarf repair: An area around the hole is sanded to expose a section of each ply in the laminate in this case as shown in Figure 10.10. Sometimes, one filler ply is added to produce a flatter surface. Taper is usually in the region of 30–60:1. Advantages of the method are that repair is only marginally thicker than the original, each repair ply overlaps the ply that gives a straighter, stronger load path, and good bonds can be achieved on the freshly exposed surfaces. However, the process is time consuming and needs a high-skill worker to achieve the work.
3. Step sanded repair: The laminate is sanded down so that a flat band of each layer is exposed, producing a stepped finish. Typical steps are 25–50 mm per layer. The advantage of the step repair is the same as taper-sanded repair, but it is extremely difficult to perform it.[1]

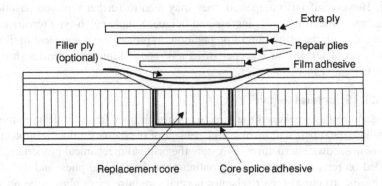

FIGURE 10.10 Process of scarf repair.[1]

Sandwich Composite Damage Assessment

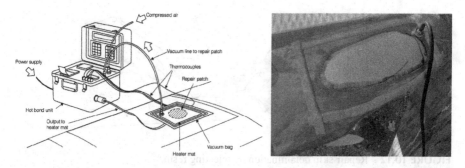

FIGURE 10.11 Vacuum bagging equipment for hand layup repair (left) and a real vacuum bagging in repair.[1]

10.3.5.4 Vacuum Bagging Pressure Curing

The repairing processes include hand layup by using dry fabrics and liquid resin or using prepreg fabrics and by vacuum infusion. After hand layup, the resin can cure in an open air condition at room temperature, in an oven at high temperature or under a vacuum bagging pressure at room or high temperature. Figure 10.11 shows a vacuum bagging equipment with a heating mat for bagging curing resin in hand layup repair and a real vacuum bagging curing repair.

10.3.5.5 Repair of Structure Face with Cracks or Surface Defects

Mask off an area approximately 75 mm larger on all sides than the area to be repaired. Then remove paint and/or raised or rough surface by sanding with 100 grit sandpaper followed by 400 grit sandpaper. Clean the surface with an acetone or other solvent-dampened clean white cotton cloth for repairing.

Cut two pieces of glass cloth so that the first will overlay the damaged area by 25 mm on all sides, and the second will be 25 mm larger still on all sides than the first. Mix an appropriate amount of the laminating adhesive and impregnate the pieces of glass cloth. Apply the two layers of impregnated glass cloth so that the first ply is centered over the damaged area, and the second ply overlaps the first on all sides. Cover the repair with a vacuum bag, apply pressure, and cure in accordance with the manufacturer's instructions. Remove the bag and sand to obtain a feather, smooth the surface, and paint as required.

10.3.5.6 Re-bonding Delaminated Skin to a Core by Resin Injection[7]

If the core material is firm and dry, re-bond the skin by injecting liquid resin between the skin and the core. If the core material is wet but still solid, re-bond the skin after the core has been thoroughly dried. This method involves drilling a pattern of holes through the skin to expose the core to air and heat and allow moisture to escape. When the core is dry, resin is injected under the skin, and the skin and core are clamped together until the resin cures. This method is useful if the delamination is small. The resin injection repairing procedures are described below:

1. Drill 5 mm diameter holes at 2.5 cm intervals, creating a pegboard-like pattern that extends several inches beyond the delaminated area as shown in

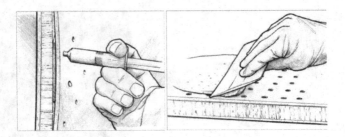

FIGURE 10.12 Repair skin delamination by injecting resin.[6]

Figure 10.12. The holes should penetrate the delaminated fiberglass skin without drilling into the core.

2. Inject the resin mixture under the skin through each of the holes by a syringe. The shortened tapered syringe tip will fit tightly in the holes. Fill any remaining voids and fair the surface with the resin.
3. Clamp the skin to the core when you are sure you have injected enough resin to bridge any gap between the skin and core until the resin cures. Clean up excess resin before it begins to gel. Allow the resin to cure thoroughly before removing clamps.

10.3.5.7 Vacuum Infusion Bag Pressure Curing Repair Technology

Although composite repair methods have been widely studied, there are still significant opportunities for less invasive, more effective, and more rapid repairs. Compared with traditional material removal and patching methods, injection methods are seen to be advantageous for both delamination in sandwich structures. In both cases, the repairs were minimally invasive and required only vacuum bagging along with low viscosity, room temperature curing resins.

Resin vacuum infusion repairs have previously been used for lightly loaded structures with small damage, delaminated monolithic structures, and sandwiches with a face-sheet disbond. Typically, an inlet hole is drilled at the center of the damage, and outlet holes are drilled at the periphery, before a very low-viscosity resin is injected to flow through the delaminated area. These repair methods have the potential to minimize weight gain and material removal, while retaining aerodynamic efficiency.

One example is to repair a delamination between skin and core of a sandwich structure. Based on the size and location of delamination in each situation, a pattern of evenly spaced 1.5 mm diameter holes were drilled through the outer skin, ensuring that outlet holes were a maximum of 30 mm from the nearest inlet across the whole delamination area. After drilling, separate strips of peel ply were placed over the inlet and outlet holes with sufficient space to then apply sealant tape between the inlet and outlet regions. Breather cloth was wrapped around the samples and over the outlet regions, connected to a vacuum port away from the sample itself. This was then encased in a vacuum bag and checked for any leaks. Low-viscosity resin was then inlet into the delaminated samples at room temperature. The vacuum bag and other auxiliary materials were removed after resin was cured completely. Clean the repaired surface for testing or reusing.

Sandwich Composite Damage Assessment

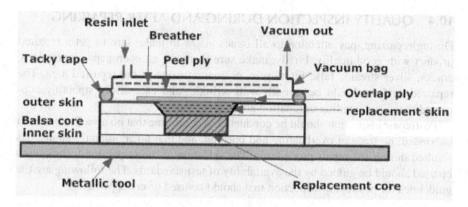

FIGURE 10.13 Repair Type B damage by vacuum infusion method.[8]

Damaged Type B sandwich panel can also be repaired by vacuum infusion after replacing core and scarfing damaged skin. One example shows that the damaged sandwich structure is abraded of extra resins at the defect corners, the outer face properly scarfed and cleaned with acetone. This outlines the procedure of repairing the damaged structure. The whole damaged structure is placed on the metallic tool, and the core of required defect length is replaced. Mild tolerance is left on both sides of the replaced core for resin fill.

The damaged scarfed upper skin is replaced with the layers of the same material with each successive layer longer than the previous layer as required for various scarf angle. Placing each successive layer 40 mm longer than the previous layer would provide the replacement skin for the damaged 3° scarf upper skin. One extra overlap layer is placed extending 20 mm beyond all damage. After building a vacuum system, the vacuum bagging procedure is continued. A liquid resin is introduced to the resin inlet. Resin flows from the inlet to outlet infusing the replaced core and skin. The repairing setup is shown in Figure 10.13.

10.3.5.8 Temperature Control During Repair Curing

A heat source such as an oven or autoclave will raise the complete part to cure temperature and so provide the most uniform temperature distribution throughout the repair. A heat source that just applies heat locally (i.e., electric heater blanket, heat lamp, and hot air blower) provides more variation in temperature distribution and is more susceptible to heat sinks. These effects are more pronounced at higher cure temperatures, and so, the risk of an unsatisfactory repair is greater; hence, as many thermocouples as possible are used and complete cure is supervised so corrective action can be taken if required.

If access is to one side only, the temperature gradient through thickness can be a problem. Attention should be paid to prevent overheat of both the repair and undamaged area. Sometime, insulation may be added under the heat blanket. Low-temperature cure materials for core plug installation should be used for surviving the repairing skin ply during curing.

10.4 QUALITY INSPECTION DURING AND AFTER REPAIRING

During repairing, pay attention to all issues below to make sure to get a repaired product with good quality. Firstly, make sure there are no open gaps, depressions, cracks, silver streaks, false interlayers, or contaminants in the repaired area. The repaired surface should be smooth with wrinkle, and excessively smoothly connected to the surrounding unrepaired area.

Postrepair inspection should be conducted to determine that no area has been left unbonded, no trace of overheating and burning, and that no additional damage has resulted during the repair procedures. The selection of the inspection technique to be used should be guided by the availability of test standards. The following are the guidelines for postrepair inspection that should be used where applicable:

1. Conduct visual inspection of the repaired area for obvious defects. Observe for signs of uncured adhesive. This can be determined by observation of the adhesive flash for lack of solidification or tackiness.
2. The degree of resin curing in the repair area should not be less than 90% of the value required by the manufacturer, which can be measured with a hardness tester.
3. Conduct a nondestructive inspection with one or more portable instruments.
4. If the repair area contains adhesive potting or core splice adhesive, an X-ray examination is recommended to ensure that all applicable areas have been filled.
5. If the repair area consists of thick laminates or multiple bondlines, the use of ultrasonic through transmission should be considered.
6. When standards for inspection are not available, the repair area may be compared to an unrepaired area of the sample construction. The inspector should be aware that the instrument readings may vary, however, due to any structural change caused by the repair.
7. Repeat the inspection scan using various instrument settings and verify the inspection results with another type of instrument if available.

REFERENCES

1. "Composite Repair," www.hexcell.com.
2. "Remote Visual Testing with an Endoscope," www.timemaxrostschutz.de/endoskopie.
3. L. Ilcewicz, L. Cheng, J. Hafenricher and C. Seaton, "Guidelines for the Development of a Critical Composite Maintenance and Repair Issues Awareness Course," U.S. Department of Transportation Federal Aviation Administration, February 2009.
4. "Damage Classification, Sandwich Structures and Solid Laminates - Aircraft Composite Honeycomb Sandwich Repairs," https://www.aircraftsystemstech.com/2019/06/damage-classification-sandwich.html.
5. R. Thomson, R. Luescher and I. Grabovac, "Repair of Damage to Marine Sandwich Structures: Part I - Static Testing," https://www.semanticscholar.org/paper/Repair-of-Damage-to-Marine-Sandwich-Structures%3A-1-Thomson-Luescher/957127cec41168e2f68f974c6b9575621fc6d024.

6. "Fiberglass Boat Repair & Maintenance," http://atlcomposites.com.au/icart/products/50/images/main/Fibreglass%20Boat%20Repair%20and%20Maintenance%20Manual.pdf.
7. R.S. Pierce, W.C. Campbell and B.G. Falzon, "Injection repair of composites for automotive and aerospace applications," *21st International Conference on Composite Materials*, Xi'an, 20–25th August 2017.
8. J. Palaniappan, S.W. Boyd, R.A. Shenoi and J. Mawella, "Repair efficiency of resin infused scarf, repair to marine sandwich structures," *ICCM-17 Proceeding*, Edinburgh, Scotland, July 27–31, 2009.

Index

3D-fabric cores xxiv
3D printing
   advantages and applicaitons 152
   alternate printing using two materials 153
   continous fiber in-nozzle impregnation 154
   continuous liquid interface production (CLIP) 155
   fused depsition molding (FDM) 152
   fused filament fabrication (FFF) 152
   introduction 151
   making honeycomb core by using rein with high aspect ratio fillers 154
   procedure by FDM method 152
   sandwich product made by dual materials 153
3D printing devleopment xx
3D printing for making sandwich composites 151

additive manufacturing (AM) for making sandwich composites 151
additives used with resin formula xxvii
adhesive and debonding/delamination modeling
   by cohesive element approach 308
   by connector element approach 311
   by surface separation approach 311
adhesives for dry lamination
   adhesive films made by EVA, EAA, LDPE, HDPE nad PP 129
   epoxy resin 128
   hot melt adhesive 128
   polyolefin film and typical physical, processing and mechanical properies 129
   polyurethane adhesive 128
   stage B epoxy film 129
   typical properites and processing condition of one epoxy film 130
   unsaturated polyester and vinyl resin with LPA 129
   urethane acrylate adhesive 128
advantages of sandwich structrues xv
air and water blast tests
   by shock tubes 222
   full scale explosion 225
   instroduction 222
air blast test full scale
   configuration and setup 226
   example 226
air shock tube test
   air pressure and velocity measurement 223
   damage progression 223
   facility 223
   maximum and permanent deformation 223

all thermoplastic sandwich composite post processes
   conductive infuse welding 180
   induction welding 180
   laser welding 181
   recycling by melting and reprocessing 181
   thermoforming 177
   ultrasonic welding 181
all-thermoplastic sandwich composites (all-TSCs)
   advantages 160
   concept and specialties 160
   construction 160
   cores - thermoplastic foams and honeycombs 162
   development 163
   expanded polypropylene bead foam board as core 164
   face materials 164
   fiberglass reinforced polypropylene board as core 163
   introduction 159
   natural fiber/pp board as core 164
   polypropylene (PP) foams as core 163
   postprocessing 177
   processing methods 169
   processing, thermoforming and recycling capacity 161
   properties of natural fiber board core 165
   skin (facing) materials with thermoplastics 160
   thermoplastics foam and honeycomb cores 160
   typical foam and honeycomb cores 163
   typical skin materials 161
all-thermoplastic sandwich compostes (all-TSCs) 159
all thermoplastic sandwich laminations
   curved mold and vacuum assistant for one step 3D forming 177
   film foaming or injection melt foaming 177
all thermoplastic sandwich post processing
   two steps heating thermoforming 179
aluminum face sheets
   mechanical and physical properties 89
aluminum honeycomb
   mechanical properties of corrugated honeycomb 46
   typical mechanical properties of flexible core 47
   typical properties 45

449

application in recreational boat building
  development 335
applications of sandwich composites 333
  acoustic barriers 399
  airplane and aerospace 347
  building and civil industries 373
  introduction xxviii, 334
  marine industry 334
  medical equipment construction 395
  radar radome construction 388
  sport and leisure products 413
  transportation 357
  wind energy industry 341
applications of sandwich composites in airplane
    and aerospace
  development history 347
applications of sandwich composites in airplane
    industry 348
  core materails for aircraft ingineering 348
  honeycomb sandwich structures for a
    aircraft 348
  metal honeycomb and facing materials for
    making flaps 350
  Nomex honeycomb, PEI and PMI cores for
    making airplanes 350
  sandwich structural components for Boeing
    and Airbus airplanes 350
  wind leading edge made from honeycomb
    sandwich structure 348
  works need to be done for making primary
    structures 351
applications of sandwich composites in Marine
    industry 334
  advantages 335
  boat hull structure of recreational boat 336
  commercial marine industry 339
  cruise liner buiilding applicaitons 340
  developments in commercial ship
    constructions 339
  examples of transitions from sandwich
    structure to solid section in boat
    building 336
  foam cored sandwich structures as buoyancy
    materials for submersibles 340
  history 334
  metal sandwich in military ship
    constructions 338
  millitary ship constructions 338
  nonmetal composite structures used
    for topside enclosures and other
    components in military ship
    constructions 339
  PVC foam cored FRP sandwich constructures
    for commercial boat building 340
  raw materials for recreational boat
    building 335
  recreational boat building 335

  recreational boat hull examples 337
  sandwich FRP construction for commercial
    boats 340
  types of commercail boats buit by sandwich
    composites 339
applications of sandwich composites in wind
    energy industiry 341
  conceptual design of wind turbine
    blade 343
  core materials for building wind turbine
    blades 343
  developments 341
  hybrid cores used for building wind turbine
    blades 343
  main requirements for wind burbine
    blades 341
  materials for building wind turbine
    blades 342
  reinforcement materials for building wind
    turbine blades 343
  structure of wind turbine blade built with
    sandwich structures 341
  vacuum asstisted resin transfer molding
    (VARTM) for building wind turbine
    blades 343
  wind burbine blade components 342
applications of sandwich composites in wind
    energy industry
  advantages of using sandwich structural webs
    of main spar 344
  buckling analysis used for designing main
    spar flange (spar cap) 344
  challenges for detecting defects and
    damages 345
  consider to make main spar web using
    sandwich structure 344
  fatigue and demage tolerance
    considerations 345
  FEA work approved to use sandwich structure
    making main spar web 344
  fiber-optic sensors used for structural health
    monitoring blades 345
  localized effects on fatigue strength of
    blades 346
  major issue - weak interfaces between core
    and facings 346
  nondestructive inspection (NID) for quality
    controlling 345
  safe-life design 345
  truss network in core for improving fatigue
    strength and damage tolerance 346
applications of sandwich structures in acoustic
    barrier constructions 399
  acustic separation for reducing sound
    transmission 405
  adding a layer of membrane meta-material for
    reducing sound transmission 401

Index

adding meta-material on both top and bottom sides of sandwich structure for reducing sound transmission 403
adding one or more membranes in honeycomb core for reducing sound transmission 405
air or sound insulation gap to core or between panels 407
behaviors of sandwich structure in low, meddle and high frequency ranges 401
coincidence sound wave 401
disadvantage of sandwich structure for blocking sound 399
double panel system with air gap for best sound proofing property 413
example of acoustic separation by using elastic material splitting the core 406
example of CSIP with different air or wool inserts 408
example of honeycomb core panel with one perforated facing for sound absorption 410
foam filled honeycomb core panel for sound insulation constructions 410
honeycomb sandwich panel with micro perforated facings 410
importances of sound insulation 399
introduce air void inserts, glass wool and layers, or connecting two panels by a wool layer 407
mass, stiffness and damping as essential parameters of reducing sound transmission 401
micro perforated panels and Helmholtz resonance absorption 410
optimizing sandwich structure for improving acoustic performance 399
perforated faceplate as sound absorbing structure 410
resonant transmission 407
sandwich structures for blocking different wave sound 401
self-coordination of wave fields 405
sound reduction index (R) 400
sound transmission loss (STL) 400
sound transmission mass law 399
sound transmission through sandwich structures 400
use damping core with perforated facings 412
applications of sandwich structures in aerospace 353
3D printed truss lattice core 354
application areas 353
example of vented truss lattice sandwich struture 354
future of aeronutic sandwich strucures 354

perforated, slotted or porous venting honeycombs 353
applications of sandwich structures in airplane and aerospace 347
applications of sandwich structures in building industry 373
advantages 373
composite sandwich panels for basement wall 378
composite structural insulated panel (C-SIP) for house construction 376
construction and advantages of C-SIP system 376
elements of SIPs building 375
foam cores for SIPs manufacturing 374
housing and building construction 373
joint designs for SIPs construction 375
light weight and fast constructing 373
PET foam core FR resistant facade cladding panels 378
pultruded C-channel used for covering top and bottom of panels 377
pultruded H-staud used for connecting sandwich composite panels 376
PUR foam cored snadwich roof panel 379
sandwich roof, wall, floor and subfloor system 373
shear connetors used for increasing shear property of sandwich panels 377
SIPs developing history 374
structrual insulated panels (SIPs) for house construction 374
studs used for increasing shear strength of SIPs 374
applications of sandwich structures in civic industry
acceptance increasing 387
advantages of SPS bridge deck system 380
aluminum honeycomb sandwich structural system for protesting highway bridge 386
balsa wood core sandwich structure for repairing existing bridge 382
bridge and dock protective systems made by sandwich structures 385
bridge building 379
challenges and issues 387
composite railway sleeper made by sandwich structures 384
core and facing materials for making SPS deck panels 380
core with high shear properties needed for 387
example and connection detail of SPS bridge deck 381
new or revised design guidelines needed for 387

applications of sandwich structures in civic industry (cont.)
  pedestrian bridge and walkway structures with sandwich bridge decks 382
  PUR foam webcore used for making sandwich bridge deck 382
  sandwich composite bumper for protecting bridge piers against ship collision 385
  sandwich panel system (SPS) for bridge deck construction 379
  structural beam made by gluing sandwich panells for bridge girder 383
applications of sandwich structures in helicopters 351
  blades made by sandwich structure 352
  developing history 353
applications of sandwich structures in medical equipment construction 395
  advantage of geometrial freedom of foam cored fiber reinforced structure 395
  advantages using PUR foam replacing PMI foam 397
  carbon composite sandwich tabletop for next generation equipment 398
  curved PMI foam cored sandwich bed 396
  example of making a sandwich cradle 396
  FEA works for designing medical equipment 398
  low X-ray and radio absorption of nonmetallic sandwich structures for medical equipment 395
  next generation radiolucent patient table 398
  PMI foam mostly used as core material for 395
  resin infusion used for making medical equipment 398
  scanning bed boards and pellets for X-ray, CT and MRI built by 395
  X-ray transmission rate of sandwich structure estimation 397
applications of sandwich structures in radar radome construction 388
  3-panel 5-panel and quasi-random radomes 389
  A, B, C and multilayers constructions for different applications 390
  advantages 391
  advantages of foam core for building radome 393
  dielectric properties of basic fiber for building radome 391
  epoxy and polyester resin for building radome 391
  fiberglass sandiwch construction for common type of radome 388
  flanges and joints for radome construction 391
  four types of sandwich constructions for radome construction 389
  low dielectric constant foam core needed for 389
  major design criteria of radome 388
  multiple pieces sandwich panels connected for large diameter radome 389
  nonmetallic honeycombs as core for building radome 392
  one sphere structure for small diameter radome 389
  pregreg lamination used mostly for building radomes 395
  radomes classifications 388
  relation of dielectric properties of core density 393
  requirements of radome 388
  right fiber reinforcement materials for radome construction 390
applications of sandwich structures in sport and leisure products 413
  3D printed core pattern for sandwich sport boards 416
  3D printing for making sport boards 415
  canoes, kayaks and paddleboard made by sandwich structures 419
  example of using wood strips as core and carbon or glass fiber composite skins for making snow boards 417
  foam cores and fiber reinforcements for making water boards 415
  laminating schedule for most water boards 415
  manufacturing process for making water boards 415
  paulownia and hard wood used as core for making ski boards 417
  polylactic acid (PLA) as raw material for 3D printing 415
  skis and snowboard made by sandwich structures 416
  soric mat core and kevlar/glass fabric foam making kayak 419
  sporting boards on the water 413
  surfboard development from wood to foam core sandwich strucures 415
  wood used for both of core and skin for making snow boards 417
applications of sandwich structures in transportation 357
  advantages 358
  advantages for bus body construction 364
  aim of rail car application 358
  aluminum honeycomb and fiberglass epoxy composite facing for bus body construction 365

Index

aluminum sandwich structures for making Formula 1 racing car 371
aspects to be paid attention 358
assembling technologies 358
automotive industry 368
balsa and foam core sandwich structure made by vacuum infusion for bus body 366
balsa and PET foam cores for making floor of Corvette sport car 370
bus body construction 364
bus boy is assemblied by two parts made by vacuum infusion 367
car chassis made by assembling sandwich panels 370
components of passenger rail cars made by sandwich structures 360
core materials for building truck body 367
cut-and-fold sandwich panel techniques used for making car chassis 371
edge and corner connection for building truck body 368
egg-crate cored metal sandwich for building conventional tanks 363
example of commodity carring tank 362
example of tilting train 362
examples of passenger rail car structures 360
extruded profile combined with foam as core for bus body construction 365
freight rail cars made by metal snadwich structure 362
honeyomb sandwich for car components 369
hybrid design combining sandwich with solid components for bus body 365
increase inner space of the rail car 359
inserts and adhesives used for assembling car chassis 372
inside components of high-speed trains 362
light weight and easy construction for truck body 367
main advantages in rail vehicles application 360
multifunctional considerations 358
Normax honeycomb used for making Formula 1 racing core 372
PET foam core and natural fiber composite sandwich for no-structural parts of cars 369
PET foam core sandwich panel made by pultrusion for semitrailer body 368
rail car application 358
reasons for not used for structural components of automobiles 369
refrigerated trailer and container made by sandwich structures 368
techniques for shaping and assembling flat sandwich panel into car chassis 370
thermoplastic sandwich structure used as bus body components 366
truck and semitrailer body application 367
vacuum infusion for making bus body stucture 366
vacuum infusion process for making front cap of fast train 361
autoclave process 136
advantage and procedure 136

Balsa wood core
affects of adhesive 25
applications 33
core format 32
density range 32
density variaion 25
development history for sandwich core material 19
end-grain core 24
fire resistance 31
fungal and moisture content 32
humidity and moistrue effections 26
impact absorption 28
lumber properties 21
making mosquito air bomber xvi
made by veneers 33
manufacturing processes 21
mechanical behaviors in different directions 23
mechanical properties at different moisture content 27
microstructure 21
milling, kiln drying, and making block 20
miscellaneous properties 28
new developments 24
processes for making end-grain core 24
product with narrow density variation 26
properties of end-grain and flat-grain 22
rigid, perforated, grooved and flexible cores 32
surface coating preventing resin absorption 29
temperature resistance 30
tree 19
tree plantation 19
venneer core mechanical properties 34
beam flexural and shear stiffness test
calculation of core shear modulus 203
calculation of transverse shear rigidity 203
calculations of panel flexural stiffness 202
coverage 201
flexural stiffness and shear rigidity by one test 204
loading configuration 201
sample 201
setup for facing modulu know 202
test setup for facing modulus unknow 202

Index

beam flexural test for core shear properties 193
  ASTM standards 193
  core shear ultimate stress 196
  core shear yield strength 197
  facing stress 198
  failure modes 196
  LVDT setup 196
  sample preparation 193
  support span determination 195
  tests setup 194
  three and four point bending tests 194
  usefule results 193
bending force works on sandwich structure xiv
biaxial composite face sheets
  physical and mechanical properties of sheets made by prepreg 117
biaxial composite sheets made by vacuum infusion
  mechanical and physical properties 117
boron and ceramic fiber reinforcement materials
  mechanical and physical properties 107

carbon fiber and synthetic fibers as face reinforcements xxvi
carbon fiber reinforcement materials
  manufacturing and properties of PAN based carbon fibers 102
  mechanical and physical properties 103
  PAN based carbon fibers 101
  pitch based carbon fibers 103
  special properties 101
  types 104
carbon foam cores
  methods for making 59
  specialties and applications 60
carbon foams
  typical properties 60
cardboard honeycomb xxiii
categories of core materials xxiii
ceramic foam cores
  spicialty and application 57
  typical properties 59
ceramica foam cores
  method for making 57
characterizaion methods for core materials 73
characterization of sandwich structures
  air water blaster test by shock tube 222
  developing history xx
  wave impact tests 218
characterizations of sandwich structures 185
  concentrated load impact tests 211
  coverage 186
  drum peel test 187
  dynamic fatigue evaluation 227
  edgewise compressive test 207
  face to core bonding tests 187
  flatwise compressive test 205

  flexural fatigue tests 227
  flexural strength and bending stiffness evaluations 192
  fracture toughness test 236
  full scale air and water blast tests 225
  introduction 186
  nondestructive evaluations (NDE) 244
  significances 186
  thermal mechanical tests 240
closed mold lamination 138
  auxiliary materials 139
  development and introduction 138
  requirements of core, facing and resin 139
  vacuum infusion 138, 139
cohesive element 310
cohesive element approach
  defined 308
  example 312
  material parameter calibration 310
cohesive zone method *see* cohesive element approach
commercial finite element software
  ABAQUS 299, 302, 303, 312, 329
  LS-DYNA 299, 302, 319
  NASTRAN 302, 328
  OPTISTRUT 325, 326
commodity plastics sheets
  PE, PP, PET and PVC plastics 93
comparisons of solid and sandwich laminates xiv
composite face sheets
  plastic matrix materials 109
composite face sheets made by vacuum infusion
  mechanical and physical properties of quasi-isotropic composites 119
composite faces properties
  biaxial laminates 116
  quasi isotropic laminates 116
  unidirectional laminate 114
composite honeycombs 48
  applications 50
  Aramid composite honeycomb 52
  expansion manufacturing processes 49
  fiberglass composite honeycomb 51
  made by carbon fiber 48
  special properties 48
  temerature resistance 49
  types and materials made by 48
composites face sheets
  reinforcement fiber materials 99
composite laminate optimization
  free-size optimization 326
  shuffling optimization 326
  size optimization 326
concentrated load impact tests 211
  actual impact energy 217
  apparatus and fixtures 213
  ASTM D7766 procedure A, B and C 216

Index

ASTM standard 211
curves of force *vs.* head displacement 215
data of forc *vs.* contact time 217
dent depth and diameter measuring 218
drop hight determination 216
drop weight and velocity determination 215
functions of different indenters 214
impact regimes
    ballistic, intermediate and low velocity impact 211
indenters
    hemispherical, conical, pyramid, and cylindrical 212
minimum drop hight 217
procedure A, B and C 212
procedures of test 216
significances 211
specimen destructive insplection after testing 216
specimen preparation 213
connection for secondary processing
connector to solid structure 289
right angle connectors 291
straignt connector of sandwich structures 289
T-joint connection 291
connector element approach
example 312
general description 311
material parameter calibration 311
consistent unit system 298, 299, 314, 320
continous fiber reinforce thermoplastic sheet (CFRTP) 127
core formats
applications of different formats 68
core materials
compressive properties 75
flatwise properties 75
maximum operatiing temperatures 82
special properties and testing methods 73
types xxii
core materials and properties 1
introduction 2
core mats 64
applications (*see* applications)
made by using polymer microsheres 65
types and specialties 65
core shear prediction under bending moment
example of calculation 258
core sheet formats
affects of format to mechanical properties 68
rigid, double cut, grooved and perforated, contoured, and plain 68
cork wood core
mechanical properties 35
pantation and harvest 34
properties and applications 36
corrugated metal sheet core(s)

metal sandwich panel made with 62
method for making 61
specialties and applications 61
crushable foam model
material model
    yield surface 303
parameters calibration
    by yield strengths under uniaxial compression, hydrostatic compression, and hydrostatic tension 303
    by yield strengths under uniaxial compression, uniaxial tension, and shear 304, 312

damage types of sandwich composites
detail damage for special products 433
gerneral clarifications of damages 432
type A, B and C damages 432
damage types of sandwich structures 432
deflection and strength of simple elements at different supports and loads 261
equations table 262
example of cantilever beam and end load 265
example of cantilever beam and uniformly distributed load 266
example of clamped support and central load 264
example of clamped support and two loads 265
example of clamped support and uniformly distributed load 265
example of simple support and central load 264
example of simple support and two loads 264
example of simple support and uniform distributed load 264
deflection prediction under bending moment
equations 258
overal deflection 259
design for minimum weight
equations and charts for optimum solutions 276
significances 274
design for minimum weight for given strength
failure mode technique 277
design principles simple element 272
basic design concept 272
choice of materials and assembly methods 272
core compression strength verification for a floor structure 273
core shear strength verification 273
design for minimum weight 274
design for minimum weight for given stiffness 274
design for minimum weight for given strength 277
design for rigidity 273

design principles simple element (*cont.*)
   environment and fire resistances 278
   environment considerations 272
   facing strength verification 273
   facings and core strengths 272
   high elastic modulus facing and thick core for increasing flexural rigidity 274
   insulation consideration 278
   main causes for failure 272
   manufacturing conditions 278
   rigidity design 272
   rigidity requirements followed by strength verification 273
   thermal, electrial and fire retardant considerations 272
   thickness of panel, facing and core considerations 278
design procedure from simple to complex structure 278
   constraint of structural member for thickness design 279
   design routine of simple element 279
   scaling up to large complex structure from simple element 280
   simple element thickness determination 278
   thickness determination simple element 279
detection and estimation of defects and damages
   battery-driven electromagnetic hammer for tap test 429
   critical damage point (CDP) 432
   damages during service time 431
   damages from production 431
   hammers used for tap testing 428
   handheld ultrasonic (UT) camera inspection 429
   moisture meter inspection 431
   personal and visual inspections 428
   personal inspection by minor destructive tests 428
   pitch-catch instrument inspection 429
   radiography instrument inspection 431
   remote visual inspection 428
   shearography instrument inspection 430
   tap testing for revealing disbond and delamination 429
   thermography instrument inspection 430
   ultrasonic phased array (UT-PA) inspection 430
   UT-PA dry-coupling roller probe inspection 430
drum peel test 187
   average peel load determining 188
   equation for calculating everage peel torque 189
   fixture and setup 187
   sample 187
   test procedure 187

dry laminating processes 130
   adhesive for bonding face and core together 128
   core and face surface preparing 130
   facing materials 126
   introduction of equipment and advantages 130
   pinch roller laminating process 131
   process for making sandwich structural insulated panel (SIPs) 131
   process using heat and pressure 131
   using 4-level vacuum powered press for making RV sandwich sidewall 131
   using hot melt adhesive and doubel belt press for making Al faces and honeycomb core panel 131
   vacuum bagging lamination for making bus wall, roof and floor 132
dynamic fatigue evaluation
   hydromat fatigue test 234
dynamic fatigue tests
   edgewise compressive fatigue test 233
   flatwise compressive fatigue test 231
dynamic fracture evaluation 227

edgewise compressive test 207
   ASTM standard 208
   failure modes 209
   Fixture 209
   material and specimen preparation 207
   significance 207
   specimen dimension recommendation 207
   specimen installation and loading 208
   ultimate strength calculation 209
edgewise damage prediction and prevention 269
   critical stress causing general buckling 269
   design procedures involving buckling 271
   designs for preventing local buckling 270
   edgewise compression failure modes 269
   general buckling under edgewise load 269
   local buckling 270
   solutions for preventing general buckling 270
end-grain balsa wood material model
   by *MAT_WOOD in LS-DYNA 317
   parameter calibration 317
engineering thermoplastics face sheets
   PA6, PEI, PPS, etc., 94
epoxy resin 27
   properties for vacuum infusion 142
   resin transfer molding (RTM) 149
   specialties and applications 111
expanded polypropylene (PP) foam
   core for thermoplastic sandwich composites 162
extruded polystyrene foam (XPS)
   core for thermoplastic sandwich composites 162

Index

fabric woven cores
    fiberglass fabric core 64
    specialties and applications 64
    three directional spacer fabrics 64
face sheet materials 85; *see also* skin materials
    characterizations of composite sheets 118
    composite faces properties 113
    fiber materials for composite face sheets 99
    instroduction 86
    made in co-curing wet lamination xxv
    metals, plastics and woods 86
    plastic matrix materials for composite sheets 109
    plastics sheets 93
    types xxiv
    wood sheet and planks 94
face tension stress prediction under bending moment
    example of calculation 257
face to core bonding strength tests
    cleavage test 192
    end-load test 191
    flatwise tensile test 189
    single cantilever test 191
facing materials
    continuous fiber reinforced thermoplastic sheet (CFRTP) 127
    FRP used for RV wall lamination 127
    metal sheet 126
    thermosetting fiber reinforced plastics (FRP) 126
facing properties by long beam test 198
    coverage 198
    effective sandwich flexural stiffness 201
    facig ultimate stress calculation 200
    failure mode acceptable 199
    sample preparation 199
    standard loading configuration 199
    support length determination 199
fatigue failure modes of sandwich composite
    core shear 326
    interface debonding 326
    tensile failure in the skin sheet 326
fiber glass 3D fabrics core
    properties 65
fiber reinforcemant materials
    development history xviiii
    leaf fibers 104
fiber reinforcement plastics (FRP) face sheets xxv
finite element analysis xxi
    key considerations 297
    key steps 297
    purpose 296
finite element mesh
    CAD (Computer-aided design) file 298
    element integration scheme 299
    element order 299

element type 299
    hourglassing control 299
    mesh quality 300
    mesh size 299
finite element preprocessor
    ANSA 299
    HyperWorks 299
finite element solver 300
    explicit solver 300, 314, 320
    implicit solver 300
    key paramters
    nonlinear geometry switch 300
    time step 300
flatwise compression test
    result calculations 76
    standard, sample and test procedure 76
flatwise compressive test for sandwich structures 205
    chord modulus determination 206
    compressive modulus determination 206
    compressive strength calculation 205
    difference from core test 205
    equipment and fixture 205
flatwise tensile test 189
    ASTM standards 76, 189
    failure modes 190
    procedure and result calculations 77
    sample and fixture 189
    sample preparation 76
    setup and test procedure 190
flexural fatigue tests 227
    ASTM standard 227
    failure modes 231
    fully reversed stress cycles 229
    loading force determination for stress controlled test 229
    loading striking frequency setting 229
    maximal displacement determination for displacement controll fatigue 229
    procedure A for three point loading system 228
    procedure B for four point bending 228
    repeated stress cycle - maximum and minimum stress are not equal 230
    repeated stress cycles 229
    setup for 3 point and 4 point loading 228
    stress amplitude 230
    stress based sinusoidal waveform 229
    stress range determination 230
    three point and four point bending 227
    two-step and block loading regimes test 231
flexural strength and bending stiffness evaluations 192
    facing properties by long beam test 198
flexural strength and bending stiffness tests
    beam flexural test for core shear properties 193
    coverage 193

foam cores
  close cell content 3
  compressive properties of PVC and PEI
    foams 83
  important of light density 2
  poly (styrene-co-acrylonitrile) (SAN) 12
  polyetherimide (PEI) 15
  polyethersulfome (PES) 17
  polyethylene terephthalate (PET) 5
  polymethacrylimide (PMI) 13
  polyurethane (PUR) 8
  polyvinyl chloride (PVC) 3
  requirements 2
  sureface engergy 3
  syntactic foam 17
  thermal conductivities at room
    temperature 80
foamed in-situ thermoformable sandwich (FITS)
    process 175
foam modeling
  by crushable foam model 302, 313
  by solid element 302, 313
force on core and facings when a bending force
    acts on a sandwich structure xiv
formats of core materials 67
formats of fiber reinforcements xxvi
fracture toughness nest
  significance 236
fracture toughness test 236
  double cantilever beam test 236
  end notched sandwich test Mode I
    and II 238
  failure modes 238
  fracture toughness calculation 238
  single cantilever beam test 237
free water drop tests
  single and repeated slamming tests 218
full-scale blast loading tests on sandwich
    strucures 225
future of aeronautic sandwich structures 354
  concept of multifunctional structure 356
  examples of multifunctional sandwich
    structures 356
  integration of multifunctionality of sandwich
    structures 354
  mechanical plus thermal, stealth,
    acoustic and vibration damping
    functions 355
  multifunctional protective skin concept 357
  multifunctions of core and outer skin 357
  next target for new structure 355

glass fiber reinforcement materials
  A-glass, C-glass, and AR-glass 100
  E-glass and ECP-glass 99
  H-glass and R-glass 99
  mechanical and phyical properties 101

quarts (silica) fibers 100
S-glass 100
types xxvi

hand layup process
  procedure with vacuum bagging
    curing 133
  resin, procedure and application
    range 133
Hashin damage model
  failure mode
    compressive matrix mode 301
    fiber compressive mode 301
    tensile fiber mode 301
    tensile matrix mode 301
  material model
    defined 301
    example 320
  material parameter calibration 301
high-performance foam core materials xviiii
high-performance sandwich composites xxviii
high-performance thermoplastics face sheets
  PAEK, PEEK and PEI 94
high temperature and fire resistant resins xxvii
history of theoretic research on sandwich
    composites xx
honeycomb core modeling
  by homogenized solid element
    example 320
    material calibration tests 306
    with *MAT_MODIFIED_HONEYCOMB
    in LS-DYNA 306, 319
  by mixed shell and solid element 307
  by shell element 305
honeycomb cores 36
  categories 37
  made by profile extrusion 38
  making bullet train from 1980s xvii
  making commercial airplane xvii
  materials for making 37
  metallic and plastic 37
  methods for making 38
  specialty of metallic honeycombs 37
  techniques for making 37
  thermoplasctics 38
hybrid core concept xxiv
hydromat fatigue test
  ASTM standard 234
  test conditions determination 235
  test fixture and sample mounting 234

I-beam and sandwich structure comparison xiii
introduction of sandwich composites xiii
in-situ core foaming
  PA12 based composite sandwich 174

J2 elastoplasticity material model 304

Index

kit cores 68
   benefits 69

lamination processes
   development xx
   dry and wet laminations introduction 126
   dry laminatoin 126
   instroduction xv
   for thermoset sandwich composites 125
lattice truss metal core 62
   applications 62
   made by 3D printing 63
   types 63
linear damage law 309

magnesium face sheets
   mechanical and physical properties 90
mat cores xxiv, 64
maximum stress and deflection of simple supported rectangular plate under uniform load
   equations and example 267
metal face materials
   steel fiber composites 87
metal face sheets
   alumnum sheets 87
   magnesium face sheets 90
   mechanical and physical properties of varous alloys 92
   miscelaneous alloys sheets 90
   special properties 86
   steel sheets 87
   titanium face sheets 90
metal fibers as facing reinforcement xxvi
metal foams
   application in automotive industry 57
   made by metal powders 56
   properties of aluminum foam 56
   properties of stainless steel foam 56
   specialty and applications 54
   types and method for making 54
metal honeycomb(s) xxiii, 43
   aluminum honeycomb made by corrugating 46
   aluminum honeycomb properties 45
   cell configurations made by expension 44
   core cell configurations made by corrugating 46
   flexible format configurations and applications 47
   made by corrugation 45
   made by expansion process 43
   methods for making 43
   types 43
   typical properties of stainless steel honeycomb 44

metallic group of facing materials xxv
mosquito airplane bomber xvi

natural fiber reinforcement materials
   bast fibers xxvi, 104
   mechanical and physical properties 104
N-CODE 328
nondestructive evaluations (NDE) 244
   basic methods 245
   hammer tap tests 246
   industrila scale inspection 249
   infrared thermography method 248
   instroduction 245
   pitch-catch swept test 246
   shearography test 248
   targets of tests 245
   ultrasonic tests 247
   visual inspection 246

optimization
   constraints 326
out-of-autoclave process
   typical vacuum laminating setup 137
   design variable 326
   objective 326
   response 326

paperboard honeycomb(s) 50
   applications 50
   developing history 50
   manufacturing processes 53
   typical properties 53
PET foam core development xviii
   manufacture and recyclling capacity xviiii
phenolic resin
   specialties and applications 111
plastic matrix materials
   epoxy resin 111
   high temperature application resins 112
   phenolic resin 111
   polyureathane resin 111
   properties of different fiber and matrix composites 113
   thermoplastic matrix 112
   unsaturated polyester 110
   vinyl esters 110
plastics face sheets 93
   commodity plastics sheets 93
   introduction 93
   thermoplastics sheets 93
plastics for making honeycomb cores xxiii
plate shear test
   procedures and result calculations 79
   sample preparation 78
   standards 78
polystyrene (PS) foam as core new development xviiii

poly (styrene-co-acrylonitrile) (SAN) foam
   heat resistance 13
   manufacturing process 13
   mechanical properties 14
   molecular structure 13
polycarbonate (PC) honeycomb
   typical properties of honeycombs made by co-extrusion 40
polyehtersulfone (PES) foam
   chemical structue 17
polyester nonwoven fabric cores
   properties 66
polyetherimide (PEI) foam 15
   fire and heat resistance 15
   mechanical properties 16
   special applications 16
   special properties 15
polyetherimide (PEI) honeycomb
   mechanical properties 41
polyethersulfone (PES)
   special properties and applications 17
polyethersulfone (PES) foam
   mechanical properties 17
polyethylene terephthalate (PET) foam 5
   core structure 7
   heat resistance 7
   manufacturing processes 6
   mechanical properties 9
   recycling capacity 5
   special properties 7
   strand foam 7
   thermoforming capacity 9
   thermoplastic foam 5
polyisocyanurate (PIR) foam 10
polymethacrylimede (PMI) foam
   mechanical properties 15
polymethacrylimide (PMI) foam
   chemical structure 14
   heat and chemical resistance 14
   manufacturing process 15
polypropylene (PP) foam 163
polypropylene (PP) honeycomb
   made by co-extrusion 40
   made by profile extrusion 39
   properties of folded honeycomb core 42
   typical properties 39
   typical properties of co-extrusion core 40
polypropylene (PP) strand foam
   core for thermoplastic sandwich composites 162
polyureathane (PUR) foam
   fiberglass reinforced 12
   manufacturing process 11
   mechanical properties 12
   molecular strcture 10
   PUR and PIR mixture 10
   web core 12

polyurethane resin
   development xviiii
   specialties and applications 112
polyvinyl chloride (PVC) foam 3
   development xviii
   manufacturing 4
   mechanical properties 5
   molicular structure 3
power law 310
prepreg compression process 137
   using sheet mold compounding (SMC) 138
prepreg lamination
   advantages 135
   autoclave curing assistance 135
   materials 135
   procedure and application range 135
pultrusion 149
   automation laminating process 149
   core materials 150
   high output and broard range products 149
   introduction and advantages 149
   processing details 150
   reinforcement fabrics 149
   specialties of resin formulations 151

quality inspection during and after repairing 446
   guidlines for post repair inspection 446

recycling of sandwich composites xxii
reinforcement fiber materials 99
   applications of different fiber formats 108
   architectural fiber and fabric forms 107
   basal fibers 100
   boron fibers 106
   carbon fibers 101
   glass fiber 99
   natural fibers 103
   SiC and alumina fibers 106
   synthetic fibers 104
repair processes for sandwich composites
   modified method for repairing type B damage 439
   modified method for repairing type C damage 440
   repair type B damage 438
   repair type C damage 439
   type A damage repairing 437
   type A repair procedure 437
repeated slamming test
   procedures 222
   sample mount 221
   servo-hydraulic slam system 220
   skin strain gauge installlation 222
   skin strain rate recording 222
resin for vacuum infusion
   properties of unsaturated polyester 141
resin for wet co-curing laminations xxvi

Index

resin transfer molding (RTM) 147
  advantages 148
  core requirements 148
  epoxy resin 149
  introduction 148
  preformed reinforcements 148
  processing parameter considerations 149
rigidity, stress, and deflection under bending moment 256
  compression stress 257
  core shear stress prediction 258
  deflection prediction 258
  face tensile stress prediction 257
  flexural rigidity predication 256
  shear rigidity prediction 257

sandwich composites
  characterizations of sandwich structures 185
  development history xvi
  definition xiii
  introduction xiii
  significance in modern industries xiii
  wet laminating process 132
sandwich composites
  all thermoplastic sandwich composites (all-TSCs) 159
  applications 333
  core materials 1
  damage assessment and repairing 425
  face sheet materials 85
  introduction xiii
  lamination processes of thermoset composites 125
  maximum operation temperature 240
  navy ship construction in Europe history xvii
  new development xxi
  special properties and testing methods for core materials 73
  structure design and mechanical property analysis 253
  structure modeling by finite element method 295
sandwich composites damage assessment and repairing 425
  damage types 426
  details of different type damages 432
  detection and estimation of defects and damages 428
  inspecting and evaluating damages 427
  introduction 426
  overall processes for sandwich composite repairing 426
  quality inspection during and after repairing 446
  repairing plan and operations 427
  repairing procedures and methods 433

sandwich composites repairing procedures and methods 433
  attention for handling materials before starting repair 436
  considerations for making repair plan 434
  damage clearing and preparation before repairing 434
  general repair process 437
  items need to be considered for choice of repair materials 435
  principles for making repair plan 434
  quality assurance factors and nondestructive inspection (NDI) methods 434
  repair philosophy 433
  repair plan design and option selection 433
  repair techniques for nonaerospace structures 434
  repair techniques used in aerospace industry 434
  repairing materials and preparation 435
sandwich core materials
  specialties 74
sandwich structure design and mechanical property analysis 253
  application conditions effect on design parameters 286
  bending moment M, shear force V and normal force N of sandwich under bending force 255
  core format effect on design parameters 283
  deflection of simple element at different support and load 261
  design principles simple element 272
  edgewise damage prediction and prevention 269
  fasteners and connections to solid structure 289
  force and deformation of core and facing under flexural load 255
  hybrid cores using for large complex design 287
  I beam comparison and extension 254
  introduction 254
  laminating method effect on design parameters 285
  load distributions 254
  maximum facing stress, core shear stress and deflection predictions of simple supported rectangular plate under uniform normal load 266
  reginal and local reinforcement used for second mounting 288
  rigidity, stress and deflection under bending moment 256
  strength of simple sandwich at different support and load 261

Index

sandwich structure design and mechanical property analysis (*cont.*)
  stress, strain and rigidity under compression 259
  stress, strain and rigidity under tension 259
sandwich structure design and mechanical analysis
  design procedure from simple to complex structure 278
  scaling up to large complex structure from simple element 280
  coupons to subcomponent to final structure of multilevel scaling 280
  laboratory-scale structure of similary scaling 281
  multilevel scaling 280
  nondimensionalization scaling methodology 282
  scaling law of similarity scaling 282
  similarity condition from small-scale models 281
  similarity theory scaling 281
sheet mold compounding (SMC) 138
single slamming test
  pressure distribution profile *vs.* slamming energy 221
  sample dead rise angel setup 220
skin/face sheet modeling
  by Hashin damage model 301, 320
  by shell element 301, 313, 320
S-N curve method for sandwich composite
  core only in shear load 328
  skin sheet only in tension load 328
solid wood and foam cores xxiii
special foam cores
  carbon foams 59
  ceramic foams 57
  metal foams 54
special repair processes for sandwich composites 441
  basic approaches for structure repair 442
  brief and short term repair 441
  face crack and defect repair 443
  patch repair for structure repair 442
  resin injection for re-bonding delaminated skin to core 443
  resin vacuum infusion repairing 444
  structural repair 441
  surface damage repair 441
  temperature control repair curing 445
  vacuum bagging pressure curing 443
  vacuum infusion for repairing type B damage 445
spray layup lamination
  materials, procedure and application range 134

spray layup process
  advantages 135
  process making honeycomb core sandwich panel 134
stainless steel honeycomb
  typical properties 44
steel face sheets
  mechanical and physical properties 88
stiffness comparison of solid sheet and sandwich at different core thickness xiv
strand foam core 7
strength and bending stiffness evaluations
  significances 192
  test for determining beam flexural and shear stiffness 201
structure modeling by finite element method 295
  basics 296
  general steps and considerations by using commercial software 297
  introduction 296
  purpose 296
surface separation approach
  example 312
  general description 311
  material parameter calibration 311
syntactic foam
  manufacturing 17
  mechanical properties 18
  properties and applications 17
  special properties and applications 18
synthetic fiber reinforcement materials
  high-modulus polypropylene fibers 105
  mechanical and physical properties 106
  para-aramid fibers 105
  UHMWPE, PBO and LCP fibers 105

testing sandwich structures 185
test methods for composite sheets 118
  compressive strength and modulus 118
  damage tolerance and impact properties 121
  flexural strength and modulus in sandwich 120
  in-plane shear strength and modulus 120
  interlamilanar shear strength 121
  introduction and standard 118
  tensile strength and modulus 118
test methods for core materials 73
  dielectric constant test 81
  fire, smoke and toxicity tests 81
  flatwise compression 75
  flatwise tesile test 76
  introduction 73
  maximum operating temperature 80
  plate shear test 78
  thermal conductivity 80

Index

thermal mechanical tests 240
    example 241
    flexural modulus *vs.* testing temperatures 242
    significance 240
    test equipment 241
    test setup 240
    test setup for one side of structure heated 244
    test with one side of structure heated 243
thermoforming
    all-thermoplastic sandwich composites 178
thermoplastic face materials
    towpregs sheets 164
thermoplastic facing materials
    commingle fiber roving process 168
    glass, carbon and synthetic fiber prepreg fabrics 165
    physical and mechanical properties of GF/PP commingled fabrics and laminates 169
    physical and mechanical properties of GF/PP sheets 167
    prepreg ocnsolidated tapes made by melt impregnation 165
    pressing formed fiber reinforced laminates made from pregreg unidirectional tapes 167
    towpres, pre-consolidated taps and conmmingled fiber roving 165
    weaved fabrics made from prepreg unidirectional tapes 167
thermoplastic sandwich compostie lamination
    fusion bonding processes 170
    introduction 169
thermoplastic sandwich composites 159
thermoplastic sandwich laminations
    3D thermoplastic sandwich made by one step forming 177
    all PP sandwich panel made by in-situ foaming process 174
    carbon fiber/PEEK skin and carbon core laminating by heat fusion 170
    a combination of different polymers 171
    compression molding 171
    diaphragm forming 176
    double belt continuous laminating 172
    example of diaphragm forming setup 176
    expaned polypropylene (EPP) foam in-situ foaming process 175
    foamed in-situ thermformable sandwich (FITS) 175
    hot melt film function 170
    in-situ core foaming 173
    making fiber reinforced PP skin/PET foam core sandwich panel 173
    melt temperatures of PP, PET and PEI heat fusion 170
    nonisothermal compression molding process 172
    pultrusion 173
thermoplastics face sheets 93
    engineering plastics 94
    high-performance sheets 94
    mechanical and physical properties of amorphous plastics 96
    physical and mechanical properties of semi-crystalline plastics 95
    semi-crystalline and amorphous plastics 94
thermoplastics honeycombs 38
    folding production 41
    format for vaccum infusion 42
    made by folding and thermoforming 40
    manufactured by co-extrusion 39
    polycarbonate (PC) 40
    polyetherimide (PEI) 40
    polyethylene terephthalate (PET) 40
    polypropylene honeycomb and properties 39
    types 38
    types made by co-extrusion 40
thermoplastics matrix materials
    types, properties and applications 113
titanium and its alloys face sheets
    mechanical and physical properties 91
types of sandwich composites xxii

underwater blast test full scale 226
    sample damage evaluation 227
    set up and example 226
uniaxial and shear stress-strain curve converting 304
unidirectional composite face sheets
    mechanical and physical properties 115
unsaturated polyester xxvii
    for vacuum infusion 141
    specialties and applcations 110

vacuum infusion 139
    auxiliary materials 143
    co-curing process 140
    consideration of core and facing combination 140
    considerations of practices, resin viscosity and documentation 146
    core gap preventing 144
    Darcy's law 139
    definition and advantages 139
    double vacuum bag (DVB) 147
    drawbacks 146
    eccentric, concentric and linear flow 147
    epoxy resin 142
    equipment setup 138
    facing reinforcement materials 140
    feed geometry of resin inlet and moving 147
    flow test 145
    leaking check 145
    materials 139

vacuum infusion (*cont.*)
    peel ply, flow media, bagging film and sealant tape as auxiliary materials 143
    pressure drop test 146
    pump condideration 145
    re-infuse for fixing dry spots 147
    requirements of core mateiral 140
    resin requirement 141
    sealant tap requirement 144
    setup for infusing a boat hull 146
    setup with right auxiliary materials 144
    vacuum gauge set up 143
    vacuum maintaining in infusion process 146
    vacuum port setup 145
    vinyl ester properties 141
vacuum-assisted resin transfer molding (VARTM) 148
vinyl ester xxvii
    specialties and applications 110
vinyl resin
    for vacuum infusion 141

water shock tube test 223
    3D digital image correlation measurement 223
    example 225
    shock tube setup 223
    underwater blast simulator 223
water wave impact tests
    free drop slamming system 219
    free drop tests 218
    post slamming techniques 218
    repeated slamming tests 220
    slamming energy determination 220
    slamming tests 218
    specemen preparation and setup 219
wave impact tests 218
    introduction 218
    significances 211
    water vave impact tests 218
weight reduction - principle using sandwich structures xxi
wet laminating process 132
    3D printing 151
    hand layup process 133
    introduction 132
    losed mold lamination 138
    out-of-autoclave (OOA) process 136
    prepreg lamination 135
    pultrusion 149
    resin transfer molding (RTM) 147
    spray layup lamination 134
    vacuum-assisted resin transfer molding (VARTM) 148
wind blade construction xviii
wood based core 18
    balsa wood 19
    cork wood core 34
wood face materials 94
    hardwood and plywood 95
    mechanical and physical properties of hardwood 97
    mechanical properties of plywood 98
    oriented strand board (OSB) 97
    softwood and its mechanical and physical properties 98
    specialties 95

Printed in the United States
by Baker & Taylor Publisher Services

Printed in the United States
by Baker & Taylor Publisher Services